AF802413

Fibrinogen
Structural Variants and Interactions
Volume 3

Fibrinogen

Structural Variants and Interactions

Volume 3

Proceedings
Workshop on Fibrinogen
Stockholm, Sweden, July 9–10, 1983

Editors
A. Henschen · B. Hessel
J. McDonagh · T. Saldeen

Walter de Gruyter · Berlin · New York 1985

Editors

Agnes Henschen, Dr.
Max-Planck-Institute of Biochemistry
8033 Martinsried
Federal Republic of Germany

Birgit Hessel Dr.
Coagulation Research Department
Karolinska Institute
Solnavägen 1
10401 Stockholm 60
Sweden

Jan McDonagh, Dr.
Dept. of Pathology
Beth Israel Hospital
Harvard Medical School
Boston, MA 02215
USA

Tom Saldeen, Dr.
Institute of Forensic Medicine
University of Uppsala
Dag Hammarskjölds Väg 17
75237 Uppsala
Sweden

Library of Congress Cataloging in Publication Data

Workshop on Fibrinogen (1983 : Stockholm, Sweden)
Fibrinogen--structural variants and interactions.
Bibliography: p.
Includes indexes.
1. Fibrinogen--Congresses. I. Henschen, A. (Agnes), 1935- . II. Title.
QP93.5.W67 1983 612'.115 85-24571
ISBN 0-89925-130-7 (U.S.)

CIP-Kurztitelaufnahme der Deutschen Bibliothek

Fibrinogen : proceedings / Workshop on Fibrinogen. –
Berlin ; New York : de Gruyter
1981 ohne Kongressbenennung
NE: Workshop on Fibrinogen
Vol. 3. Structural variants and interactions : Stockholm, Sweden,
July 9–10, 1983. – 1985.
ISBN 3-11-009806-7

Copyright © 1985 by Walter de Gruyter & Co., Berlin 30.
All rights reserved, including those of translation into foreign languages. No part of this book may be reproduced in any form – by photoprint, microfilm or any other means nor transmitted nor translated into a machine language without written permission from the publisher. Printing: Gerike GmbH, Berlin. – Binding: Dieter Mikolai, Berlin. – Printed in Germany.

PREFACE

This volume contains the contributions to two symposia on fibrinogen. The first symposium, which was entitled "Abnormal Fibrinogens as Probes of Structure-Function Relationships", was held at Stockholmsmässan (Stockholm International Fairs) in Älvsjö outside Stockholm on July 8th, 1983, as a part of the IX. International Congress on Thrombosis and Haemostasis. The second symposium, which was entitled "Fibrinogen-Structure and Function", was held at the Berzelius Laboratory of the Karolinska Institute in Stockholm on July 9th and 10th, 1983, as a Satellite Symposium of the IX. International Congress. Approximately 150 scientists from all parts of the world participated in the symposia.

Several European Fibrinogen Workshops had already been held during the previous 10 years, some of them in Martinsried outside München, the most recent one in Noordwijkerhout, The Netherlands in May 1982. In Stockholm many scientists from outside Europe participated for the first time.

The purpose of the present symposia was to review and summarize recent progress in fibrinogen research and to enable scientists interested in the various aspects of fibrinogen to meet and discuss. Obviously, the titles of the contributions reflect which research areas are regarded as being currently of special interest. Thus, many presentations dealt with structural-functional variants of fibrinogen and with the interaction between fibrinogen and fibrin, various proteolytic enzymes or various cell types. At the Satellite Symposium an introductory lecture was given by the President of the main congress, Professor Birger Blombäck.

The organizers of the symposia wish to express their sincere gratitude to the Executive Committee of the IX. International

Congress for support and advice and to several members of the Staff of the Blood Coagulation Research Department at the Karolinska Institute for invaluable help with the local organisation. Most generous financial support was given by the companies Pharmacia, Kabi, Serono and Imco. The excellent secreterial help of Mrs. T. Nieberle and the cooperation of Dr. R. Weber and the publisher, Walter de Gruyter, are gratefully acknowledged.

This fibrinogen conference was held in a building named after J. Berzelius, who was the first professor of chemistry at the Karolinska Medico-Surgical Institute and who 145 years ago to the day, i.e. July 10th, invented the name "protein" and used it for the first time for fibrin! The organizers hope that the conference participants will remember the exceptionally hot summer days for the many important scientific contributions made and perhaps also for the delightful banquet that was given in the Stockholm City Hall, with the more personal contributions by professors John Shainoff, Staffan Magnusson and Jack Hawiger, dealing with such, so far unpublished, topics as how to recognize enemies at meetings, slide projection in the dark and which molecules are ugly and which are beautiful.

München, Stockholm, Boston and Uppsala, October 1985

A. Henschen
B. Hessel
J. McDonagh
T. Saldeen

Earlier publications in this series:

Fibrinogen
Recent Biochemical and Medical Aspects
Volume 1
Editors: A. Henschen, H. Graeff and F. Lottspeich
Walter de Gruyter, Berlin • New York, 1982

Fibrinogen
Structure, Functional Aspects and Metabolism
Volume 2
Editors: F. Haverkate, A. Henschen, W. Nieuwenhuizen and P.W. Straub
Walter de Gruyter, Berlin • New York, 1983

Forthcoming publication:

Fibrinogen
Fibrin Formation and Fibrinolysis
Volume 4
Editors: D.A. Lane, A. Henschen and M.K. Jasani
Walter de Gruyter, Berlin • New York, 1986

CONTENTS

I. A HISTORICAL NOTE

II. GENE ANALYSIS

III. FIBRINOGEN-FIBRIN CONVERSION

VIII. INTERACTION WITH PLASMINOGEN AND ITS ACTIVATOR

IX. INTERACTION WITH CELLS

I. A HISTORICAL NOTE

TWO THOUSAND YEARS OF FIBRINOGEN RESEARCH AND EVIDENCE FOR FIBRIN BEING THE FIRST PROTEIN

Agnes Henschen
Max-Planck-Institute for Biochemistry, 8033 Martinsried/ Munich, FRG

Blood coagulation has attracted the interest of scientists, physicians, philosophers and even theologists for several thousand years. Already very long time ago blood and blood flow were considered as essential for life. A few examples of the central role of blood clotting and fibrin(ogen) in science and philosophy will be presented.

About 2300 years ago

Ever since Hippokrates and Aristoteles and until less than 150 years ago all life processes were believed to take place in the blood, and for a long time period all living and dead things were thought to be composed of the four elements: air, fire, earth and water. According to ancient medical theory (the humoral pathology) the body contains four fluids (humors) and the right proportion or mixing of them constitutes health, but an improper distribution constitutes disease. The four humors were all believed to be contained in the clotted blood and were also regarded as closely related to the four elements, the four temperaments, the four ages, a.s.o. (Fig. 1). The four components observed in the clotted blood were: the light red, oxygen-rich part of the clot - "blood", the dark red, oxygen-deficient part - "black bile", the yellowish

Fibrinogen - Structural Variants and Interactions
© 1985 Walter de Gruyter & Co., Berlin · New York - Printed in Germany

Components of clotted blood	Red blood cells, oxidised form	Serum	Red blood cells, reduced form	Fibrin
Colours	Red	Yellow	Black	White
Humors	Blood	Yellow bile	Black bile	Phlegm
Elements	Air	Fire	Earth	Water
Tempera-ments	Sanguinic	Choleric	Melancholic	Phlegmatic
Ages	Child	Adolescent	Adult	Old
Seasons	Spring	Summer	Fall	Winter
Organs	Heart	Liver,Stomach	Milt	Brain,Bladder

Fig. 1. The four humors, elements, temperaments, a.s.o. and their relation to the components of clotted blood.

serum - "yellow bile" and the white layer on the surface of the clot - "phlegm". It is obvious that the "phlegm" corresponds to fibrin in modern nomenclature. Already at the time of Hippokrates it was known that many diseases were accompanied by increased amounts of "phlegm", and for thousand of years to follow the main medical diagnostic method consisted in the analysis of the patient's clotted blood, the most important therapeutic method being the attempts to remove the excess of "phlegm".

About 1300 years ago

In many religions blood is regarded as having special, philosophically significant properties. Thus, the Koran contains the following text, which is a part of Mohammed's first revelation: "Recite thou, in the name of thy Lord who created; created man from clots of blood".

About 150 years ago

Relatively pure preparations of the white, fibrous material from blood clots, now named fibrin, and of its precursor from unclotted blood, soon after named fibrinogen, were analysed for chemical composition and other properties by several scientists, among them Berzelius, Mulder and Virchow. It occurred to Berzelius that fibrin, albumin and "probably the coloured material", i.e. hemoglobin, contained the same kind of oxidised basic component, and as it seemed less convenient to name this component only after fibrin, he proposed the word protein, derived from primary or principal (Fig. 2). Thus, fibrin was the first substance designated as a protein!

Nowadays

Our understanding of blood coagulation and fibrin(ogen) is steadily increasing. At present we have reached the stage where the complete covalent structures of fibrinogen (Fig.s 3 and 4) and many other hemostasis-related proteins are known. For many of these proteins at least models for their spatial structures and for structure-function relationships have been conceived; their various interactions have been studied in great detail. Which kind and degree of comprehension will be achieved in the future?

Fig. 2. (next page) Excerpt from one of Berzelius' letters to G.J. Mulder in Rotterdam, the original letter being considerably longer. From: Jac. Berzelius Bref, II:2 (Söderbaum, H.G., ed.), Almquist & Wiksell, Uppsala, 1916.

Stockholm, le 10 juillet 1838.

Monsieur et ami,

Lorsqu'on jouit d'une correspondance aussi intéressante que la votre, et qu'on diffère de répondre, il faut avoir des raisons valables. Voici les miennes. J'ai pendant les dernières 3 semaines de juin et la première de juillet pris les eaux de Spa, et pendant ce temps-là j'ai renoncé aux occupations scientifiques pour la promenade et le mouvement dans l'air libre, ce dont j'avais furieusement besoin....

Je reviens à présent à vos travaux sur la fibrine. Il est sans doute de la plus haute importance que la substance organique, séparée du soufre et du phosphore, se trouve être la même pour la fibrine et l'albumine. Mais en même temps il devient nécessaire d'étudier ses propriétés à l'état isolé et de déterminer son poids atomique......je présume que l'oxyde organique, qui est la base de la fibrine et de l'albumine (et auquel il faut donner un nom particulier p.ex. protéine) est composé d'un radical ternaire, combiné avec de l'oxygène dans quelqu'un de ces rapports simples que la nature inorganique nous présente....

Le nom protéine que je vous propose pour l'oxyde organique de la fibrine et de l'albumine, je voulais le dériver de πρωτειος, parce qu'il parait être la substance primitive ou principale de la nutrition animale que les plantes préparent pour les herbivores, et que ceux-ci fournissent ensuite aux carnassiers. En dérivant le nom du mot grec pour la fibre, il serait moins convenable, puisque l'oxyde organique est la base aussi de l'albumine et probablement de la matière colorante, ainsi que d'autres encore.

Mais il est temps de finir mon griffonage. J'attents avec bien de l'intérêt d'avoir de vos nouvelles.

Votre dévoué

J. Berzelius

Human fibrinogen Aα-chain:

1 A D S G E G D F L A E G G G V R↓G P R V V E R H Q S A C K D S D W P F C S D E D
41 W N Y K C P S G C R M K G L I D E V N Q D F T N R I N K L K N S L F E Y Q K N N
81 K D S H S L T T N I M E I L R G D F S S A N N R D N T Y N R V S E D L R S R I E
121 V L K R K V I E K V Q H I Q L L Q K N V R A Q L V D M K R L E V D I D I K I R S
161 C R G S C S R A L A R E V D L K D Y E D Q Q K Q L E Q V I A K D L L P S R D R Q
201 H L P L I K M K P V P D L V P G N F K S Q L Q K V P P E W K A L T D M P Q M R M
241 E L E R P G G N E I T R G G S T S Y G T G S E T E S P R N P S S A G S W N S G S
281 S G P G S T G N R N P G S S G T G G T A T W K P G S S G P G S T G S W N S G S S
321 G T G S T G N Q N P G S P R P G S T G T W N P G S S E R G S A G H W T S E S S V
361 S G S T G Q W H S E S G S F R P D S P G S G N A R P N N P D W G T F E E V S G N
401 V S P G T R R E Y H T E K L V T S K G D K E L R T G K E K V T S G S T T T T R R
441 S C S K T V T K T V I G P D G H K E V T K E V V T S E D G S D C P E A M D L G T
481 L S G I G T L D G F R H R H P D E A A F F D T A S T G K T F P G F F S P M L G E
521 F V S E T E S R G S E S G I F T N T K E S S S H H P G I A E F P S R G K S S S Y
561 S K Q F T S S T S Y N R G D S T F E S K S Y K M A D E A G S E A D H E G T H S T
601 K R G H A K S R P V

Human fibrinogen Bβ-chain:

1 Z G V N D N E E G F F S A R↓G H R P L D K K R E E A P S L R P A P P P I S G G G
41 Y R A R P A K A A A T Q K K V E R K A P D A G G C L H A D P D L G V L C P T G C
81 Q L Q E A L L Q Q E R P I R N S V D E L N N N V E A V S Q T S S S S F Q Y M Y L
121 L K D L W Q K R Q K Q V K D N E N V V N E Y S S E L E K H Q L Y I D E T V N S N
161 I P T N L R V L R S I L E N L R S K I Q K L E S D V S A Q M E Y C R T P C T V S
201 C N I P V V S G K E C E E I I R K G G E T S E M Y L I Q P D S S V K P Y R V Y C
241 D M N T E N G G W T V I Q N R Q D G S V D F G R K W D P Y K Q G F G N V A T N T
281 D G K N Y C G L P G E Y W L G N D K I S Q L T R M G P T E L L I E M E D W K G D
321 K V K A H Y G G F T V Q N E A N K Y Q I S V N K Y R G T A G N A L M D G A S Q L
361 M G E N^{+}R T M T I H N G M F F S T Y D R D N D G W L T S D P R K Q C S K E D G G
401 G W W Y N R C H A A N P N G R Y Y W G G Q Y T W D M A K H G T D D G V V W M N W
441 K G S W Y S M R K M S M K I R P F F P Q Q

Human fibrinogen γ-chain:

1 Y V A T R D N C C I L D E R F G S Y C P T T C G I A D F L S T Y Q T K V D K D L
41 Q S L E D I L H Q V E N^{+}K T S E V K Q L I K A I Q L T Y N P D E S S K P N M I D
81 A A T L K S R K M L E E I M K Y E A S I L T H D S S I R Y L Q E I Y N S N N Q K
121 I V N L K E K V A Q L E A Q C Q E P C K D T V Q I H D I T G K D C Q D I A N K G
161 A K Q S G L Y F I K P L K A N Q Q F L V Y C E I D G S G N G W T V F Q K R L D G
201 S V D F K K N W I Q Y K E G F G H L S P T G T T E F W L G N E K I H L I S T Q S
241 A I P Y A L R V E L E D W N G R T S T A D Y A M F K V G P E A D K Y R L T Y A Y
281 F A G G D A G D A F D G F D F G D D P S D K F F T S H N G M Q F S T W D N D N D
321 K F E G N C A E Q D G S G W W M N K C H A G H L N G V Y Y Q G G T Y S K A S T P
361 N G Y D N G I I W A T W K T R W Y S M K K T T M K I I P F N R L T I G E G Q Q H
401 H L G G A K Q A G D V

Fig. 3. (previous page) Amino acid sequences of human fibrinogen Aα-, Bβ- and γ-chains, sequences according to A. Henschen, F. Lottspeich, M. Kehl and C. Southan, Ann.N.Y.Acad.Sci, 408, 28-43 (1983); M.W. Rixon, W.-Y. Chan, E.W. Davie and D.W. Chung, Biochemistry 22, 3237-3244 (1983); D.W. Chung, B.G. Que, M.W. Rixon, M. Mace and E.W. Davie, Biochemistry 22, 3244-3250; D.W. Chung, W.-Y. Chan and E.W. Davie, Biochemistry 22, 3250-3256 (1983). Arrows and crosses indicate thrombin cleavage sites and carbohydrate side chains, respectively.

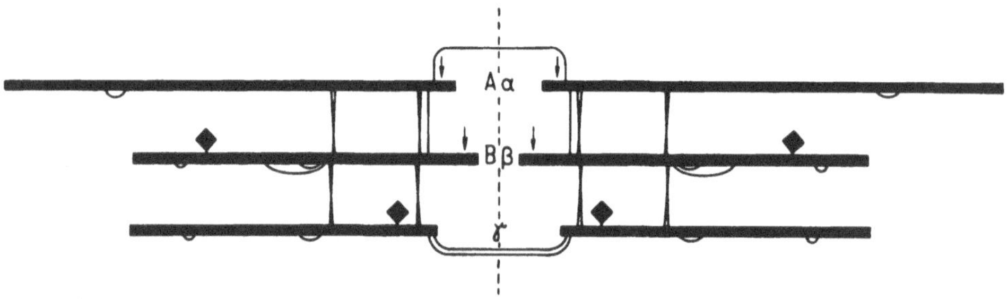

Fig. 4. Primary structure model of human fibrinogen, the chains being aligned according to homology, the N-termini with the thrombin cleavage sites (↓) in the middle; the thin lines represent disulfide bridges, the diamonds carbohydrate side chains and the broken line a center of symmetry.

II. GENE ANALYSIS

The Structure and Evolution of Fibrinogen: The Coiled Coil Region

Dana M. Fowlkes, Neal T. Mullis
Department of Pathology, University of North Carolina
Chapel Hill, North Carolina

Claudette M. Comeau and Gerald R. Crabtree
Laboratory of Pathology, The National Cancer Institute
Bethesda, Maryland

Introduction

The central event in the vertebrate coagulation system is the conversion of soluble fibrinogen into an insoluble fibrin gel. This is appropriately followed by the subsequent dissolution of the fibrin gel. A mechanistic explanation of how these processes occur depends on an understanding of fibrinogen's structure.

One approach to understanding structure-function relationships involves the comparison of functionally similar proteins such as fibrinogen between various species. The primary structure of human fibrinogen has been determined directly at the amino acid level (1, 2) and indirectly through nucleic acid sequencing of the appropriate mRNA's (3, 4, 5, 6). We have undertaken the structural analysis of the rat fibrinogen genes. The isolation of cDNA clones for all three rat genes has been carried out (13). In addition, a full-length cDNA for the rat gamma gene has been sequenced (7). Using cDNA as probes, phage clones from a rat

Fibrinogen - Structural Variants and Interactions
© 1985 Walter de Gruyter & Co., Berlin · New York - Printed in Germany

genomic DNA library of all three rat fibrinogen chains have been isolated and mapped (8). Exons which code for the intermediate disulfide knot and the first half of the putative "coiled coil" region have been sequenced. Thus, the amino acid sequences for those regions of the rat fibrinogen chains can be surmised.

Using the primary sequence of human fibrinogen Doolittle et al. (11) built elaborate models of the coiled coil region of the human fibrinogen molecule. Using these models they sought to determine the potential for helix formation in the part of the structure bound by the two sets of cysteine rings found in each half of the fibrinogen monomer. When the "helical-net" approach of McLachlan et al. (12) was applied, a non-polar face could be seen for each of the chains. This led them to propose a parallel three-stranded rope that would begin with a coiled coil, would be interrupted in its central portion, and then would resume the coiled coil structure. The key to this configuration is the alternating pattern of the polar and non-polar amino acids that occur within the coiled region. The feasibility of such a structure was confirmed by the building of three dimensional models.

The comparison of the rat and human amino acid sequence in the region of the proposed "coiled coil" shows a large number of differences between the two species. However, the differences are such that the alternating arrangement of polar and non-polar amino acids in that region of fibrinogen molecule is preserved. Thus, the pattern Doolittle (11) found in the human molecule is almost certainly of structural significance. Moreover, the region that interrupts the coiled coil in the human is further interrupted in the rat fibrinogen molecule by the presence of three prolines in the rat gamma chain.

Materials and Methods

The appropriate Hind III DNA fragments were subcloned from Charon 4A phage isolates that have been described (8) into the plasmid pBR322. To determine the sequence of these DNA fragments the plasmid DNA was further subcloned into the M13 vectors mp-8 and mp-9.

The DNA sequence of M13 subclones was determined by the method of Sanger (9). A 15-mer from New England Biolabs was used for all reactions. However, whenever ambiguities were apparent a 15-mer (obtained from Bethesda Research Labs) having another sequence was used. Sequencing reactions were carried out with the Klenow fragment of DNA polymerase from several suppliers. Reactions were performed at 20°C or 37°C and were run on denaturing 6% acrylamide gels. Sequences were determined for both strands of DNA for 80% of the regions reported here. Sequences for the gamma region have been previously reported (7). Sequence analyses were carried out using the program of Queen and Korn (10) and microcomputer programs from the laboratories of Marshall Edgell and Clyde Hutchinson.

Results

The nucleic acid sequence of those exons that code for the intermediate disulfide knot and nearby amino acids is shown in Figure 1. The analysis of the rat genomic sequences revealed a splice junction corresponding to a position three amino acids proximal to the first cysteine of the intermediate disulfide knot. That exon ends approximately 60 amino acids downstream in all three genes. In Figure 1 the sequences have been aligned to show homology between these three exons. This required the insertion of spaces in the alpha and gamma sequences because the beta chain has an

```
       1       10           20        30        40        50
ALPHA  TTCATTTTCAG...AACCACAAATGCCCTTCAGGCTGCAGGATGAAAGGGTTGATTGATG
       ||  ||| |||           ||| ||| |||| ||   | ||          |
BETA   TTGGTTTACAG...GGAGTGCTATGTCCTACAGGGTGTGAGTTGCGTCAAACTTTGCTAA
       1       10           20        30        40        50
       |   ||| |||   ||       || || ||    ||||   |     |     ||
GAMMA  TGACTTTGCAG...GGTAGTTACTGCCCAACCACTTGTGGCATCTCAGACTTCCTGAATT
       1       10           20        30        40        50
       |   ||| |||          | |||||  |    ||  | ||   ||  ||  |  ||
ALPHA  TTCATTTTCAG...AACCACAAATGCCCTTCAGGCTGCAGGATGAAAGGGTTGATTGATG
       1       10           20        30        40        50

        60        70        80        90       100       110
ALPHA  AAGCCAATCAGGACTTTACAACCAGAATCAACAAGCTCAAAAACTCACTATTTGATTTTC
       |   | |   |    | | || ||| ||     || | || | |    ||      | |
BETA   ACCACGAAAGGCCAATCAAAAACAGTATTGCTGAGTTAAATAGCAACATAAACTCTGGTT
        60        70        80        90       100       110
          || |||      |  | |      |        |  | | || | || |    | |
GAMMA  CTTACCAAACCGACGTGGACACTGACCTCCAGACTCTGGAAAACATCTTACAACGAGCTG
        60        70        80        90       100       110
           | |    ||| |    ||     || | |  ||  |||||    ||        |
ALPHA  AAGCCAATCAGGACTTTACAACCAGAATCAACAAGCTCAAAAACTCACTATTTGATTTTC
        60        70        80        90       100       110

       120       130          140       150       160       170
ALPHA  AAAAGAACAACAAG...GATTCTAATTCACTGACCAGGAATATCATGGAGTATTTGAGAG
          ||| |     |      | | | |  |||||   |   |       ||     |
BETA   CTGAGACCTCTTCGGTCACCTTTCAGTACCTGACTCTGCTGAAAGACATGTGGAAAAAGA
       120       130       140       150       160       170
          | |   |  |           || | |||| |      |   |  | |   | ||
GAMMA  AAAACAGAACCACA...GAAGCCAAGGAACTGATTAAAGCCATCCAGGTTTACTACAACC
       120       130          140       150       160       170
       |||| |  | ||     ||  | ||   |||||  |     |||  ||  || |  |
ALPHA  AAAAGAACAACAAG...GATTCTAATTCACTGACCAGGAATATCATGGAGTATTTGAGAG
       120       130          140       150       160       170

          180       190          200       210       220       230
ALPHA  GGGACTTCGCTAACGCCAACA...GTAAGTGGGACATATTTAGTGCTTGGACTTTCTAAC
        | |                    |  |  | |       | | | |    ||  || |
BETA   AGCAGGCACGAGTTAAAG...GTAGCCATCGTGGTGGTGGTGGGGTTCTACCTAACTCAT
       180       190          200       210       220       230
        | |   ||     ||           |           |||| ||| |  | | | | |
GAMMA  CGGACCAACCCCCAAAGCC...GTTTGAGGAAACCCACTGTGGAGTTGTTACCNTCCCCT
          180       190          200       210       220       230
        ||||    |               | ||     |   | | | | |   || |||
ALPHA  GGGACTTCGCTAACGCCAACA...GTAAGTGGGACATATTTAGTGCTTGGACTTTCTAAC
          180       190          200       210       220       230
```

Figure 1. The genomic sequence of the rat fibrinogen exons coding for the intermediate disulfide knot and the first half of the "coiled coil" region. The sequences have been aligned optimally to show the maximum homology between the alpha, beta and gamma rat fibrinogen genes. The tickings flag homologous nucleotides.

```
                 AsnHisLysCysProSerGlyCysArgMetLysGlyLeuIleAspGluAlaAsnGlnAsp

RAT    ALPHA     AACCACAAATGCCCTTCAGGCTGCAGGATGAAAGGGTTGATTGATGAAGCCAATCAGGAC
                 1        10        20        30        40        50        60
                 ||| ||||||||||||| ||||||||||||||||||||||||||||||| |||||| || 
HUMAN ALPHA      AACTACAAATGCCCTTCTGGCTGCAGGATGAAAGGGTTGATTGATGAAGTCAATCAAGAT
                 1        10        20        30        40        50        60

                 AsnTyrLysCysProSerGlyCysArgMetLysGlyLeuIleAspGluValAsnGlnAsp


                 PheThrThrArgIleAsnLysLeuLysAsnSerLeuPheAspPheGlnLysAsnAsnLys

RAT    ALPHA     TTTACAACCAGAATCAACAAGCTCAAAAACTCACTATTTGATTTTCAAAAGAACAACAAG
                         70        80        90       100       110       120
                 ||||||| |||||| || ||||||||||| ||||||||||| | ||| |||||||| |||
HUMAN ALPHA      TTTACAAACAGAATAAATAAGCTCAAAAATTCACTATTTGAATATCAGAAGAACAATAAG
                         70        80        90       100       110       120

                 PheThrAsnArgIleAsnLysLeuLysAsnSerLeuPheGluTyrGlnLysAsnAsnLys


                    AspSerAsnSerLeuThrArgAsnIleMetGluTyrLeuArgGlyAspPheAlaAsn

RAT    ALPHA     ///GATTCTAATTCACTGACCAGGAATATCATGGAGTATTTGAGAGGGGACTTCGCTAAC
                        130       140       150       160       170       180
                    |||||| ||||  ||||||  ||||| |||||   ||||||||| || ||  |    
HUMAN ALPHA      ///GATTCTCATTCGTTGACCACTAATATAATGGAAATTTTGAGAGGCGATTTTTCCTCA
                        130       140       150       160       170       180

                    AspSerHisSerLeuThrThrAsnIleMetGluIleLeuArgGlyAspPheSerSer


                 AlaAsn<<<<<<<<<<<<<<<<<<<<<<<<<<<<<<<<<<<<<<<<<<<<<<<<<<<<<<

RAT    ALPHA     GCCAAC<<<<<<<<<<<<<<<<<<<<<<<<<<<<<<<<<<<<<<<<<<<<<<<<<<<<<<

                 |||||
HUMAN ALPHA      GCCAAT<<<<<<<<<<<<<<<<<<<<<<<<<<<<<<<<<<<<<<<<<<<<<<<<<<<<<<

                 AlaAsn<<<<<<<<<<<<<<<<<<<<<<<<<<<<<<<<<<<<<<<<<<<<<<<<<<<<<<
```

Figure 2. The nucleotide and amino acid sequences coding for the intermediate disulfide knot and the first half of the "coiled coil" region of the rat and human alpha fibrinogen chain (ref. 3, 4). The sequences have been aligned to show maximal homologies.

```
                GlyValLeuCysProThrGlyCysGluLeuArgGlnThrLeuLeuAsnHisGluArgPro

RAT   BETA      GGAGTGCTATGTCCTACAGGGTGTGAGTTGCGTCAAACTTTGCTAAACCACGAAAGGCCA
                1        10        20        30        40        50        60
                || ||| | ||||||||||| ||| ||||||   |  |||||||| | || |||||||||
HUMAN BETA      GGGGTGTTGTGTCCTACAGGATGTCAGTTGCAAGAGGCTTTGCTACAACAGGAAAGGCCA
                1        10        20        30        40        50        60

                GlyValLeuCysProThrGlyCysGlnLeuGlnGluAlaLeuLeuGlnGlnGluArgPro


                IleLysAsnSerIleAlaGluLeuAsnSerAsnIleAsnSerGlySerGluThrSerSer

RAT   BETA      ATCAAAAACAGTATTGCTGAGTTAAATAGCAACATAAACTCTGGTTCTGAGACCTCTTCG
                         70        80        90       100       110       120
                |||| ||| ||| ||| ||||||||||| |||  |  |  ||| |||  ||||||| ||
HUMAN BETA      ATCAGAAATAGTGTTGATGAGTTAAATAACAATGTGGAAGCTGTTTCCCAGACCTCCTCT
                         70        80        90       100       110       120

                IleArgAsnSerValAspGluLeuAsnAsnAsnValGluAlaValSerGlnThrSerSer


                ValThrPheGlnTyrLeuThrLeuLeuLysAspMetTrpLysLysLysGlnAlaArgVal

RAT   BETA      GTCACCTTTCAGTACCTGACTCTGCTGAAAGACATGTGGAAAAAGAAGCAGGCACGAGTT
                        130       140       150       160       170       180
                    ||||||||||| ||  | ||||||||||| ||||| |||||| ||||   | |||
HUMAN BETA      TCTTCCTTTCAGTACATGTATTTGCTGAAAGACCTGTGGCAAAAGAGGCAGAAGCAAGTA
                        130       140       150       160       170       180

                SerSerPheGlnTyrMetTyrLeuLeuLysAspLeuTrpGlnLysArgGlnLysGlnVal


                Lys<<<<<<<<<<<<<<<<<<<<<<<<<<<<<<<<<<<<<<<<<<<<<<<<<<<<<<<<<

RAT   BETA      AAA<<<<<<<<<<<<<<<<<<<<<<<<<<<<<<<<<<<<<<<<<<<<<<<<<<<<<<<<<

                |||
HUMAN BETA      AAA<<<<<<<<<<<<<<<<<<<<<<<<<<<<<<<<<<<<<<<<<<<<<<<<<<<<<<<<<

                Lys<<<<<<<<<<<<<<<<<<<<<<<<<<<<<<<<<<<<<<<<<<<<<<<<<<<<<<<<<
```

Figure 3. The nucleotide and amino acid sequences coding for the intermediate disulfide knot and the first half of the "coiled coil" region of the rat and human beta fibrinogen chain (ref. 5). The sequences have been aligned to show maximal homologies.

```
               GlySerTyrCysProThrThrCysGlyIleSerAspPheLeuAsnSerTyrGlnThrAsp

RAT    GAMMA   GGTAGTTACTGCCCAACCACTTGTGGCATCTCAGACTTCCTGAATTCTTACCAAACCGAC
               1       10        20        30        40        50        60
               |||||||| || ||||| || ||||||||  |||| ||||||  | |||| |||||| | 
HUMAN  GAMMA   GGTAGTTATTGTCCAACTACCTGTGGCATTGCAGATTTCCTGTCTACTTATCAAACCAAA
               1       10        20        30        40        50        60

               GlySerTyrCysProThrThrCysGlyIleAlaAspPheLeuSerThrTyrGlnThrLys


               ValAspThrAspLeuGlnThrLeuGluAsnIleLeuGlnArgAlaGluAsnArgThrThr

RAT    GAMMA   GTGGACACTGACCTCCAGACTCTGGAAAACATCTTACAACGAGCTGAAAACAGAACCACA
                      · 70        80        90       100       110       120
               || ||||  || || ||| || ||||| |||||||||| | || |||||||| |||  ||
HUMAN  GAMMA   GTAGACAAGGATCTACAGTCTTTGGAAGACATCTTACATCAAGTTGAAAACAAAACATCA
                        70        80        90       100       110       120

               ValAspLysAspLeuGlnSerLeuGluAspIleLeuHisGlnValGluAsnLysThrSer


                  GluAlaLysGluLeuIleLysAlaIleGlnValTyrTyrAsnProAspGlnProPro

RAT    GAMMA   ///GAAGCCAAGGAACTGATTAAAGCCATCCAGGTTTACTACAACCCGGACCAACCCCCA
                      130       140       150       160       170       180
                  |||| |||  | ||||| |||||| |||||  |    || || || ||  || |  ||
HUMAN  GAMMA   ///GAAGTCAAACAGCTGATAAAAGCAATCCAACTCACTTATAATCCTGATGAATCATCA
                      130       140       150       160       170       180

                  GluValLysGlnLeuIleLysAlaIleGlnLeuThrTyrAsnProAspGluSerSer


               LysAla<<<<<<<<<<<<<<<<<<<<<<<<<<<<<<<<<<<<<<<<<<<<<<<<<<<<<<

RAT    GAMMA   AAAGCC<<<<<<<<<<<<<<<<<<<<<<<<<<<<<<<<<<<<<<<<<<<<<<<<<<<<<<

               ||| |
HUMAN  GAMMA   AAACCA<<<<<<<<<<<<<<<<<<<<<<<<<<<<<<<<<<<<<<<<<<<<<<<<<<<<<<

               LysPro<<<<<<<<<<<<<<<<<<<<<<<<<<<<<<<<<<<<<<<<<<<<<<<<<<<<<<
```

Figure 4. The nucleotide and amino acid sequences coding for the intermediate disulfide knot and the first half of the "coiled coil" region of the rat and human gamma fibrinogen chain (ref. 6). The sequences have been aligned to show maximal homologies.

amino acid whose counterpart is not seen in alpha or gamma. This is the only region of the fibrinogen genes where the same intron-exon arrangement occurs in all three chains (unpublished data). That is, other exons do not have homologous coding information to their counterparts. For instance, while the exon coding for the third disulfide knot in beta and gamma also contains the coding information for the second half of the coiled coil region, that is not the case for the alpha chain. Thus, this exon coding for the intermediate disulfide knot is the best conserved based on intron-exon structure. Yet one can readily see that there is little nucleic acid homology. There is 31% homology between alpha and beta, 31% between beta and gamma and 37% between gamma and alpha. This is only slightly greater than what would be expected by chance. However, there is 36% homology in the first position of codons, 46% in the second position but only 25% in the third or neutral position.

Figures 2, 3 and 4 show the inferred amino acid sequence of the intermediate disulfide knot in the rat compared with its human counterpart. An analysis of these differences are shown in Table 1.

TABLE 1

Summary of the differences between human and rat amino acid sequences within the exons coding for the first half of the coiled coil region. The apparent changes from rat to human are shown.

	Non-polar to non-polar	Polar to polar	Non-polar to polar	Polar to non-polar
alpha:	3 of 18	5 of 43	1	1
beta:	4 of 20	13 of 41	3	4
gamma:	3 of 20	14 of 41	2	1
total:	10 of 58	31 of 125	6	6

Discussion

The comparison of the rat and human fibrinogen sequence for that portion that includes the intermediate disulfide knot demonstrates the importance of certain portions of the primary sequence. One of the most noticeable features is the preservation of the Cys-Pro-Y-Z-Cys structure of the intermediate disulfide knot. The selective pressures to maintain the relative positions of the cysteine pairs and the proline is obvious. The important function of the cysteines to form interchain bridges is a simple explanation for their preservation. However, the function of the prolines is not as easy to appreciate. Is it possible that they function in a helix-altering capacity in order to assure that the cysteines have the opportunity to form the appropriate interchain bridges? Without the prolines between the cysteine pairs, the structure in the region of those cysteines might remain internalized as do the adjacent non-polar residues within the coiled coil and thus fail to establish interchain bridges.

Another feature that seems to be well preserved is the arrangement of polar and non-polar amino acid residues. It is remarkable that there are 52 amino acid differences between the human and rat fibrinogen molecules in this region and yet there is a striking conservation of the polar and non-polar amino acid composition. Relative to human fibrinogen the periodicity of non-polar and polar residues is well maintained in the rat fibrinogen, suggesting to us that this region of the fibrinogen gene is under little constraint to maintain any particular nucleic acid composition. Apparently the only constraint on the nucleic acids is that required to maintain the locations of the polar and non-polar residues. This is highlighted by the lack of conservation between the three genes in the third nucleotide position of homologous genes versus the high degree of conservation (47% between all three genes in the rat) in the second codon position. The most likely explanation for this is the relative importance of the second nucleotide position in determining the polar versus non-polar nature of an amino acid residue. A thymidine (uridine in the coding mRNA) in the second codon position always codes for a non-polar amino acid. In the genetic code there are forty codons which use adenine, cytosine or guanine in the second codon position for polar amino acids yet only five codons use these bases in the second position to code for non-polar amino acids. Thus, one would predict that if functional constraints are limiting the evolution of the fibrinogen molecule that there would be certain biases in range of differences seen between different species. This appears to be the case. For instance, there are relatively few non-polar to non-polar amino acids substitutions shown here between the rat and human genes when compared to the number of polar to polar substitutions. Being teleological, one can see that fewer nucleic acid substitutions are tolerable to codons for non-polar residues than for polar ones. That appears to be because uridine is so frequently used in the second codon position for non-polar amino acids.

One question that has raised so much interest among those studying fibrinogen is the ancestral relationship of the fibrinogen chains to each other. It has been proposed that the fibrinogen genes may have arisen from a common ancestor 1000 to 1500 million years ago (1, 14). The natural rate of nucleic acid changes in the neutral positions of genes is too high to permit meaningful observations from the nucleic acid sequence data. For instance, the homology in the neutral codon position between the three rat chains shown in Figure 1 is only 25%, i.e., that which one could expect to occur by chance alone. However, this must not be interpreted as evidence that the three genes are unrelated. Assuming that they did arise from a common ancestral precursor, there has been adequate time for natural mutations to decrease their homology to background levels. From the data we presented here, it would appear that similarities between the three genes may be either the result of a convergent form of evolution or from the sharing of common ancestors. The functional need for cysteine bridges and alpha helical structures may indeed be responsible for the similarities in alpha, beta and gamma fibrinogen genes.

References

1. Henschen, A., Lottspeich, F., Kehl, Southan, C.: Ann. N.Y. Acad. Sci. 408, 28-43 (1983).
2. Doolittle, R.F., Watt, K.W.K., Cottrell, B.A., Strong, D.D., Riley, M.: Nature 280, 464-468 (1979).
3. Kant, J.A., Lord, S.T., Crabtree, G.R.: Proc. Natl. Acad. Sci. USA 80, 3953-3957 (1983).
4. Rixon, M.W., Chan, W-Y., Davie, E.W., Chung, D.W.: Biochemistry 22, 3237-3244 (1983).
5. Chung, D.W., Que, B.G., Rixon, M.W., Mace, M., Davie, E.W.: Biochemistry 22, 3244-3250 (1983).
6. Chung, D.W., Chan, W-Y., Davie, E.W.: Biochemistry 22, 3250-3256 (1983).
7. Crabtree, G.R., Kant, J.A.: Cell 31, 159-166 (1982).
8. Kant, J.A., Crabtree, G.R.: J. Biol. Chem. 258, 4666-4667 (1983).
9. Sanger, F., Nicklen, S., Coulson, A.R.: Proc. Natl. Acad. Sci. USA 74, 5463-5467 (1977).
10. Queen, C.L., Korn, L.J.: Methods Enzymol. 65, 595-609 (1980).
11. Doolittle, R.F., Goldbaum, D.M., Doolittle, L.R.: J. Mol. Biol. 120, 311-325 (1978).
12. McLachlan, A.D., Stewart, M., Smillie, L.B.: J. Mol. Biol. 98, 281-291 (1975).

13. Crabtree, G.R., Kant, J.A., Fornance, A.J., Rauch, C.A., Fowlkes, D.M.: Ann. N.Y. Acad. Sci. 408, 457-466 (1983).

14. Doolittle, R.F.: Ann. N.Y. Acad. Sci. 408, 13-26 (1983).

ABSENCE OF GROSS DEFECT OF FIBRINOGEN GENES IN ONE PATIENT WITH CONGENITAL AFIBRINOGENEMIA

Gilles Courtois, Georges Uzan, Zahra Assouline, Gérard Marguerie.
INSERM U. 143 - Hôpital Bicêtre - 94270 le Kremlin Bicêtre - France.
and

Axel Kahn
INSERM U. 129 - CHU Cochin - 75674 Paris Cedex 14 - France.

INTRODUCTION

Fibrinogen is involved in qualitative and quantitative congenital disorders associated with bleeding disorders. Afibrinogenemia and hypofibrinogenemia usually describe total or partial quantitative deficiencies whereas dysfibrinogenemia usually represents patients with normal plasma level of functionaly deficient molecules. The hereditary pattern for dysfribrinogenemia and hypofibrinogenemia seems to be autosomal dominant whereas autosomal intermediate and recessive patterns have been reported for afibrinogenemia. The mechanisms involved in these inherited disorders are unknown. Various amino acid substitutions have been reported in dysfibrinogenemia and this may result most likely from point mutations. Hypofibrinogenemia has been associated with an abnormal intracisternal storage and may result from deficient hepatic secretion. Afibrinogenemia is a rare but not exceptional disease in which the plasma fibrinogen level is often undetectable with radioimmunometric assays. Since the first observation of Rabe and Salomon in 1920, more than 100 cases have been reported but the molecular basis for the lack of synthesis of fibrinogen in these patients is unknown. In the recent past years, considerable progress have been made on the molecular genetic of fibrinogen. The three chains of the molecule are synthesized under the direction of three different mRNAs, transcripted from three coodinated genes (1-4) linked into a portion of chromosome 4 with the following sequence 5'-γ - Aα - Bβ -3' (5-6).

Fibrinogen - Structural Variants and Interactions
© 1985 Walter de Gruyter & Co., Berlin · New York - Printed in Germany

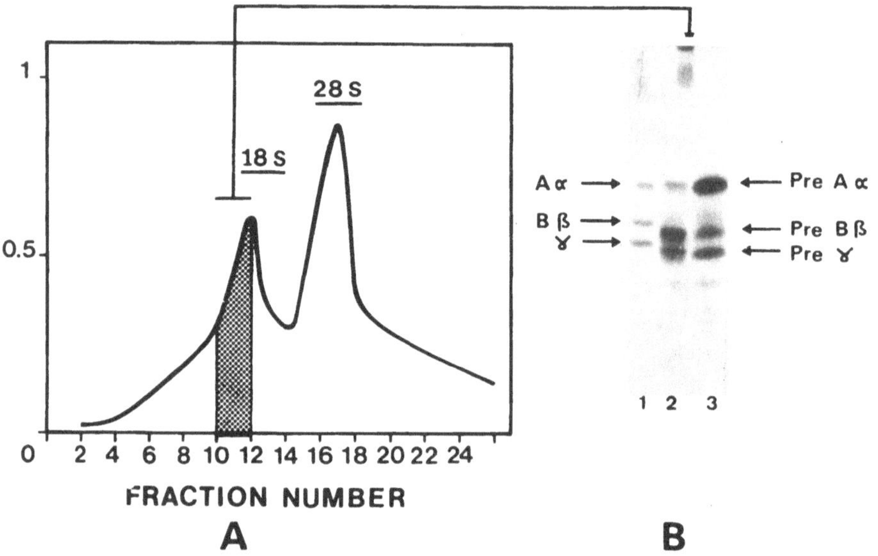

Fig. 1 Sucrose density gradient fractionation of human liver poly A^+ mRNA

Panel A : Sedimentation profile

Panel B : Fractions 10-12 were translated in a cell free system in the presence of ^{35}S-methionine. The synthesized material was immunopurified and analyzed on 5 to 20% polyacrylamide gels in the presence of 2% SDS and 2 % 2-mercaptoéthanol. Lane 1 :Standard (^{14}C)-labeled human fibrinogen. Lane 2 : Fibrinogen encoded by fractions 10-12. Lane 3 : Fibrinogen encoded by a total mRNA preparation.

With the use of recombinant DNA it is now feasible to analyse the structure and the expression of the fibrinogen genes in patients with afibrinogenemia and to examine wether this is due to a major deletion or insertion, an abnormal RNA processing and/or metabolism or a translational defect.

In our group we have isolated human Aα , Bβ and γ cDNAs and we have recently initiated a study of the DNA from different patients with afibrinogenemia. This paper describes a preliminary restriction endonuclease analysis of the DNA of one of these patients.

cDNA CLONES FOR Aα,Bβ AND γ CHAINS OF HUMAN FIBRINOGEN

Poly A^+ mRNA were isolated from a total human liver RNA preparation by chromatography on oligo dT-cellulose and fractionated by sucrose gradient sedimentation in the presence of 20. mM methylmercury hydroxide (Fig. 1A). Different fractions of this gradient were translated in a cell free reticulocyte lysate system and the neosynthesized material was analyzed on polyacrylamide gels after immunoaffinity purification (4-7). Those fractions containing RNAs encoding for the preAα , preBβ and preγ chains were selected (Fig. 1B) and reverse transcripted to make single strand ^{32}P-cDNA hybridization probes enriched in fibrinogen sequences. These probes were used to screen an adult human liver cDNA library (8-10). Different negative probes were used for screening. The high molecular weight fractions from the sucrose gradient (18-20 S) and a preparation of human muscle mRNA were also reverse transcripted to make negative ^{32}P-cDNA probes. In addition a rat albumin cDNA was used for evaluation of non specific hybridization. Twenty clones out of 7,000 hybridized with ss^{32}P-cDNA prepared with enriched preparation of fibrinogen mRNA. Of these isolates, three were recognized to have inserts with coding sequences for the Aα, Bβ and γ chains (Fig. 2). The specificity of these inserts were based on the capacity of the probes to select mRNAs of the correct size which translated to yield the predicted preforms of the Aα, Bβ and γ chains. Restriction mapping of these inserts and map fitting with known cDNA for human fibrinogen chains allowed the identification of the coding sequences (Table I).

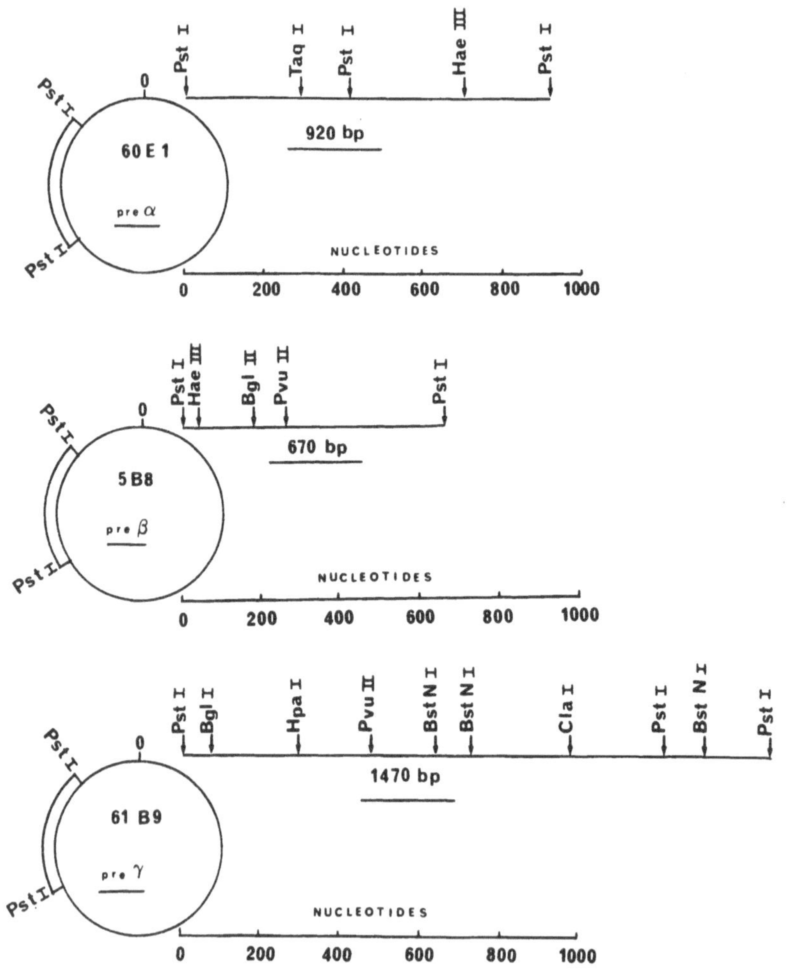

Fig. 2 Restriction maps of cDNA for Aα, Bβ and γ chains of human fibrinogen.

TABLE I

Coding sequences inserted in human Aα, Bβ and γ cDNA clones

Hybridization Probes	Chains	Sequence
60E1	pre Aα	GLY 404 to PRO 625
5B9	pre Bβ	ALA 277 to GLU 461
61B9	pre γ	MET -26 to VAL 411

ANALYSIS OF NORMAL AND AFIBRINOGENEMIC GENES

The recombinants were used as probes to analyse normal genomic DNA digested with various restriction enzymes. Single band patterns were observed, representing unique sequence consistent with the existence of a single copy of fibrinogen genes (Fig. 3). Analysis of the DNA from several normal individuals did not reveal restriction fragment polymorphism.

The DNA of one patient with congenital afibrinogenemia was analyzed with the same procedure. Absence of fibrinogen in the plasma of this patient has been documented for many years. Using a radioimmunoassay, the plasma level was less than 1.5 ug/ml. The DNA was isolated from peripheral white blood cells and digested with endonucleases. The pattern of digestion products obtained were indistiguishable from those obtained for normal DNA. These results indicated that fibrinogen genes, for Aα, Bβ and γ chains were present in the DNA of this patient who do not produce fibrinogen, and did not exhibit a major deletion. Further studies are now required to explain the lack of expression of the molecule in this patient. In the futur, genomic probes will be used to further characterize the gene defects present in afibrinogenemia.

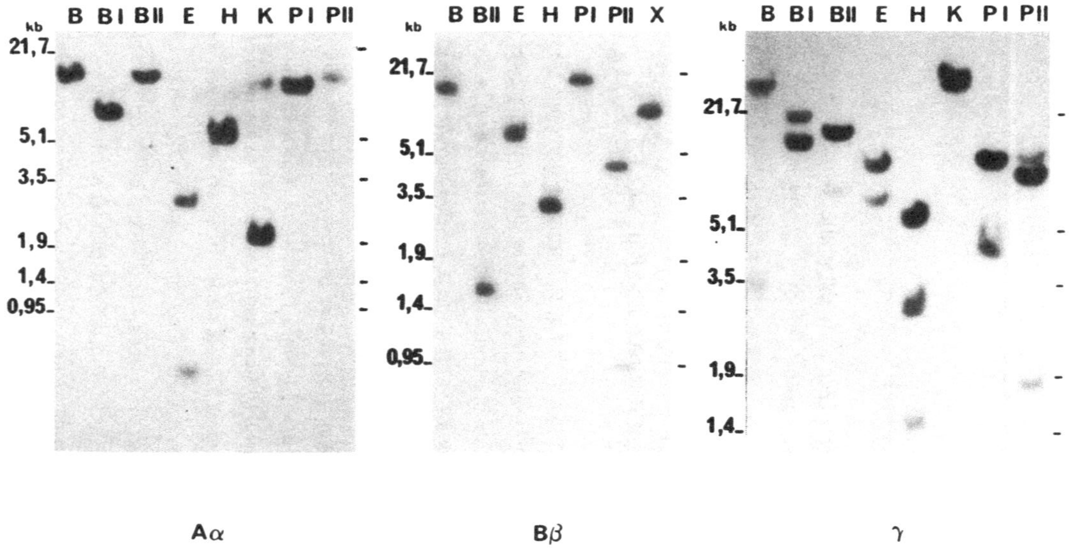

Fig.3 : Southern blot hybridization of human DNA digested with Bam HI (B), Bgl I (BI), Bgl II (BII), Eco RI (E), Hind III (H), Kpn I (K), Pst I (PI), Pvu II (PII), Xba I (X). The three filters were hybridized with nick translated purified inserts from Aα, Bβ and γ cDNAs.

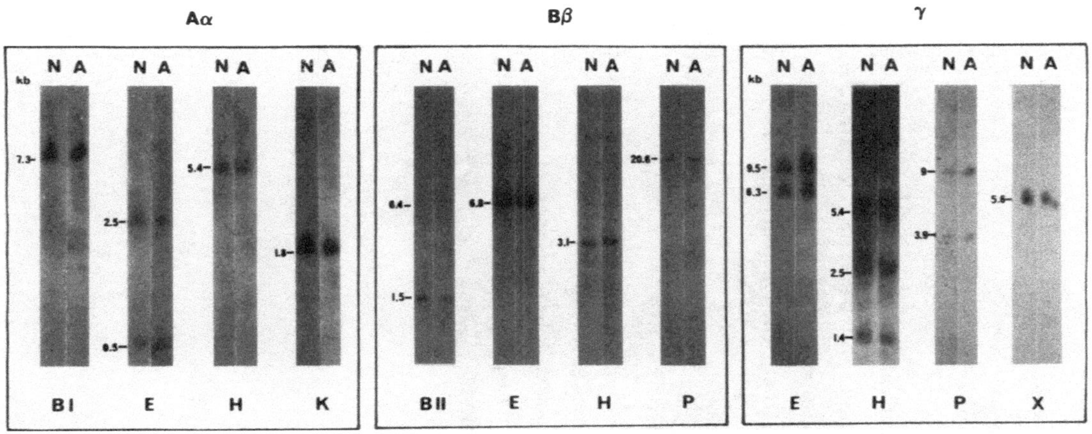

Fig.4 : Southern Blot analysis of normal (N) and afibrinogenemic (A) DNA digested with various restriction enzymes

References

1. Nickerson, J.M. and Fuller, G., Proc. Nat. Acad. Sci. USA, 78, 303-307, (1981).

2. Crabtree, G.R. and Kant, A. J., J. Biol. Chem. 256, 9718-9723, (1981).

3. Crabtree, G.R. and Kant, A. J., J. Biol. Chem, 257, 7277-7279, (1982).

4. Uzan, G., Besmond, C., Kahn A. and Marguerie, G., Biochemistry Int. 4, 271-274, (1982).

5. Kant, A.J. and Crabtree, G.R., J. Biol. Chem. 258, 4666-4667, (1983).

6. Henry, I., Uzan, G., Weil, D., Nicolas, H. Kaplan, J.C., Marguerie, G., Kahn, A., and Junien, C.: Am. J. Hum. Gen. 36, 760-768 (1984).

7. Kahn, A., Cottreau, D., Daegelen, D., and Dreyfus, J.C., Eur. J. Biochem. 116,7-12, (1981).

8. Sala-Trépat, J.M., Savage, M. and Bonner, J., Biochem. Biophys. Act. 519, 173-193, (1978).

9. Sala-Trépat, J.M., Sargent, T.D., Sell, S. and Bonner, J., Proc. Nat. Acad. Sci. USA, 76, 695-699, (1981)

10. Uzan, G., Courtois, G., Besmond, C., Frain, M., Sala-Trepat, J., Kahn, A. and Marguerie, G.,: Biochem. Biophys. Res. Comm. 120, 376-383 (1984).

III. FIBRINOGEN-FIBRIN CONVERSION

FIBRINOGEN TO FIBRIN - AN OVERVIEW

Birger Blombäck
Karolinska Institutet, Stockholm, Sweden and The New York Blood Center, New York, N.Y.

From having in the ancient medicine been an obscure constituent of phlegm and crusta phlogistica, fibrinogen emerged in the latter part of the 19th century as the first plasma protein to be isolated in relatively pure form. Denis de Commercy in France and Olof Hammarsten in Sweden are credited with this accomplishment (1-4). After the pioners's work it would take another 60 years before the conceptual contours of this protein started to rize from enigma. This review will cover the trail of research leading to our present appreciation of fibrin formation.

The early hydrodynamic and spectroscopic studies performed on fibrinogen and fibrin formation provided information on the basis of which size and shape of the fibrinogen molecule could be deduced and also allowed for proposals of how individual fibrinogen units were aligned in polymerizing fibrinogen (5-16). Work employing electron microscopy, X-ray crystallography and light-scattering techniques (17-23) essentially confirmed the concept that fibrinogen is an asymmetric rod-like particle. The current view is that the molecule in its dehydrated form, is 45 nm long and is composed of three nodular domains, one in the center and one at each end of the molecule. The hydrated molecule has a "sausage"-like appearance having the same length as the dehydrated one but does not clearly show the nodular domains. Its width in its thickest portion is about 9-10 nm.

Chemical studies of the fibrinogen molecule showed that it was composed of three pairs of polypeptide chains (24). The chains were held together by disulfide bridges and could be isolated after reduction or sulfitolysis (25-26). The next crucial information on the structure of fibrinogen came from structural analysis of the so called NH_2-terminal disulfide knot, N-DSK (27-29). The identification of symmetrical disulfide bonds in this structure unequivocally demonstrated that the fibrinogen molecule, like the immunoglobulins, consisted of two identical halves, joined along a twofold symmetry axis. By now the complete primary structure of the fibrinogen molecule has been successfully delineated (cf. 30).

These discoveries set the stage for conceptualization at the molecular level of fibrin formation as a process in which

Fibrinogen - Structural Variants and Interactions
© 1985 Walter de Gruyter & Co., Berlin · New York - Printed in Germany

activated divalent molecules interact in a half-staggered fashion to form the polymers earlier proposed by Ferry. It was known that fibrin formation required preliminary release of fibrinopeptides from the NH_2-terminal end of the molecule and that this release was closely associated with polymerization and gelation (31-35). Thus, fibrinogen needed activation in order for subsequent reactions to take place. It was also recognized at this time that polymers which were formed on release of fibrinopeptide A (FPA) were different from those formed when fibrinopeptide B (FPB) in addition to FPA was released (36). The first point mutation (Aα 19 Arg ⟶ Ser) in an abnormal fibrinogen was also known and this mutation in N-DSK was associated with impaired gelation (37). These findings led to the notion that following activation of fibrinogen in the presence of thrombin, two sets of polymerization domains are operative in polymer formation (28). One set becomes functional on release of FPA by unfolding of a polymerization domain (domain A) in N-DSK. This domain subsequently interacts with a complementary domain (domain a) in a neighbouring fibrinogen molecule. On release of FPB a similar set of interacting domains, i.e. B and b, were assumed to become operative. Evidence for these notions were obtained subsequent to the demonstration that fibrinogen coupled to Sepharose on activation with thrombin acquired affinity for fibrinogen (38). It was now discovered that N-DSK, or some of its counterparts obtained by digestion of fibrinogen with plasmin, on activation with thrombin acquired affinity for conjugated fibrinogen. Furthermore, fragment D, obtained by plasmic digestion of fibrinogen, was adsorbed to activated fibrinogen conjugates but not to unactivated ones (29, 39-41). Similarly, fragment D in solution was adsorbed to conjugated N-DSK after the latter had been activated by thrombin and fragment D (from cross linked fibrin) had affinity for activated N-DSK (or counterparts) in solution. Although the initial studies (39) only provided evidence for the existence of the A:a set of sites, support for the B:b set of polymerization sites was soon to be obtained (41-43). It was also found that cross linking of fibrin in a dramatic way changed the affinity of cross linked fragment D (fragment D dimer) for activated N-DSK or its congeners (41). The interaction of the two sets of complementary polymerization sites, which I have described, provides for formation of linear or possibly branched fibrin polymers.

Whatever polymers are formed, they eventually crosslink in the presence of factor XIIIa (cf. 44). They also participate in interpolymer cross linking, which results in formation of the three dimensional net work structure, the fibrin gel. The mechanical and spectroscopic properties of fibrin gels have been studied by several investigators. Ferry and Morrison (45) described properties of gels ranging between the extremes, transparent and opaque. In the transparent gels, the polymer strands are thin and the meshwork fine, whereas in the opaque gels thick polymer strands are encompassing wide liquid spaces. In general one can say that the turbidity of a fibrin gel reflects on the strand thickness in the gel matrix and that the ionic strength and pH are important factors influencing the turbidity (46).

It was believed that fibrin gels have a rather disordered structure since they were thought to be formed as the result of random association of polymers of varying length (cf. 45). This view is no longer tenable since recent studies have shown that fibrin gels, formed in the absence of shear stress, display a considerable regularity. The first indication of this came from liquid and particle permeation studies of fibrin gels. These studies showed that the pores in the fibrin matrix not only are rather uniform but that the size of these pores are, under otherwise identical conditions, controlled by the enzymes used for gelation. Thus, the pore size is inversely related to the enzyme concentration (47-49). This fact suggests that the activation and polymerization steps preceding gelformation are important determinants for organization of the fibrin matrix in the gel. Activation of fibrinogen is effectuated, as I have already mentioned, by release of FPA and FPB. In normal fibrinogen the initial rate of the release of FPA is much faster than that of FPB (33,50). The subsequent polymerization event is a fast reaction and stochiometrically related to release of FPA (33,35,49). This means that any set of polymerization sites exposed on release of FPB must be operative at the polymer level. Interaction between such sites may lead to bond formation between the polymers formed subsequent to release of FPA.

How is it possible that ordered net work structures are formed as a consequence of interaction between polymers? We assume that activation starts at initiation points in the solution phase (49). The number of initiation points is dependent on the enzyme concentration. Activated fibrinogen molecules at the initiation points interact and thereby form polymers, which grow in different directions in space from the initiation points. When they get in touch with polymers growing from other initiation points, an infinite net work structure is formed at the time of gelation. The size of the liquid spaces in the net work will be smaller at high enzyme concentrations than at low. We further assume that the initial net work (established at the gel point) creates a frame-work and that all subsequent activation and polymerization reactions occur in proximity to the strands present in this framework. Obviously, the strands will grow in thickness as the network matures with time. It is also obvious that the strands formed at high enzyme concentration must be thinner than those formed at low enzyme concentration. Changes in substrate concentration below and around K_m will necessarily affect the activation step and thereby the number of initiation points. Therefore, on increasing the substrate concentration we expect that in the beginning the pore size of the gels will decrease at the expence of strand width. At substrate concentrations much above K_m, the number of initiation points would only be slightly affected, but the size of the pores would be somewhat decreased as a consequence of the increasing strand width.

FPA is released at a higher initial rate than FPB. Therefore, in the formation of the initial polymers, A:a interactions play the dominant role. The release of FPB will predominantly occur on the polymers being formed. We envisage that the release of FPB also occurs at the initiation points and provides for interaction between the growing polymers. The latter interactions may result in a change in spatial direction of the growing polymers.

This model for fibrin gel formation takes into account most of what is known about this process from other studies. It also takes into account the well-known fact that only partial release of FPB occurs during gelation. In the model release of FPB only serves the purpose of cross linking, at relatively few points, polymer units being formed by A:a interaction.

Most attempts to visualize fibrin structure by electron microscopy, yield pictures showing strand bundles crossing each other in an irregular fashion. These results are not surprising in view of the fact that the preparation of specimens for such studies involves removal of water. As a result the gel matrix will collapse and the inherent order be disguised. Light microscopy (49) of intact fibrin gels gives on the other hand an impression of a rather ordered lattice structure composed of straight strands of uniform width (Fig. 1a and b). The regularity is even more impressive when observed in scanning electron microscopy (SEM) (49). In such studies the gels are quickly frozen in liquid nitrogen and subsequently lyophilized. During this process rifts appear in the gel matrix. When the surfaces of these rifts are studied by SEM, ordered strand structures like those shown in Fig. 2 are observed. The SEM micrographs of fibrin gels suggest that the lattice structures originating from initiation points have different orientations in different parts or domains of the volume element (compare upper and lower part of Fig. 2b). Within each domain the axial dimension of the lattice is determined by the enzyme-substrate parameters I have mentioned above, i.e. the pore size is similar in all domains, but the orientation of pores is different. As the lattice grows it will come in close contact with similar lattices originating from other domains. An interlattice zone or a "grain boundary" is laid down in the process. The binding forces of such "grain boundaries", like in steel, are expected to be weaker than the intra-lattice binding forces. Consequently, the grain boundaries constitute presumptive crack lines in the gel matrix. The rifts observed during freeze-drying most likely occur along these grain boundaries (e.g. crack line in Fig. 2b). The ordering of the lattices is not a fine crystalline ordering and, therefore, oblique cuts through such lattices will not clearly show the

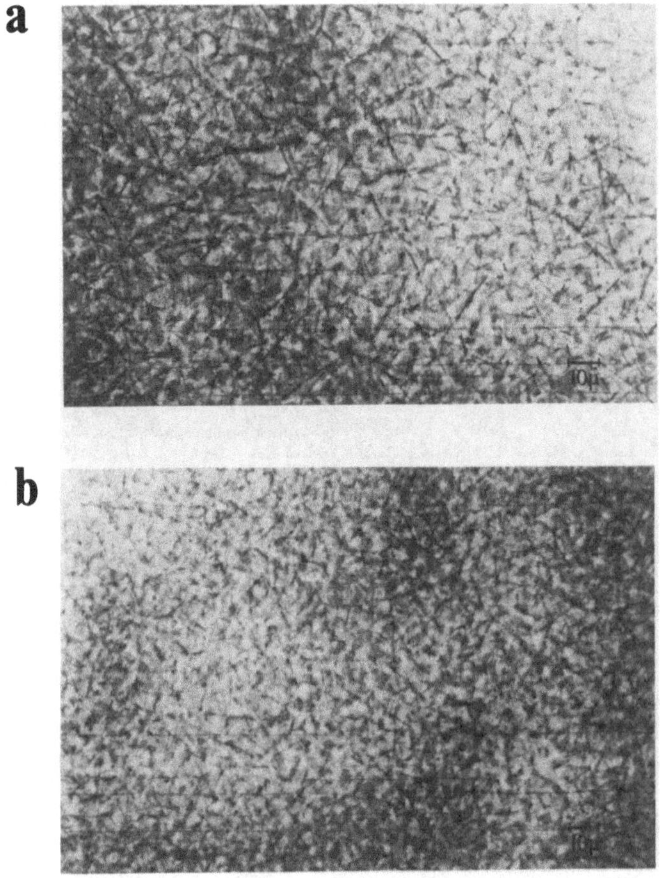

Fig. 1

<u>a and b: Light-microscopy</u>. Gel formation (47) in 0.04 M Tris-imidazole buffer, pH 7.4, containing 20 mM $CaCl_2$, 2.24 g/l of fibrinogen and Trasylol (Bayer AG) 5KIE/ml. Thrombin concentrations in panel a 0.01 NIH units/ml and in panel b 0.15 NIH units/ml. Ionic strength: 0.21. Temperature: 22°C. The gels were formed in microchambers constructed on microscopic slides by glueing two plastic bars across the slide and subsequently attaching a cover glass on the top of the bars. The chambers were filled by capillary force and placed in a moist chamber until gelation was complete (about 2 hrs). Prior to light microscopy, the open sides of the chamber were closed with high vacuum grease. Clotting time in panels a and b was 350 and 62 s, respectively. (Adopted from Blombäck et al. Ref. 49.)

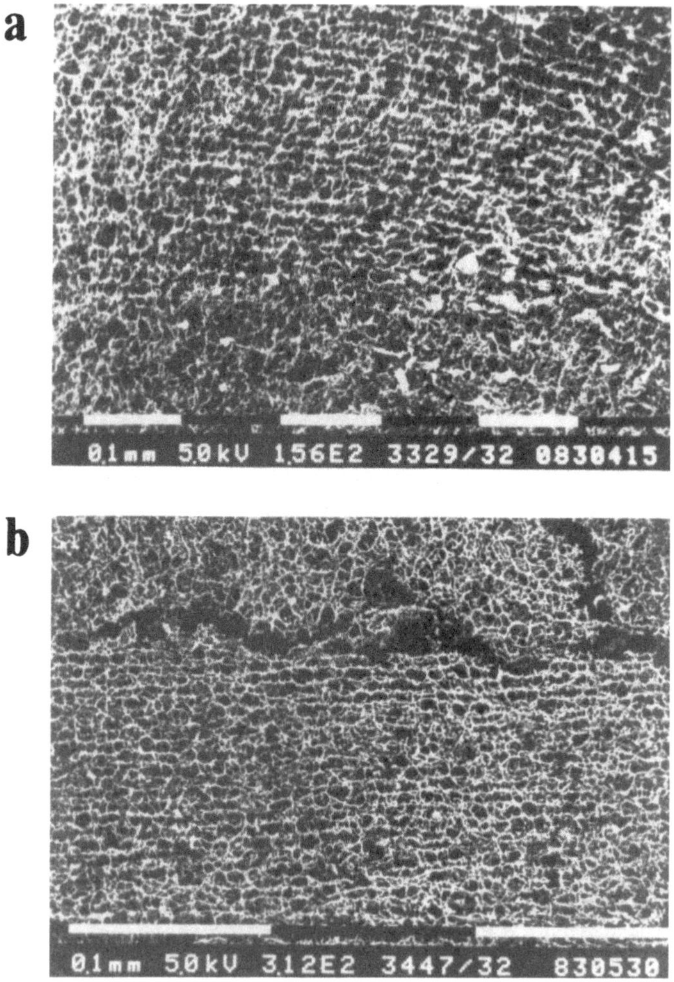

Fig. 2

a and b: Scanning electron microscopy (SEM). a: Gel formation (47) in Tris-imidazole buffer, pH 7.4, 20 mM $CaCl_2$, 2.15 g/l fibrinogen and 5 g/l of Dextran (20 Kd weight average). Thrombin: 0.037 NIH units/ml. Clotting time: 108 s. Ionic strength: 0.21. Temperature: 23°C. Gels fixed with glutaraldehyde (48). Permeability coefficient (Ks): 5.4×10^{-9} cm^2 (47). The gels were freeze-dried (at about 0.001 mm Hg) and rift surfaces of gold sputtered specimens observed by SEM. b: Rifts between domains supposedly represent "grain boundaries". The conditions for gel formation were the same as in a, except that 15 g/l of Dextran 250 Kd was used. Clotting time: 36 s. Ks: 7.3×10^{-9} cm^2. White bar indicates 0.1 mm. (Adopted from Blombäck et al. Ref. 49).

underlying ordering (Fig. 2b, upper p?t). Only right-angle, or near-right angle cuts will reveal the ordered net-work (Fig. 2a and 2b, lower part).

What role does the fibrinogen-fibrin system play in physiology? There is evidence that it plays a role in haemostatic processes. Thus, it is known that lack of fibrinogen is associated with haemorrhagic tendencies. This is especially the case when other supportive haemostatic functions, like platelet aggregation and/or adhesion, are also perturbed. The current view is that platelet aggregates supported by fibrin gels formed at sites of lesions in the vasculature are primarily responsible for physiological haemostasis.

There is no doubt that the fibrinogen-fibrin system plays a role in the formation of pathological thrombin in thromboembolic diseases afflicting the venous and very likely also the arterial system. A thrombus or an embolus is a fibrin gel into which the cellullar elements of the blood are trapped. The quantity and quality of the gel is determining the sequel of a thrombotic process. A gel with small pores does not synerize easily and is therefore space filling. It is also brittle, because of the fine strands, and may therefore easily give rize to embolization. A gel formed at low enzyme concentration is expected to have wider pores, thicker strands and to synerize easily. It is stronger and more plastic - i.e. deformable. It is therefore less space filling and may not be prone to embolization.

A physiological role of the fibrinogen-thrombin system for compartmentalization of tissues is also possible, as has been suggested by Copley (51). An ordered fibrin structure may in this sense act as an effective screen or scaffold regulating permeation processes of various kind.

Acknowledgements

This work was supported by grants from the National Institutes of Health, Bethesda, Maryland (HL 27279-04) and the Swedish Medical Research Council (B 85-13X-02475-18 B). I thank Niklas Bark and Mrs Birgitta Strimme for excellent help in the preparation of this manuscript.

References

1. Denis, P.-S.: Mèmoire sur le sang, J.B. Bailliere et fils, Paris (1859).

2. Hammarsten, O.: Nova acta reg. soc. scient. Ups. Ser. III, 10, 1-130 (1876).

3. Hammarsten, O.: Pflueg. Arch. Ges. Physiol. 19, 563-622 (1879).

4. Hammarsten, O.: Pflueg. Arch. Ges. Physol. 22, 431 (1880).

5. Mihalyi, E.: Acta Chem. Chem. Scand. 4, 317-333 (1950).

6. Bailey, K., Astbury, W.T. and Rudall, K.M.: Nature 151, 716-717 (1943).

7. Oncley, J.L., Scatchard, G. and Brown, A.: J. Phys. & Colloid. Chem. 51, 184-198 (1947).

8. Caspary, E.A. and Kekwick, R.A.: Biochem.J. 67, 41-48 (1957).

9. Katz, S., Gutfreund, K., Shulman, S. and Ferry, J.D.: J. Amer. chem. Soc. 74, 5706-5715 (1952).

10. Edsall, J.T., Foster, J.F. and Scheinberg, H.: J.Amer. chem. Soc. 69, 2731-2738 (1947).

11. Hocking, C.S., Laskowski Jr., M. and Scheraga, H.A.: J. Amer. chem. Soc. 74, 775-778 (1952).

12. Steiner, R.F. and Laki, K.: Arch. Biochem. 34, 24-37 (1951).

13. Ferry, J.D., Shulman, S., Gutfreund, K. and Katz, S.: J. Amer. chem. Soc. 74, 5709-5715 (1952).

14. Backus, J.K., Laskowski Jr., M., Scheraga, H.A. and Nims, L.F.: Arch. Biochem. 41, 354-366 (1952).

15. Casassa, E.F.: J. Amer. Chem. Soc. 78, 3980-3985 (1956).

16. Ferry, J.D.: Proc. Natl. Acad. Sci 38, 566-569 (1952).

17. Hall, C.E. and Slayter, H.S.: J. Biophys. Biochem. Cytol. 5, 11-15 (1959).

18. Krakow, W., Endres, G.F., Siegel, B.M. and Scheraga, H.A.: J. Mol. Biol. 71, 95-103 (1972).

19. Fowler, W.E. and Erickson, H.P.: J. Mol. Biol. 134, 241-249 (1979).

20. Bachman, L., Schmitt-Fumian, W.W.,Hammel, R. and Lederer, K.: Die Makromomolekulare Chemie 176, 2603-2618 (1975).

21. Stryer, L, Cohen, C. and Langridge, R: Nature 197, 793-794 (1963).

22. Cohen, C., Weisel, J.W., Phillips Jr., G.N., Stauffacher, C.V., Fillers, J.P. and Daub, E.: Ann. New York Acad. Sci. 408, 194-213 (1983).

23. Hantgan, R., McDonald, J. and Hermans, J.: Ann. New York Acad. Sci. 408, 344-366 (1983).

24. Blombäck, B. and Yamashina, I.: Arkiv Kemi 12, 299-319 (1958).

25. Henschen, A.: Arkiv Kemi 22, 1-28 (1963).

26. Henschen, A.: Arkiv Kemi 22, 375-396 (1964).

27. Blombäck, B., Blombäck, M., Henschen, A., Hessel, B., Iwanaga, S. and Woods, K.R.: Nature 218, 130-134 (1968).

28. Blombäck, B. and Blombäck, M.: Ann. New York Acad. Sci. 202, 77-97 (1972).

29. Blombäck, B., Hogg, D.H., Gårdlund, B., Hessel, B. and Kudryk, B.: Thrombosis Research 8, suppl. II, 329-346 (1976).

30. Henschen, A., Lottspeich, F., Kehl, M. and Southan, C.: Ann. New York Acad. Sci. 408, 28-43 (1983).

31. Bailey,K., Bettelheim, F.R., Lorand, L. and Middlebrook, W.R.: Nature 167, 233-234 (1951).

32. Bailey, K. and Bettelheim, F.R.: Brit. Med. Bull. 11, 50-53 (1955).

33. Blombäck, B. and Vestermark, A.: Arkiv Kemi 12, 173-182 (1958).

34. Blombäck, B.: Arkiv Kemi 12, 321-335 (1958).

35. Blombäck, B. and Laurent, T.C.: Arkiv Kemi 12, 137-146 (1958).

36. Laurent, T.C. and Blombäck, B.: Acta Chem. Scand. 12, 1875-1877 (1958).

37. Blombäck, M., Blombäck, B., Mammen, E.F. and Prasad, A.S.: Nature 218, 134-137 (1968).

38. Heene, D.L. and Matthias, F.R.: Thrombosis Research 2, 137-154 (1973).

39. Kudryk, B.J., Collen, D., Woods, K.R. and Blombäck, B.: J. Biol. Chem. 249, 3322-3325 (1974).

40. York, L. and Blombäck, B.: Thrombosis Research 8, 607-618 (1976).

41. Olexa, S.A., Budzynski, A.Z.: Proc. Natl. Acad. Sci. 77, 1374-1378 (1980).

42. Blombäck, B., Hessel, B., Hogg, D.H. and Therkildsen, L.: Nature 257, 501-505 (1978).

43. Shainoff, J.R. and Dardik, B.N.: Science 204, 200-202 (1979).

44. Lorand, L.: Ann. New York Acad. Sci. 408, 226-232 (1983).

45. Ferry, J.D. and Morrison, P.R.: J. Amer. chem. Soc. 69, 388-400 (1947).

46. Carr, Jr., M.E. and Hermans, J.: Macromolecules 11, 46-50 (1978).

47. Blombäck, B. and Okada, M.:Thrombosis Research 25, 51-70 (1982).

48. Okada, M. and Blombäck, B.: Ann. New York Acad. Sci. 408, 233-253 (1983).

49. Blombäck, B., Okada, M., Forslind, B. and Larsson, U.: Biorheology 21, 93-104 (1984).

50. Bilezikian, S.B., Nossel, H.L., Butler, V.P., Jr. and Canfield, R.E.: J. Clin. Invest. 56, 438-445 (1975).

51. Copley, A.L.: Clin. Hemorrheol.1, 9-72 (1981).

FIBRIN-SPECIFIC MONOCLONAL ANTIBODIES ARE ELICITED BY IMMUNIZATION WITH A SYNTHETIC FIBRIN-LIKE PEPTIDE

Gary R. Matsueda, Kwan Y. Hui, and Edgar Haber

Harvard Medical School and the Cellular and Molecular Research Laboratory, Massachusetts General Hospital, Boston, MA 02114

Introduction

Thrombus detection remains an important clinical problem. The use of antibodies as agents for the in vivo localization of fibrin deposits has been hampered by antifibrin cross-reactivity with fibrinogen, the circulating fibrin-precursor. Blood clots when thrombin cleaves two pairs of small peptides from fibrinogen to yield fibrin monomers. These fibrin monomers aggregate spontaneously to form an insoluble gel which is then covalently stabilized by Factor XIIIa. Despite the dramatic gelation process, fibrin retains 98% of the original covalent structure of fibrinogen. Thus it is understandable that antifibrin sera cross-react strongly with fibrinogen, and that there has been only one isolated report of a fibrin-specific serum (1).

Results and Discussion

Rather than examining antibodies elicited by the entire human fibrin molecule with all of the epitopes it shares with fibrinogen, we chose to focus the immune response on a synthetic fibrin-unique peptide. We anticipated that antibodies which recognized a synthetic fibrin-unique epitope might bind to fibrin exclusive of fibrinogen. Arbitrarily, the NH_2-terminal

Fibrinogen - Structural Variants and Interactions
© 1985 Walter de Gruyter & Co., Berlin · New York - Printed in Germany

heptapeptide of fibrin's β-chain was synthesized to serve as a fibrin-unique antigen rationalizing that the amino terminus is exposed only following thrombin cleavage. In addition to the first seven amino acids of the β-chain, as shown in Figure 1, Cys was placed at the carboxy-terminus to permit unidirectional attachment of the synthetic peptide to maleimidobenzoylated(MB)-keyhole limpet hemocyanin (KLH). It was assumed that after reacting the β-peptide with MB-KLH, all peptide amino-termini would be uniformly oriented away from the protein carrier in a manner most analogous to fibrin β-chains.

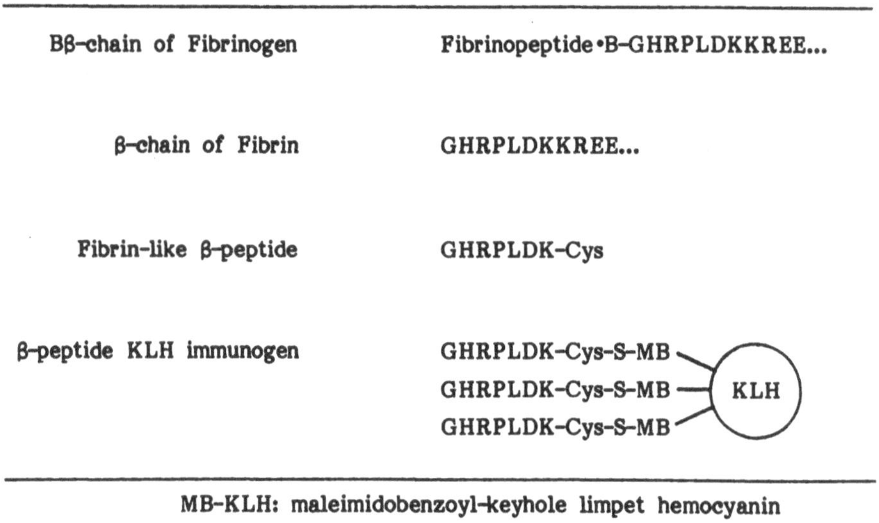

Figure 1. Structure of fibrin-like β-peptide.

Synthesis of the fibrin-like β-peptide.

Using the Merrifield solid-phase method (2), β(Gly^1-Lys^7) was assembled on Cys(Dmb)-OCH_2-resin in a stepwise fashion. A sample of the completed peptidyl-resin, Gly-His(Tos)-Arg(Tos)-Pro-Leu-Asp(Bzl)-Lys(2-ClZ)-Cys(3,4-Dmb)-OCH2-resin, was sequenced by solid-phase Edman degradation (3). The resulting sequencing data (i) confirmed the amino acid sequence and (ii) indicated that the average level of amino acid deletions, i.e., synthesis error, was less than 1% per cycle. Following anhydrous hydrogen fluoride treatment, the fully deprotected peptide gave the following amino acid composition upon acid hydrolysis: Asp, 1.05[1]; Pro, 1.01[1]; Gly, 0.98[1]; Leu, 1.01[1]; His, 0.90[1]; Lys, 1.03[1]; and Arg, 1.02[1]. The peptide was then coupled via its COOH-terminal Cys residue to MB-KLH (4).

Preparation and selection of antifibrin monoclonal antibodies

Three female BALB/c mice were immunized and boosted at 4 weeks with 30 μg of the β-peptide-KLH conjugate. In the plasma samples of each mouse, antifibrin immunoreactivity was detected. When the plasma samples were retested in the presence of 4 mg/ml of human fibrinogen, a significant fraction remained bound to the human fibrin monomer.

The spleen cells of the mouse which provided the highest fibrin-specific immunoreactivity were fused with Sp2/0 myeloma cells (5) to yield hybridomas. After 11 days in primary culture, 13 wells contained cells which had multiplied to become confluent. Aliquots of culture medium from those wells were withdrawn and mixed with heparin and phenylmethylsulfonyl fluoride. Thirty minutes after mixing an equal volume of human fibrinogen (8 mg/ml) with the heparinized aliquot, the mixture was tested on fibrin monomer-coated microtiter wells (6). Antifibrin immunoreactivity was detected in 3 of 13 wells; 7 of the remaining wells contained anti-KLH activity; none contained both activities. After 20 days in primary culture, 12 of 72 confluent wells contained antifibrin immunoreactivity which was not inhibitable by fibrinogen. After cloning by the limiting dilution method, three hybridomas continued to produce antifibrin immunoreactivity. The isoelectric points of two monoclonal antibodies, 59D8 and 64C5, were 6.4 and 7.6, respectively. The isotype of all three monoclonal antibodies was Ig(gamma$_1$-kappa).

Specific binding to clotted fibrin

Because clotted fibrin is mechanically difficult to manipulate in a quantitative manner, 9.5-mm discs of filter paper were used as solid supports. On each paper disc, human fibrinogen (40 µg in 20 µl, Kabi L grade) was clotted by adding human thrombin (Sigma, 1.25 U in 25 µl) in the presence of 20 mM $CaCl_2$ and Trasylol (Mobay, 0.05 KIU/ul). After 2 hr, the fibrin-bearing discs were washed exhaustively and then stored in 10 mM KH_2PO_4 containing 0.15 M NaCl (PBSA) at 4°C. The clotting efficiency was determined to be 80% by absorbance measurements. Fibrin discs stained strongly with a Coomassie brilliant blue dye solution, but only if thrombin had been used in the clotting mixture.

As shown in Figure 2, each monoclonal antibody bound to human fibrin-coated discs. This confirmed antibody recognition of aggregated fibrin monomers. When human fibrinogen (4 mg/ml) as inhibitor was tested, the binding of each monoclonal antibody was unaffected. By comparison, a control antidigoxin antibody (7), Dig 26-10, did not bind to fibrin discs (2390 $\pm$ 20 cpm).

Since the hybridomas were selected on the basis of fibrin monomer specificity, peptide inhibition of binding by the monoclonal antibodies was of interest. As shown in Figure 3, the synthetic β-fibrin peptide inhibited each monoclonal antibody, whereas the control α-fibrin peptide inhibitor, GPRVVER-Cys, did not. These results confirmed that the fibrin-specific monoclonal antibodies were elicited against the synthetic fibrin-like antigen and were not fortuitously selected by radioimmunoassay screening procedure (which is described above).

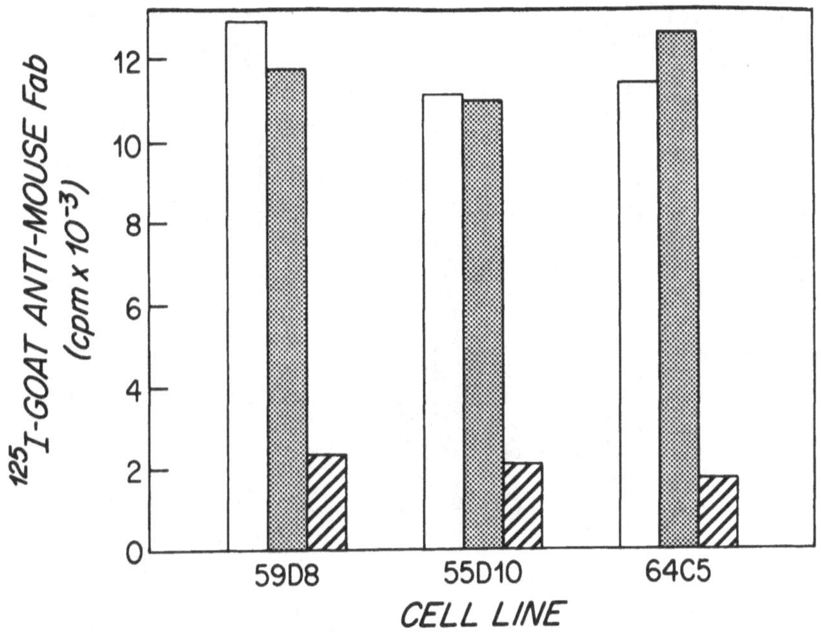

Figure 2. Effect of fibrinogen on the direct binding of antibodies 59D8, 55D10 and 64C5 to fibrin-coated discs. Heparinized culture supernatant (50 µl) was assayed on human fibrin discs in the absence (unfilled bar) and presence (stippled bar) of 4 mg/ml of fibrinogen (Kabi, grade L). Culture medium (hatched bar) was used as negative control. Specifically bound antibody was detected with 10^5 cpm of ^{125}I-goat anti(mouse Fab fragment). Each bar represents the mean of two determinations.

Species specificity.

Using the fibrin-disc assay, the species specificity of anti-fibrin monoclonal antibody 59D8 was examined. As compared to human fibrin discs, the 59D8 monoclonal antibody cross-reacted strongly with canine and rabbit clotted-fibrin discs (>75%), partially with bovine, porcine and ovine fibrin discs (>25%), and not at all with chicken fibrin discs.

Ex vivo circulatory model for human fibrin detection

Based upon the results of the species specificity study, an *ex vivo* circulatory model for human fibrin detection was devised.

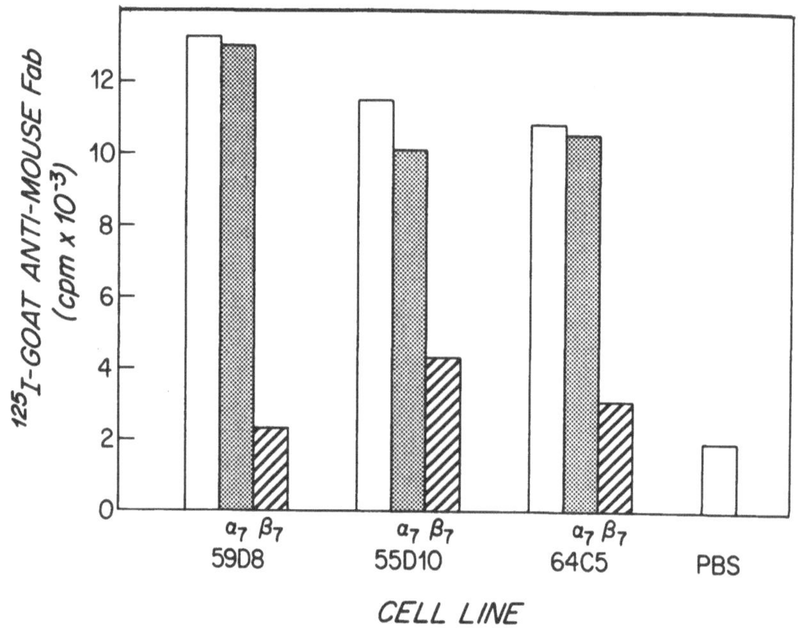

Figure 3. Effect of fibrin-like peptides on direct binding of antibodies 59D8, 55D10, and 64C5 to fibrin-coated discs. Each monoclonal was assayed in the presence of GPRVVER-Cys (α_7) as control inhibitor (stippled bar), and the β-peptide (hatched bar), GHRPLDK-Cys (β_7). Maximum binding (unfilled bar) was determined in the absence of either inhibitor peptide. Each bar represents the mean of two determinations.

The chicken was chosen as the experimental animal in order to avoid cross-reactivity with spontaneously formed thrombi in the experimental model. An extracorporeal chamber was attached via cannulae to a carotid artery and the contralateral jugular vein. The chamber was maintained at 37°C. At the beginning of the experiment, 50 ug of monoclonal antibody was injected into the jugular vein via the efferent cannula. Blood was then permitted to circulate through the chamber for 15 min. Ex vivo circulation was momentarily interrupted to permit insertion of two human fibrin discs, each rolled into an S-shaped configuration. After a predetermined time, the circulation was again interrupted to remove the fibrin discs, which were immediately washed with an excess of PBSA containing 5 KIU/ml Trasylol. The bound monoclonal antibody was measured by in vitro incubation of the fibrin discs for 1 hr with ^{125}I-goat anti[mouse $(Fab')_2$] antibody.

The results showed that (i) within 10 min intravenously injected 64C5 or 59D8 monoclonal antibody exceeded 50% binding to human fibrin discs as compared to the *in vitro* control fibrin discs; and (ii) fibrin discs exposed to circulating chicken blood for 1 hr did not differ from unexposed fibrin discs in their binding of antibody 64C5 or 59D8. In other *ex vivo* experiments, it was demonstrated that (i) circulating monoclonal antibody failed to bind to filter paper discs which lacked human fibrin; (ii) under the same circumstances, fibrin discs did not bind ^{125}I-goat anti[mouse $(Fab')_2$] antibody when the monoclonal was not injected; and (iii) human fibrin discs could be exposed to circulating blood of anesthetized chickens for 2 hr without loss of antibody recognition.

Conclusion

By focusing the immune response on a small fibrin-like peptide, we were able to detect and select murine antibodies specific for human fibrin. Since this strategy yielded fibrin-specific rabbit antibodies (8), it is likely to be applicable in other experimental systems as well. The uniqueness of this approach is emphasized by the report of Qureshi *et al.* (9), who could not detect fibrin-specific antibody when thrombin-treated N-DSK(N-DSK-T) was used as immunogen. This suggests that even when the N-DSK-T fragment is used, other epitopes, as in clotted fibrin itself, remain more immunodominant than the NH_2-termini of the β-chains.

In summary, fibrin-specific monoclonal antibodies were elicited by a synthetic heptapeptide which corresponded to the NH_2-terminus of human fibrin β-chain. The resulting monoclonal antibodies, 64C5, 55D10 and 59D8, recognized human fibrin monomer as well as fibrin polymer in the presence of plasma concentrations of fibrinogen in both in vitro and ex vivo experimental models. The potential for applying these antibodies in clinical diagnosis is of considerable interest.

References

1. Bosnjakovic, V.B., Horvat, J., Cvoric, J.: Lancet, 2, 452-455 (1977).
2. Merrifield, R.B.: J. Am. Chem. Soc., 85, 2149-2153 (1962).
3. Laursen, R.A.: Eur. J. Biochem. 20, 89-102 (1971).
4. Kitagawa, T., Aikawa, T.: J. Biochem. (Japan), 79, 233-236 (1976).
5. Köhler, G., Milstein, C.: Nature (London) 256, 496-497 (1975).
6. Klinman, N.R., Pickard, A.R., Sigal, N.H., Gearhart, P.J., Metcalf, E.S., Pierce, S.K.: Ann. Immunol. (Paris), 127C, 489-502 (1976).
7. Mudgett-Hunter, M., Margolies, M.N., Ju, A., Haber, E.: J. Immunol. 129, 1165-1172 (1982).
8. Pacella, B.L. Jr., Hui, K.Y., Haber, E., Matsueda, G.R.: Mol. Immunol. 20, 521-527 (1983).
9. Qureshi, G.D., Butler, V.P. Jr., Harfenist, E.J., Canfield, R.E., Nossel, H.L.: Thromb. Res. 6, 357-374 (1975).

ENHANCEMENT OF FIBRIN POLYMERIZATION BY ACTIVE SITE-INHIBITED THROMBIN

Marek Kaminski and Jan McDonagh

Department of Pathology, Harvard Medical School and Beth Israel Hospital, Boston, Massachusetts 02215, USA

Introduction

The assembly of a three-dimensional fibrin gel structure involves a series of complex biochemical reactions. The process is initiated when thrombin cleaves the A peptides from fibrinogen and then proceeds through multiple polymerization steps with concomitant release of the B peptides. Polymerization proceeds through a linear association reaction, resulting in the formation of protofibrils, and is followed by lateral association of protofibrils and branchpoint formation (1). The overall process of gelation is kinetically determined, so that any condition which alters the reaction rate of one of the steps in the process can have a marked effect on the structure of the final products (2). Consequently, a number of potential mechanisms exist for controlling the structure of the fibrin gel. Such mechanisms are important because alterations in the gel structure result in changes in the functional properties of the fibrin gel. For example, fibrin gels in which the degree of lateral association is high have increased mechanical rigidity (3).

In _in vitro_ test systems, the gel structure is regulated by the concentrations of fibrinogen, thrombin, and calcium ions and also by the pH and ionic strength of the reaction. Fibrinogen has three high affinity Ca^{2+} binding sites (4) and the bound or structural Ca^{2+} helps to maintain the

Fibrinogen - Structural Variants and Interactions
© 1985 Walter de Gruyter & Co., Berlin · New York - Printed in Germany

structural integrity of the molecule and the appropriate configuration of its polymerization domains. In addition, however, the free Ca^{2+} in the clotting system also regulates the rate of polymerization (5). In this report we describe the effect of active site-inhibited thrombin on the process of fibrin formation and compare this to the effect of Ca^{2+}. Data are presented which show that active site-inhibited thrombin binds to fibrinogen and fibrin monomer and increases the rate of gelation. This effect is similar to the enhancing effect of Ca^{2+} on fibrin polymerization.

Experimental Procedures

Thrombin preparation. Purified human α-thrombin (2.6 mg/ml, ~8000 NIH units/ml) was kindly provided by Dr. John Fenton. It was trace labeled with ^{125}I by the solid phase lactoperoxidase procedure. Functionally active enzyme was separated from radiolytically damaged material by affinity chromatography on fibrin monomer Sepharose, with elution of active thrombin from the column with 50 mM $CaCl_2$ or 1.0 M NaCl, followed by dialysis. This procedure resulted in the preparation of labeled thrombin with high specific clotting activity (~3 NIH units/μg protein). Details of the methods for preparation of fibrinogen Sepharose and its conversion to fibrin monomer Sepharose and for labeling and purification of thrombin have been described (6). Labeled thrombin was diluted with unlabeled material to the appropriate radioactivity. This mixture is referred to as ^{125}I-thrombin.

Thrombin Inhibition. D-phenylalanyl-prolyl-arginyl chloromethyl ketone (PPACK), a thrombin specific inhibitor, was obtained from Calbiochem-Behring. PPACK irreversibly inhibits the proteolytic activity of thrombin by alkylation of histidine in the active center (7). Two ml of thrombin and 100 μl of PPACK (5 mg/ml in buffer) were mixed at an enzyme:inhibitor molar ratio of 1:6 (weight ratio of 10:1) and then diluted with 1.9 ml of saline. Inhibited thrombin was then dialyzed extensively against saline and then against 0.05 M Tris-0.1 M NaCl, pH 7.35. 60 μg of PPACK inhibited thrombin had no enzyme activity with respect to fibrinogen. ^{125}I-thrombin was reacted with PPACK under essentially the same conditions as unlabeled thrombin.

Analysis of inhibited thrombin interaction with fibrinogen and fibrin. PPACK-thrombin was examined for its ability to bind to fibrinogen and fibrin monomer and for its effect on the clotting of fibrinogen by active thrombin. Human fibrinogen was purified from cryoprecipitate depleted Cohn Fraction I or from plasma by $(NH_4)_2SO_4$ precipitation (8). Thrombin clotting times were measured in a test system containing 20 μl fibrinogen (24 mg/ml), 330 μl buffer, and 50 μl active thrombin. Specific changes in the test system with respect to Ca^{2+} concentration, PPACK-thrombin concentration, and ionic strength were made by appropriate alterations in the buffer. In some experiments Ca^{2+} concentration was increased, but constant ionic strength was maintained. The clotting time was measured as the time from the addition of thrombin to the first appearance of the gel.

Fibrinopeptide release by active thrombin in the presence or absence of PPACK-thrombin was quantitated by reaction of peptide-bound arginine with phenanthrenequinone (9). Fibrinogen (1.2 mg) was activated with 50 μl of thrombin in 350 μl of 0.05 M Tris-0.1 M NaCl, pH 7.35 (control), thrombin plus buffer containing PPACK-thrombin, or thrombin plus Ca^{2+} buffer (at constant ionic strength of 0.15). Fibrinopeptide cleavage was stopped by addition of 50 μl 20% trichloroacetic acid to the test system. Under these conditions the clotting time of the control sample was ~170 seconds. Fibrin polymerization was monitored by following the change in turbidity at 350 nm in a Beckman D-8 spectrophotometer. The test system contained 50 μl of fibrinogen (24 mg/ml) and 20 μl active thrombin (~2 units/ml) diluted with 780 μl of the same reaction buffers used for the fibrinopeptide assay. Change in optical density at 350 nm was recorded continuously at a chart speed of 2 cm/min.

Binding of inhibited thrombin to fibrin monomer was studied with ^{125}I-thrombin. Ten µg of ^{125}I-thrombin were reacted with 10 µl PPACK (1 mg/ml) in 0.05 M Tris-0.1 M NaCl, pH 7.35. This mixture was applied to a column containing 5 ml of fibrin monomer Sepharose (13.9 mg of fibrin per ml of settled gel, equilibrated with the same Tris-NaCl buffer). After sample application the column was washed with approximately 4 column volumes of buffer. Elution of PPACK(^{125}I-)-thrombin was achieved with buffer containing 40 mM $CaCl_2$-20 mM NaCl-10 mM Tris, pH 7.35. The column flow rate was 3 ml/hr, and the fraction volume was 15 drops. Radioactivity of each fraction was monitored in a γ-spectrometer.

Results

We have previously demonstrated that thrombin binds to fibrin monomer Sepharose with high affinity and that it can be specifically desorbed and eluted by addition of Ca^{2+} to the elution buffer (6). Figure 1 shows that the same results were obtained when the active center histidine residue was alkylated by reaction with PPACK. PPACK(^{125}I-)-thrombin was adsorbed to fibrin monomer in the absence of divalent metal ions and at physiologic pH and ionic strength. Prolonged washing of the column with Tris-NaCl did not result in desorption. However, elution with Ca^{2+} containing buffer at the same pH and ionic strength resulted in rapid desorption and elution of inhibited thrombin. The same results could also be obtained by simple addition of $CaCl_2$ to the Tris-saline buffer. The binding affinity of PPACK-thrombin for fibrin monomer was similar to that for active thrombin. Analogous experiments showed that PPACK-thrombin also bound to fibrinogen Sepharose and could be eluted by Ca^{2+}.

Similar studies were also performed to measure the effect of phenylmethylsulfonyl fluoride (PMSF) inhibition of thrombin on the binding of thrombin to fibrin monomer and fibrinogen (10). PMSF also inhibits the proteolytic activity of thrombin by reaction with the active center serine. PMSF-

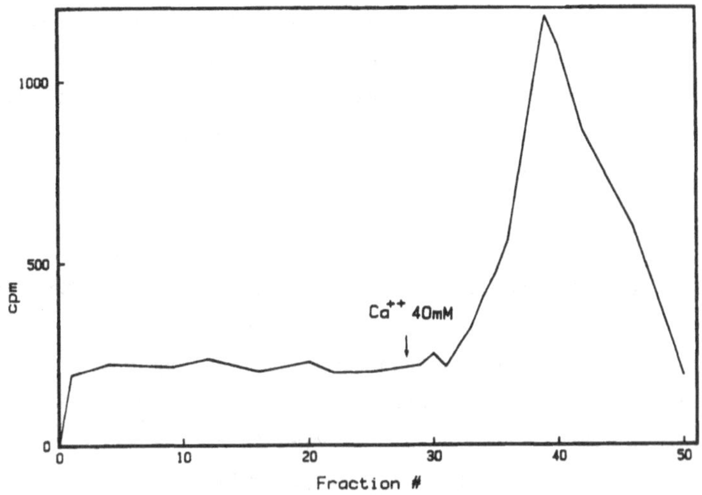

Figure 1. Elution of PPACK(^{125}I-)-thrombin from fibrin monomer Sepharose. 10 μg of test material were applied to 5 ml of gel. The column was washed with 0.05 M Tris-0.1 M NaCl, pH 7.35; at fraction 28 the buffer was changed to 0.01 M Tris-0.02 M NaCl-0.04 M $CaCl_2$, pH 7.35.

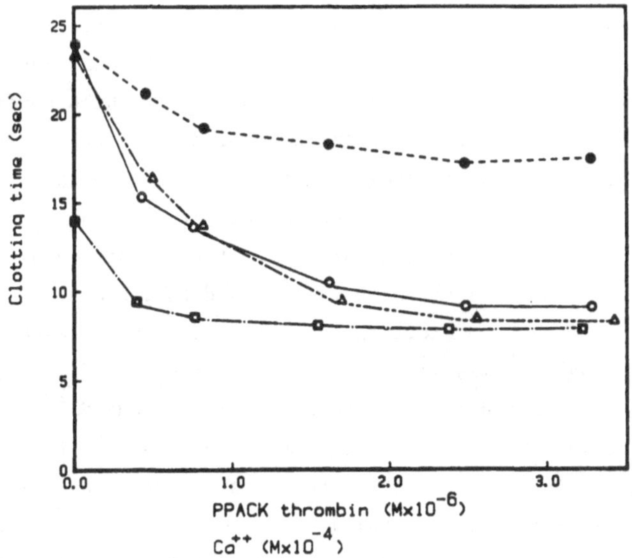

Figure 2. Effects of PPACK-thrombin and Ca^{2+} on the clotting time. ○: increasing Ca^{2+} with constant ionic strength; ●: increasing Ca^{2+} and increasing ionic strength; △: increasing PPACK-thrombin, constant ionic strength; □: constant PPACK-thrombin and increasing Ca^{2+} (μ= 0.15).

(^{125}I-)-thrombin behaved in an identical manner with PPACK (^{125}I-)-thrombin with respect both to binding to fibrinogen and fibrin monomer and to desorption by Ca^{2+}. These results clearly indicate that a functional active center is not necessary for the initial binding of thrombin to fibrinogen and fibrin monomer.

Initial attempts to study the effect of inhibited thrombin on the activation of fibrinogen by active thrombin produced results that might at first be considered paradoxical. PPACK-thrombin alone did not clot fibrinogen. However in the presence of PPACK-thrombin the clotting time with active thrombin was significantly shortened. As seen in Figure 2, in a test system containing 3.5 μM fibrinogen and ~15 nM thrombin, addition of PPACK-thrombin resulted in a dose dependent decrease in the clotting time. At 2.5-3.4 μM PPACK-thrombin the clotting time was decreased by 65%. Further increase in the PPACK-thrombin concentration did not have any additional effect. Hence, the maximum effect of PPACK-thrombin on the clotting time was observed at an equimolar ratio of fibrinogen to inhibited thrombin. A similar effect on the clotting time was also observed by addition of Ca^{2+} if the ionic strength was held constant at 0.15. Under these conditions Ca^{2+} also decreased the clotting time in the same dose dependent manner, which again reached a plateau at one-third the control time when the Ca^{2+} concentration was about 25 mM. The marked effect of Ca^{2+} on the clotting time was observed only when the ionic strength of the test system was held constant at 0.15. When ionic strength was allowed to increase by addition of $CaCl_2$ to 0.05 M Tris-0.1 M NaCl, the effect of Ca^{2+} with respect to decreasing the clotting time was significantly less pronounced. Furthermore, at constant, physiologic ionic strength PPACK-thrombin and Ca^{2+} acted to decrease the clotting time to the plateau level more rapidly than did either acting alone, indicating that the effects of both reagents were being observed. In a test system containing 0.86 μM PPACK-thrombin and 8 mM $CaCl_2$, the clotting time was shortened to one-third the control value. Results similar to these were also obtained when PMSF-thrombin was used in place of PPACK-thrombin in the clotting assay.

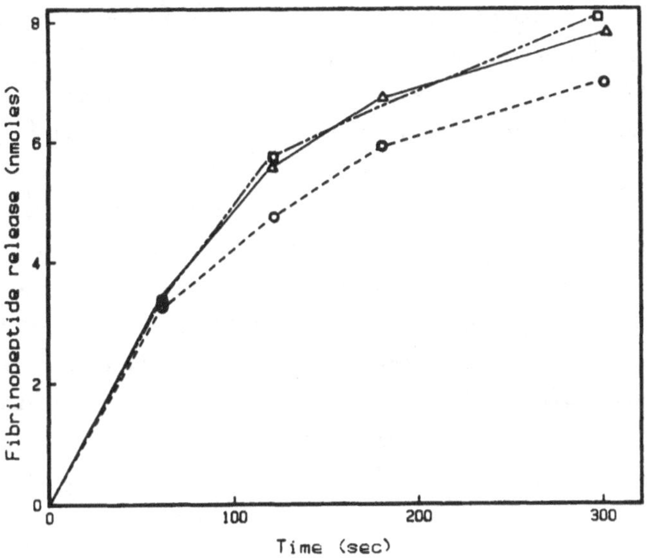

Figure 3. Fibrinopeptide release in the presence of PPACK-thrombin. △: control; □: with 31 mM $CaCl_2$; O: with 3.6 μM inhibited thrombin.

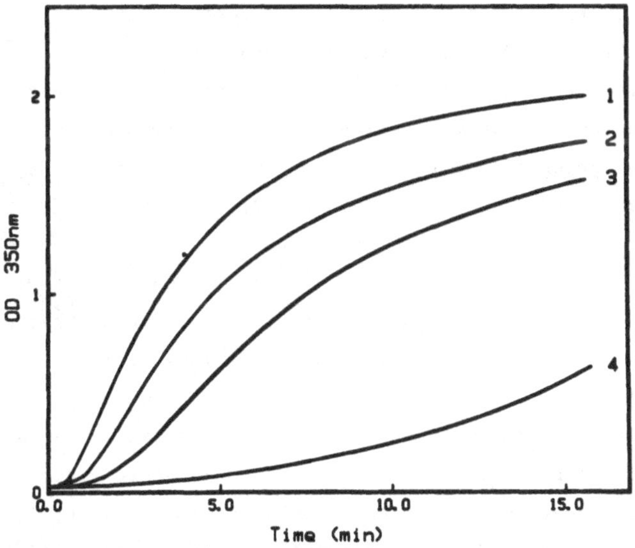

Figure 4. Fibrin polymerization with varying Ca^{2+} concentration and constant ionic strength of 0.15. 1: 36.7 mM $CaCl_2$; 2: 18.3 mM $CaCl_2$; 3: 4.6 mM $CaCl_2$; 4: no $CaCl_2$ in the buffer (control).

Further studies were done to investigate the mechanism for accelerated fibrin formation in the presence of inhibited thrombin. Total fibrinopeptide release by active thrombin was measured in the presence of Ca^{2+} and PPACK-thrombin at concentrations where shortening of the clotting time was maximal. Figure 3 shows the expected result for Ca^{2+}. As has previously been found (5,11), Ca^{2+} does not affect the rate of fibrinopeptide release by thrombin, and maximal fibrinopeptide cleavage is obtained in Ca^{2+} containing systems. Similarly, in a test system containing 1.8 nM active thrombin and 3.6 μM PPACK-thrombin, the initial rate of fibrinopeptide release was not significantly different from the control. Also, after incubation of the clotting system with PPACK-thrombin for two hours with high thrombin, 3.8 moles of fibrinopeptides per mole of fibrinogen were detected in the supernatant fluid.

The effect of inhibited thrombin on the assembly of fibrin monomers into a gel was studied by measuring the change in turbidity which occurs with gelation. This was compared to the known effect of Ca^{2+} on the acceleration of polymerization and also to the combined effect of both inhibited thrombin and Ca^{2+}. Figure 4 describes the acceleratory effect of Ca^{2+} on fibrin polymerization. With increasing Ca^{2+} concentration both a decrease in the lag time and an increase in the aggregation rate were observed. In this test system the aggregation rate increased linearly with the Ca^{2+} in solution over the concentration tested (4.6 to 37.5 mM $CaCl_2$).

Similar effects on the lag time and aggregation rate were observed when polymerization was monitored in the presence of inhibited thrombin (Figure 5). Aggregation rate increased and the lag phase decreased as the concentration of PPACK-thrombin increased. In this system, as also in measurements of the clotting time, the effect approached maximal as the molar ratio of fibrinogen to inhibited thrombin became close to 1 (10.1:9.2 in curve 1 of Figure 5). Consequently, the effect of inhibited thrombin on increasing aggregation is sufficient to account for the decrease observed in the clotting time. Again, as was observed in the clotting time, the effects of both Ca^{2+} and inhibited thrombin were observed when both

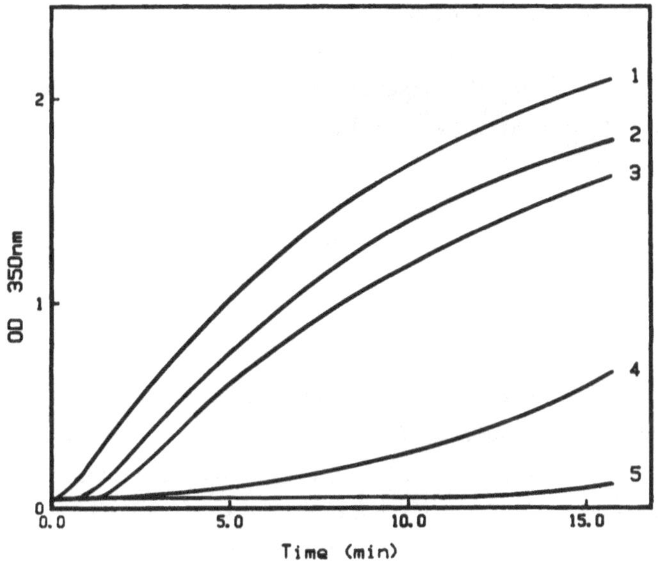

Figure 5. Fibrin polymerization with varying PPACK-thrombin concentration. PPACK-thrombin = 1: 3.8 μM; 2: 1.9 μM; 3: 0.95 μM; 4: none; 5: 3.8 μM with no active thrombin in the test system.

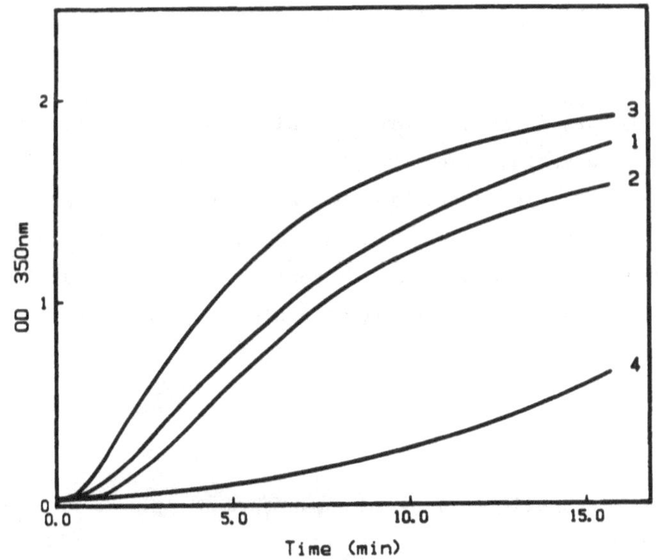

Figure 6. Combined effect of PPACK-thrombin and Ca^{2+} on fibrin polymerization. 1: 4.3 mM $CaCl_2$; 2: 1.9 μM PPACK-thrombin; 3: 4.3 mM $CaCl_2$ + 1.9 μM PPACK-thrombin; 4: control (no $CaCl_2$ or PPACK-thrombin).

were present in the test system. In a system containing 4.5 mM $CaCl_2$ and a molar ratio of fibrinogen:PPACK-thrombin of 2:1 (Figure 6, line 1) the increase in aggregation was the same as that observed with equimolar fibrinogen:PPACK-thrombin or high Ca^{2+} acting alone.

Discussion

The results presented here further define the interactions of thrombin with fibrinogen and fibrin monomer. The binding of active thrombin to fibrin monomer is reversible and saturable (6,10); there is a single class of high affinity sites with a binding affinity of $4.9 \times 10^5\ M^{-1}$ (6). Active site-inhibited thrombin binds in a similar manner, irrespective of whether histidine-365[1] or serine-527 in the active center is derivatized. These results clearly show that a functional active center is not required for thrombin binding to fibrinogen or fibrin monomer and indicate that it is the extended binding domains which are necessary for formation of the enzyme:substrate complex.

The observation that inhibited thrombins also accelerate the clotting of fibrinogen is interesting, even though the mechanism is not fully apparent. It is clear that inhibited thrombin shortens the clotting time by acceleration of the polymerization reaction, since there is no effect of inhibited thrombin on the initial rate of fibrinopeptide cleavage by active thrombin. The effect on aggregation is maximal at equimolar concentrations of fibrinogen and inhibited thrombin.

---

[1]Numbering of amino acid residues in human thrombin is based on the numbering of residues in prothrombin (13).

Change in turbidity was used as an index of polymerization in these experiments. Since turbidity is a function of the size of molecules in solution, this method cannot completely differentiate the various stages of aggregation; but it does reflect accurately the overall process. The initial rate of aggregation probably reflects a composite of linear and lateral association, while the maximum increase in turbidity may relate more to the thickness of the fibers and the formation of branchpoints in the fiber network. It has been shown, for example, that turbidity begins to increase when about 40% of fibrinopeptide A has been cleaved (14). It is also clear that aggregation rate and maximum turbidity are a function of the rate of production of fibrin monomers (15). Changes in turbidity have also been used to examine the effects of other molecules, such as Ca^{2+} and dextran, on fibrin polymerization.

In the studies described here PPACK-thrombin resulted in an increase in both the initial rate of aggregation and in the maximum turbidity at 15 minutes. The increase in maximum turbidity was directly proportional to the concentration of PPACK-thrombin. This enhancing effect appears to be similar to that seen with dextran, where both aggregation rate and maximal turbidity were also increased, and the maximal increase in turbidity squared was proportional to the dextran concentration (15). It is well known that Ca^{2+} also has a similar enhancing effect on polymerization (5, 11,15,16), although the mechanism may be different. In these experiments the initial rate of aggregation, rather than the maximum increase in turbidity, was proportional to the Ca^{2+} concentration. Also, the effects of PPACK-thrombin and Ca^{2+} were found to be additive at concentrations below the maximal level of enhancement.

Although dextran, Ca^{2+}, and active site-inhibited thrombin all have the effect of increasing fibrin polymerization, the mechanisms may not necessarily be the same. The specific mechanisms are not yet known, but there is sufficient information for some reasonable hypotheses. Any mechanism which increases the attractive forces between fibrin molecules will result in enhancement of polymerization. The effect of Ca^{2+} could be due to alteration of net charge, salt bridge formation, or a conformational change in fibrin monomer. We have previously found that the thrombin:fibrin monomer complex dissociates as fibrin monomer polymerizes, indicating that the thrombin binding domain is altered by polymerization (6). Similarly, occupation of the thrombin binding domain might have a conformational effect on a polymerization domain. Since the effect of inhibited thrombin approaches maximal when one molecule of inhibited thrombin is bound per molecule of fibrin monomer, it seems probable that this complex formation produces an alteration in a polymerization domain which enhances the contact between fibrin molecules. Further studies on the conformation of fibrin monomer and fibrin monomer:inhibited thrombin will be required to test this hypothesis.

Acknowledgements

We wish to thank Dr. John W. Fenton II, New York State Department of Health, Albany, for a generous supply of purified human α-thrombin. This research was supported in part by Grant HL 29511 from the U.S. National Institutes of Health.

References

1. Hermans, J., McDonagh, J.: Sem. Thromb. Hemostas. 8, 11-24 (1982).

2. Hantgan, R.R., Hermans, J.: J. Biol. Chem. 254, 11272-11281 (1979).

3. Shen, L.L., Hermans, J., McDonagh, J., McDonagh, R.P.: Amer. J. Physiol. 232, H629-H633 (1977).

4. Nieuwenhuizen, W., van Ruijven-Vermeer, I.A.M., Nooijen, W.J., Vermond, A., Havercate, F.: Thromb. Res. 22, 653-657 (1981).

5. Hardy, J.J., Carrell, N.A., McDonagh, J.: Ann. N.Y. Acad. Sci. 408, 279-287 (1983).

6. Kaminski, M., McDonagh, J.: J. Biol. Chem. 258, 10530-10535 (1983).

7. Shaw, E.N., Kettner, C.A.: Thromb. Res. 14, 969-973 (1979).

8. Gralnick, H.R., Givelber, H.M., Shainoff, J.R., Finlayson, J.S.: J. Clin. Invest. 50, 1819-1830 (1971).

9. Yamada, S., Itano, H.A.: Biochim. Biophys. Acta 130, 538-540 (1966).

10. Kaminski, M., McDonagh, J.: Circulation 66, II-294 (1982).

11. Boyer, M.H., Shainoff, J.R., Ratnoff, O.D.: Blood 39, 382-387 (1972).

12. Liu, C.Y., Nossel, H.L., Kaplan, K.L.: J. Biol. Chem. 254, 10421-10425 (1979).

13. Mann, K.G., Elion, J., Butkowski, R.J., Downing, M., Nesheim, M.E.: In: Methods in Enzymology (Lorand, L., ed.) 80, 286-302 (1981).

14. Abildgaard, U.: Scand. J. Clin. Lab. Invest. 18, 518-524 (1966).

15. Muzaffar, T.Z., Youngson, G.G., Bryce, W.A.J., Dhall, D.P.: Thromb. Diath. Haemorrhag. 28, 244-256 (1972).

16. Endres, G.F., Scheraga, H.A.: Arch. Biochem. Biophys. 153, 266-278 (1972).

PEPTIDES RELEASED FROM HUMAN FIBRINOGEN BY THROMBIC ENZYMES

Job Harenberg, Jan Xaver de Vries and Stephan Waibel
Departments of Internal Medicine and Clinical Pharmacology, Medical University Clinic, Bergheimerstr.58, 6900 Heidelberg, FRG

Introduction

Thrombin releases the fibrinopeptides A, AP, AY and B from human fibrinogen during the coagulation process (1). Enzymes from snake venoms with procoagulant activities also split fibrinopeptides from fibrinogen (2-5). Ancrod and reptilase, the procoagulant fractions of the snake venoms of Agkistrodon rhodostoma and Botrops atrox, respectively, release peptides only from the α-chain of fibrinogen, i.e. A peptides (2-4,11, 12) and peptides of higher molecular weight (6-12).

Recently, a rapid, high-performance liquid chromatography-based method for the detection of the fibrinopeptides has been developed (13). We report here on the release of human fibrinopeptides from the α- and β-chain of fibrinogen by thrombin, Arvin, reptilase and Southern Copperhead venom using this highly sensitive detection method.

Materials and Methods

Apparatus: Liquid chromatography was carried out with the following components: an automatic injector WISP Model 710 B, two Model 6000A pumps controlled by a Model 660 solvent programmer (all from Waters, FRG), a Model SDP-2A UV detector, and a Model C-R1B chromatographic printer data processor (Shimadzu Europe GmbH, FRG).

Fibrinogen - Structural Variants and Interactions
© 1985 Walter de Gruyter & Co., Berlin · New York - Printed in Germany

Reagents: Fibrinogen (Forschungsfibrinogen) was purchased from Kabi (Munich, FRG), Bovine thrombin (Behringwerke, Marburg,FRG, 30 NIH/ml solution), Arvin (Knoll, Ludwigshafen, FRG, 70 IU/ml solution), reptilase (Merz and Dade, Munich, FRG, 20 BU/ml solution) and snake venom enzyme from Southern Copperhead (Sigma Fine Chemicals, 0.5 mg/ml) were used as enzymes. Distilled water was purified with inorganic cartridges (Millipore GmbH,Neu-Isenburg, FRG). Acetonitrile (Merck, Darmstadt, FRG) for spectroscopy was used without further purification. All other substances were analytical grade.

Enzyme digestion: Fibrinogen was dissolved (5 mg/ml) in 0.15 M ammonium acetate, pH 8.5. Thrombin (2.0 NIH/ml fibrinogen solution), Arvin (1.0 IU), reptilase (4.0 BU), and Southern Copperhead venom (0.5 mg/ml fibrinogen solution) were dissolved in sterile water and incubated with the substrate at 37°C for 0, 0.1, 0.2, 0.5, 1, 2, and 20 hours. Hydrolysis was terminated by a 5 min incubation at 100°C. The clot was then sedimented by centrifugation at 1800 g for 10 min. The supernatant was analysed by high performance liquid chromatography within 24 h.

Chromatography: A reversed phase C18 column was used for the separation (Lichrosorb RP-18 with 5 μm particles, 25 mm x 4 mm I.D., Hibar RT 250-4, Merck, Darmstadt, FRG). Hydrolysates were analysed using solvent gradient elution with 0.025 M ammonium acetate, pH 6.0 as solvent A and 0.05 M ammonium acetate, pH 6.0, —acetonitrile (1:1, v/v) as solvent B. Solvents were filtered and degassed before use. A linear gradient from 12 to 28% solvent B in 30 min with a total flow rate of 1.5 ml/min and 10 min for reequilibration were used. All separations were run at room temperature. The detector was set at a wavelength of 210 nm and a sensitivity of 0.04 AUFS. Quantitation was carried out by peak area integration. After a series of analyses the columns were washed with water and methanol.

Results

The separation by the high performance liquid chromatography method of the human fibrinopeptides enzymatically released is shown in Fig. 1. The peptides are eluted from the column in the following sequence: AP, A, AY, Des-Arg B, and B. They were identified by amino acid analyses.

Fibrinopeptide A is released from the α-chain of fibrinogen by thrombin (Fig.2), reptilase (Fig.3) and Southern Copperhead enzyme (Fig.5) to about 62 - 68% of total A peptide liberation after 20 h. Arvin, however, released only 43% FPA of total A peptides (Fig.4). The snake venom enzyme from Southern Copperhead liberates FPA more slowly and relatively little peptide is detected within the first 30 min. After incubation for 2 h with this enzyme most of FPA is released (Fig.5).

Release of the AP peptide by both thrombin and the different snake venom enzymes was 10 - 20% of total peptides. No differences in the maximal liberation of this peptide were found (Fig. 2-5). Cleavage of peptide AY by thrombin and reptilase was about 13% of the FPA, but Arvin and the Southern Copperhead venom seemed to release 27 - 30% of this peptide from human fibrinogen (Fig. 2-5).

Fibrinopeptide B is rapidly released from the β-chain of fibrinogen by thrombin. The velocity of this cleavage is slower than that of FPA (Fig.2,6). Both Arvin and reptilase release FPB from fibrinogen in our experiments (Fig.6). The Southern Copperhead enzyme releases FPB faster than FPA (Fig.5 and 6) with maximum cleavage rate at 0.1 and 1 h for FPB and 1 and 2 h for FPA.

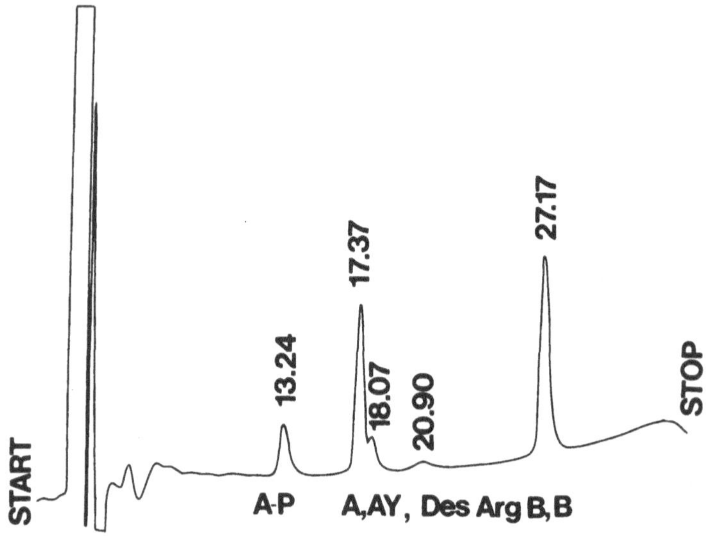

Fig. 1. Chromatogram of fibrinopeptides after incubation of human fibrinogen with thrombin for 2 h obtained by the high performance liquid chromatography method. Ordinate: absorption at 210 nm, abscissa: elution time in min. 13.24 = FPA-P, 17.37 = FPA, 18.07 = FPA-Y, 20.90 = Des-Arg B, and 27.17 = FPB.

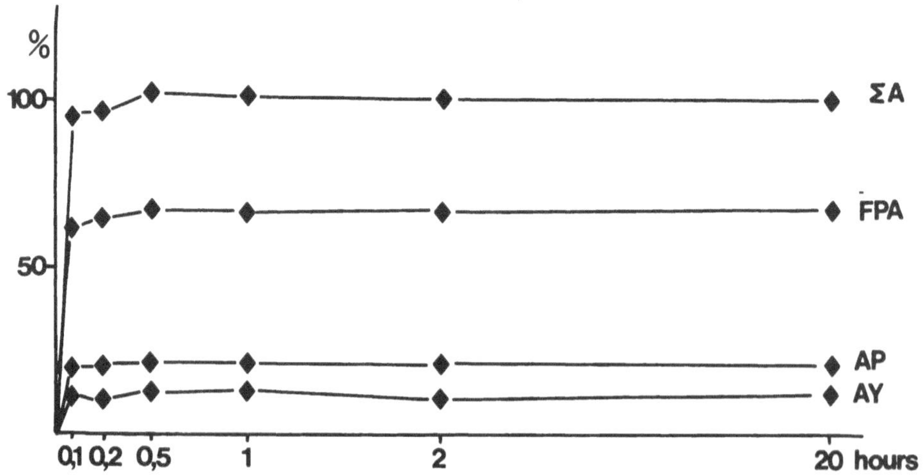

Fig. 2. Human fibrinopeptide A release by thrombin; peak areas of the total and of the individual peptides plotted as a function of time.

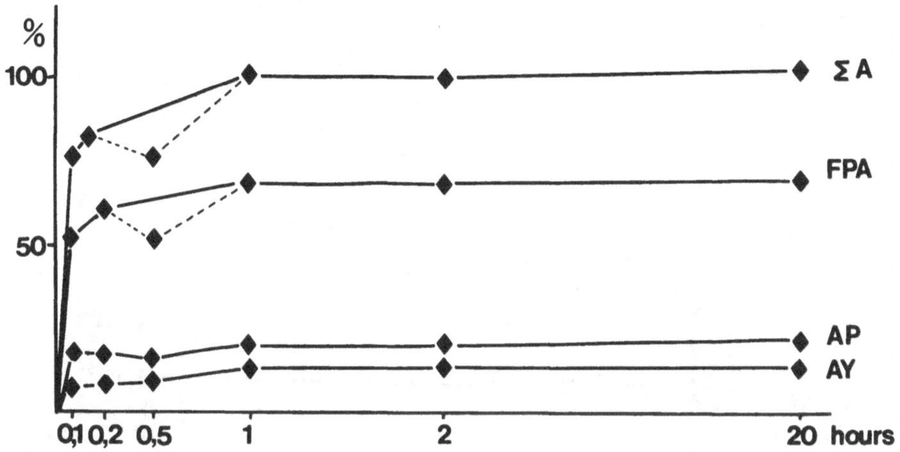

Fig. 3. Human fibrinopeptide A release by reptilase; peak areas of the total and of the individual fibrinopeptides plotted as a function of time.

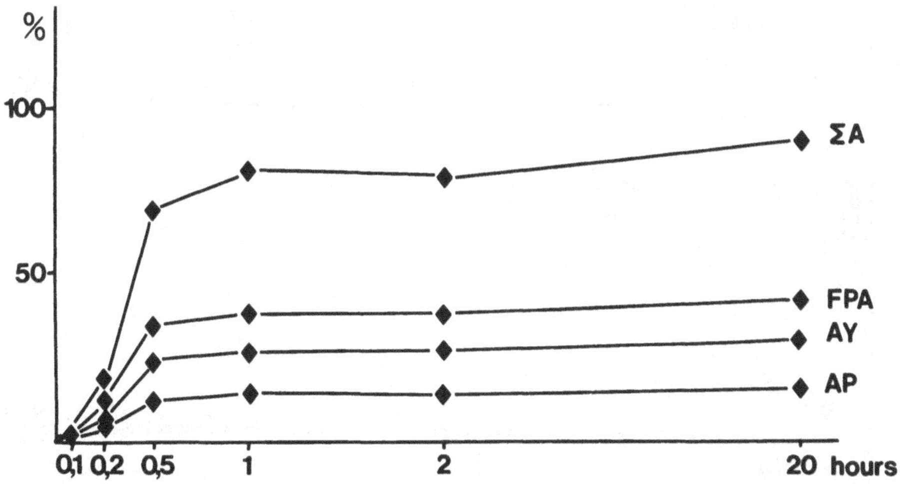

Fig. 4. Human fibrinopeptide A release by Arvin; peak areas of the total and of the individual peptides plotted as a function of time.

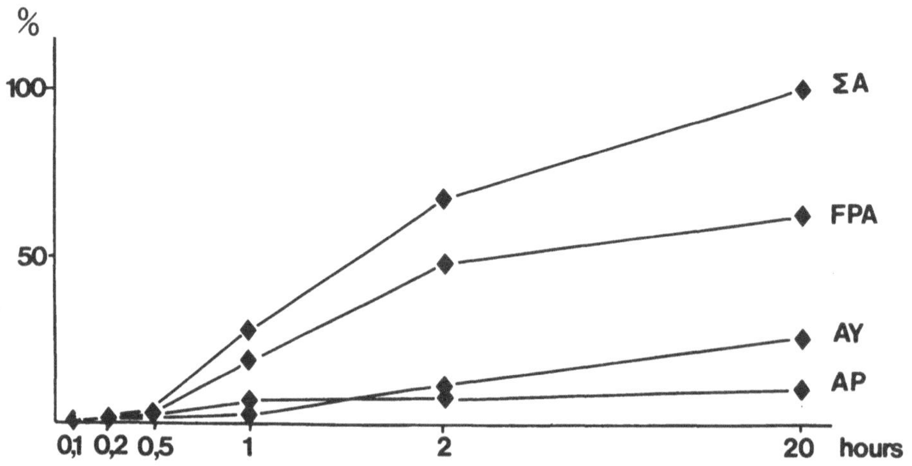

Fig. 5. Human fibrinopeptide A release by the coagulant fraction of the snake venom of Southern Copperhead, the peak area of the total and of the individual peptides plotted as a function of time.

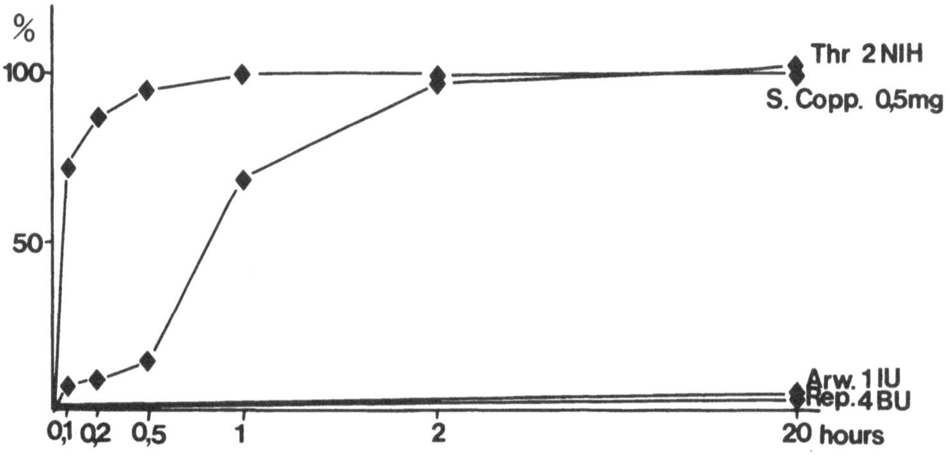

Fig. 6. The human fibrinopeptide B release by thrombin, Arvin, reptilase and Southern Copperhead venom shown as a function time.

Discussion

In the present investigation, high performance liquid chromatography was used to analyse the release of fibrinopeptides from human fibrinogen by thrombin and procoagulant fractions of snake venoms from Bothrops atrox, Agkistrodon rhodostoma and contortrix.The release of the human fibrinopeptides from native fibrinogen has been described in detail by Blombäck et al. (1) and the results have been confirmed by the data from high-performance liquid chromatography studies (13,14). The present results are consistent with these findings.

Hessel et al. have shown that peptides from the α-chain of fibrinogen are released by Arvin (6). Our data demonstrate that peptides from the β-chain are also found after an adequate equilibration period. The cleavage of human fibrinogen by reptilase, the coagulant fraction of the snake venom of Botrops atrox, occurs in the α-chain (2,3). In the present study, fibrinopeptide B also appeared after incubation with human fibrinogen in vitro. The liberation of B peptides from fibrinogen has been assumed, but had not been demonstrated so far using electrophoretic techniques (11).

The procoagulant fraction of the snake venom Southern Copperhead, Agkistrodon contortrix, releases FPB faster than FPA (5). Our results confirm these findings and that this enzyme also liberates AY and AP peptide from the α-chain.

The results of the present experiments indicate that the different enzymes investigated here have different affinities to the α- and β-chain of fibrinogen. Accordingly, this affinity decreases for the α-chain from thrombin to reptilase, Arvin and Southern Copperhead venom and for the β-chain from thrombin and Southern Copperhead venom to Arvin and Reptilase. Since the amino acid bonds of the α- and β-chain of fibrinogen which are split are both arginyl-glycine the affinity has to be determined also by other affinity centers on the α- and the β-chain, respectively. This would explain that the coagulant fraction of the snake venom enzyme of Southern Copperhead splits first the β-,then the α-chain of fibrinogen. Thus, the high performance liquid chromatography of fibrinopeptides may be useful in determining the affinity of enzymes for Arg-Gly bonds and the surrounding structures.

References

1. Blombäck, B., Blombäck, M., Edman, P. and Hessel, B.: Biochim. Biophys. Acta 115, 371, 1966.
2. Blombäck, B., Blombäck, M. and Nilsson, J.M.: Thrombos. Diathes. Haemorrh. 1, 76, 1957.
3. Bilezikian, S.B., Nossel, H.L., Butler, V.P. and Canfield, R.E.jr.: J. Clin. Invest. 56, 438, 1975.
4. Ewart, M.R., Hatton, M.W.C., Basfort, J.M. and Dodgson, K.S.: Biochem. J. 118, 603, 1970.
5. Herzig, R.H., Ratnoff, O.D. and Shainoff, J.R.: J. Lab. Clin. Med. 76, 451, 1970.
6. Hessel, B. and Blombäck, B.: FEBS Letters 18, 318, 1971.
7. Stocker, K.: In: Handbook of experimental pharmacology (Markwardt, F., ed.) Springer, Berlin, pp.451, 1978.
8. Zajdel, M., Wegrzynowicz, Z., Sawecka, J. and Kopec, M.: Thromb. Res. 6, 337, 1975.
9. Edgar, W. and Prentice, C.R.M.: Thromb. Res. 2, 85, 1973.
10. McKillop, C., Edgar, W., Forbes, C.D. and Prentice, C.R.M.: Nature 255, 638, 1975.
11. Rath, N.F. and Wuppermann, T. : Blut 36, 9, 1978.
12. Asbeck, F., Lechler, E., Martin, M. and van de Loo, J.: Haemostasis 3, 340, 1974.
13. Kehl, M., Lottspeich, F. and Henschen, A.: Hoppe-Seyler's Z. Physiol. Chem. 362, 1661, 1981.
14. Martinelli, R.A. and Scheraga, H.A.: Anal. Biochem. 96, 246, 1979.

THE ANALYSIS OF FIBRINOPEPTIDE RELEASE FROM S-CARBOXYMETHYLATED FIBRINOGEN CHAINS USING HIGH-PERFORMANCE LIQUID CHROMATOGRAPHY

Christopher Southan and Agnes Henschen

Max-Planck-Institut für Biochemie, D-8033 Martinsried/München, Federal Republic of Germany

Introduction

For the analysis of fibrinopeptide release from purified fibrinogen in vitro high-performance liquid chromatography (HPLC) has become the method of choice. A variety of HPLC techniques have now been developed and some of them have already been employed for an extensive kinetic characterisation of fibrinopeptide release, both from normal fibrinogen (1,2,3) and genetic variants (4,5). Up to now such measurements have been limited to the study of native fibrinogen. A similar type of analysis for S-carboxymethylated (Scm) fibrinogen (i.e. fully dissociated polypeptide chains) is of considerable theoretical interest but is associated with experimental difficulties because of the relatively low solubility of Scm material at physiological pH. The HPLC system of Kehl et al. (6) has already been used to analyse the fibrinopeptides released from Scm samples of fibrinogen Bicêtre (7) and London III (4) after extensive thrombin digestion. However, for these specific elution conditions the digest sample needs to be de-proteinated for the removal of Scm fibrin chains before application to the HPLC column.

The investigations reported here had two aims; the first was to establish a system for the simultaneous analysis of fibrinopeptides and fibrin(ogen) chains directly from an enzyme digest, the second was the application of such a technique for the kinetic analysis of fibrinopeptide release from Scm fibrinogen. Both these methods were specifically

Fibrinogen - Structural Variants and Interactions
© 1985 Walter de Gruyter & Co., Berlin · New York - Printed in Germany

developed for eventual application to the examination of genetically abnormal fibrinogens. In particular the characterisation of fibrinopeptide release from dissociated chain material could provide a useful method of discriminating between structural defects that exert a direct effect on the thrombin cleavage reaction and those defects where the alteration of fibrinopeptide release is influenced by indirect interactions which depend on the native conformation of the molecule.

Materials and Methods

Enzyme Digests. The S-carboxymethylation of fibrinogen and the preparation of solutions were performed according to Henschen & Edman (8). The freshly freeze-dried sample was dissolved to 10 mg/ml in distilled water brought to pH 10 by the addition of a drop of triethylamine (TEA). Preparations that remained insoluble were dissolved in alkaline urea and dialysed against H_2O/TEA, pH 9.0, for 16 hours at 4°C. A subsequent dilution of 1:1 with a 0.2M sodium bicarbonate solution and a careful titration with dilute acetic acid brought the pH down to 8.5. Before the addition of enzyme, free tryptophan (Trp) was added to the solution as an internal standard to facilitate the quantitative assessment of fibrinopeptide release. For complete peptide chain cleavage thrombin (Behringwerke, GFR) was added to a final concentration of 10 NIH u/ml and left at 22°C for 4 hours. For Reptilase digestion the contents of one bottle (Boehringer Haemodiagnostica, GFR) were dissolved in 1.0 ml, and an aliquot of 10 µl was added to 200 µl of Scm fibrinogen. The incubation conditions were the same as given for the thrombin-treated samples. Reactions were stopped by the addition of acetic acid to pH 3. These samples could be directly applied to the HPLC for analysis using the Vydac/TFA system.

HPLC Conditions. Three different eluent/column combinations were used. The gradient profiles employed are indicated on the chromatograms in Results. Instrumentation and operating conditions are fully described in the references given. For the separation of fibrinogen chains Vydac TPRP columns were used with a 0.1% trifluoroacetic acid (TFA)/acetonitrile eluent as described by Kehl et al. (9). During the development of a simultaneous assay for fibrinopeptides and fibrin(ogen) chains the Vydac/TFA system described above was used for the analysis of fibrinopeptides. The peaks were identified by re-chromatography using the 0.025 M ammonium acetate/acetonitrile system with a Merck, Hibar, C18 column, as described by Kehl et al.(6). For the measurement of the time-course of fibrinopeptide release the TFA/acetonitrile eluent was used with the Hibar, C18 column. The peak areas, including the internal standard, were corrected for differences in sample volume and converted to % of total release for each fibrinopeptide using a computer programm.

Results

Preliminary Separations. The two systems used for the analysis of fibrinopeptides are shown in Fig. 1 and 2 and the separation of Scm fibrinogen chains in Fig. 3. By collection of the already identified

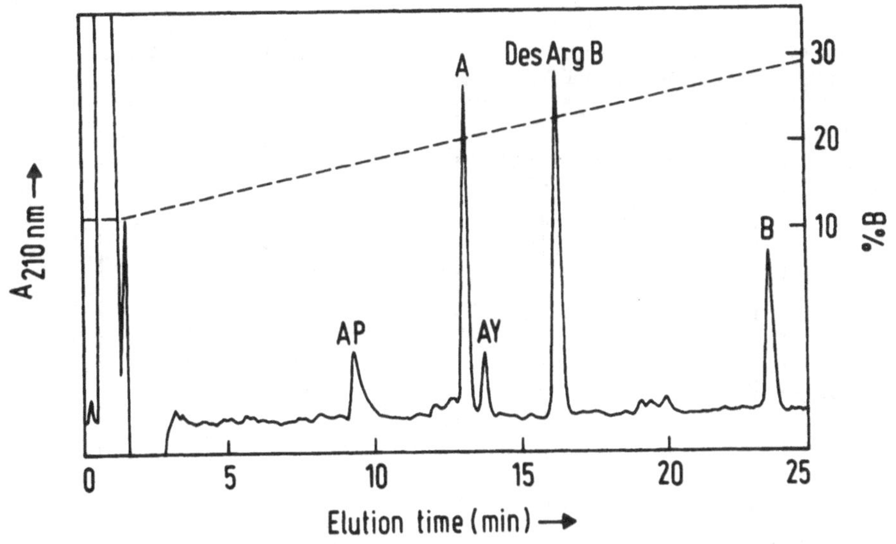

Figure 1: HPLC separation of fibrinopeptides from a 16 hour thrombin digest using the 0.025 M ammonium acetate/acetonitrile system (6) The absorbtion units full-scale (AUFS) was 0.2 and the sample loading equivalent to the peptides released from approximately 1 mg of fibrinogen.

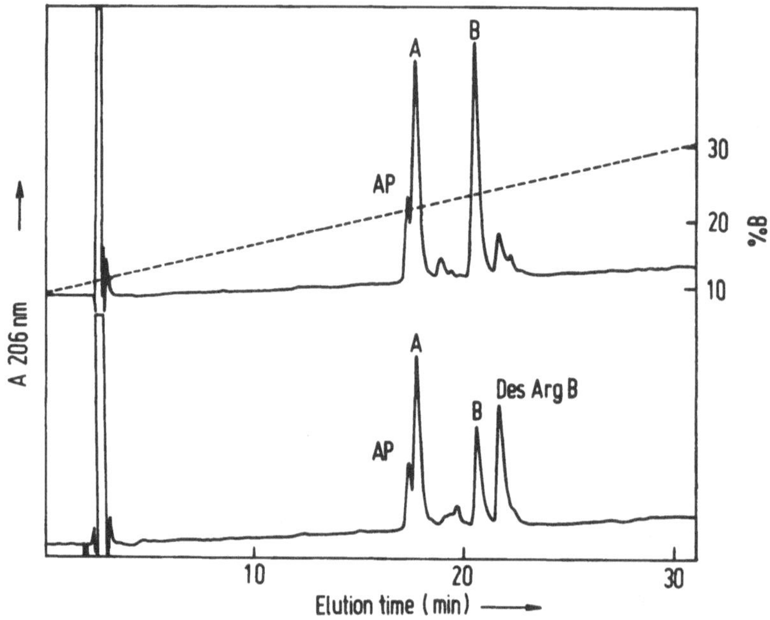

Figure 2: HPLC separation of fibrinopeptides from a 2 hour (upper curve) and a 16 hour (lower curve) thrombin digest using a Vydac/TFA system. The peaks represent the amount of peptide released from approximately 2 mg of fibrinogen at an AUFS of 0.5.

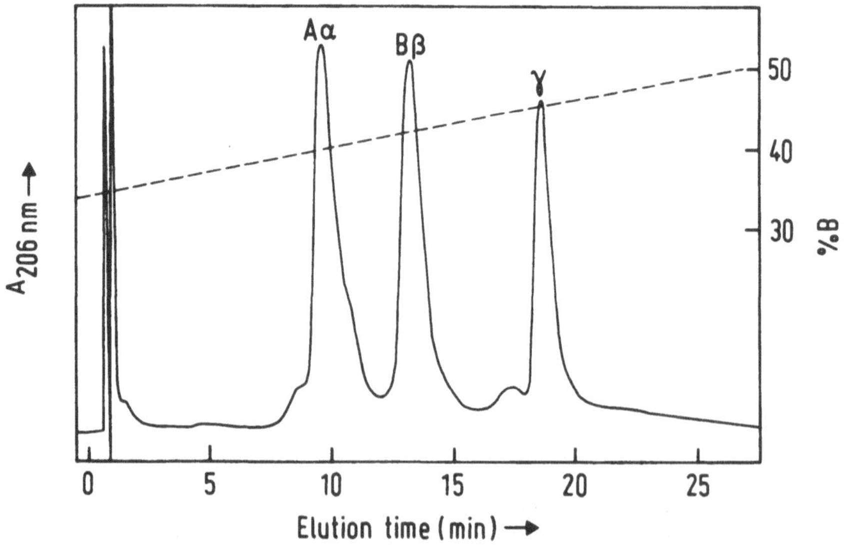

Figure 3: HPLC separation of Scm fibrinogen chains using a Vydac/TFA system according to Kehl et al. (7). The sample loading was approximately 0.5 mg of Scm fibrinogen (dialysed against 0.5 % TFA) and the AUFS was 1.0.

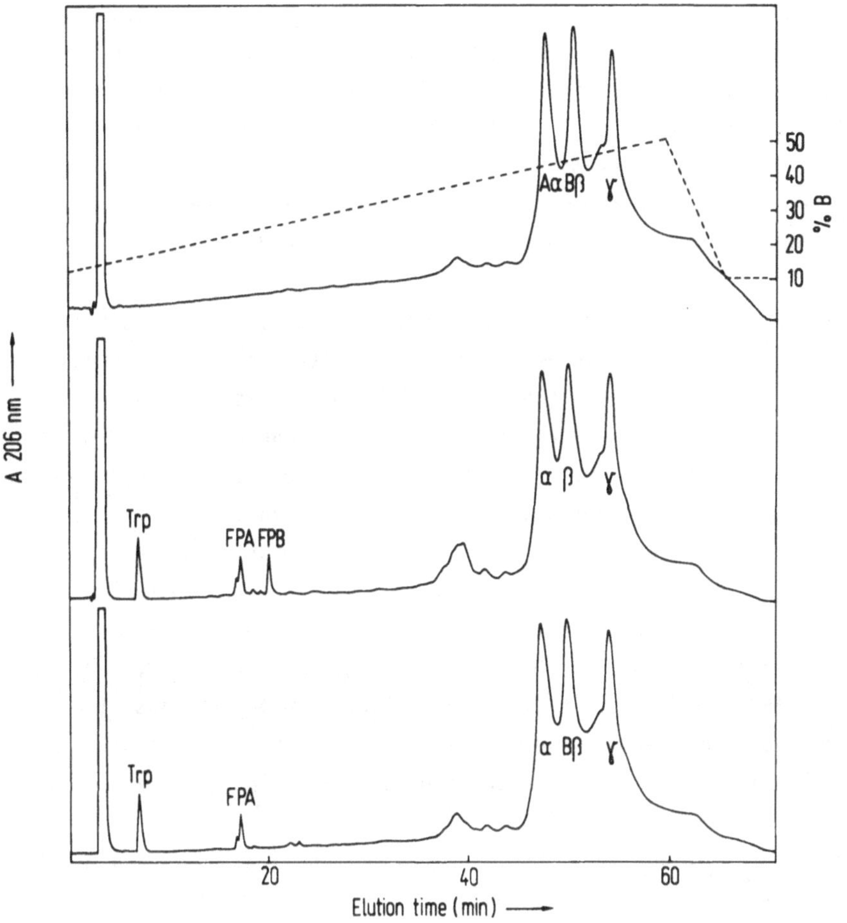

Figure 4: HPLC separations of untreated Scm fibrinogen (upper curve) and of Scm fibrinogen digested with thrombin (centre) and Reptilase (bottom) using a Vydac/TFA system. The small peak eluting at 35 % B is an additional cleavage product of thrombin or a contaminating proteinase.

fibrinopeptides eluted under the conditions shown in Fig. 1 and re-chromatography of them under the conditions given in Fig. 2 it was possible to establish the differences in elution order.

Combined Analysis of Scm Chains and Fibrinopeptides. The fibrinogen chains shown in Fig. 3 are eluted by the same TFA/acetonitrile solvent system used to separate the fibrinopeptides in Fig. 2. It was thus clear that both fibrinopeptides and chain material present in a thrombin digest of Scm fibrinogen might be separable by an extended elution gradient. The results of eluting undigested Scm fibrinogen and preparations treated with thrombin and Reptilase using an extended gradient are shown in Fig.4.

The chromatograms in Fig. 4 show that all the cleavage products from the digestion mixture can be adequately resolved in a single gradient elution. It is remarkable that peptide components as small as 14 residues (FPB) and as large as 610 residues (Aα-chain) can be recorded on the same run, in spite of the large absorbance difference at 206 nm. The inclusion of the internal standard (Trp) was essential for assessing the extent of cleavage because the comparative peak areas between the fibrinopeptides and Scm fibrin(ogen) chains varied considerably with changes in sample volume and different UV detectors. The resolution of Scm fibrin chains could be improved by decreasing the gradient from 0.66%/min to 0.5%/min (as used in Fig.3). With Reptilase treatments a peak was sometimes observed eluting prior to FPA but this was established to be a non-peptide contaminant of the enzyme preparation.

Individual Chain Analysis. From the results obtained it was clear that the same elution system could also be used to examine the cleavage products of purified chains (i.e. chains previously isolated from Scm fibrinogen by preparative HPLC or ion exchange chromatography). An example of this application is shown in Fig. 5 for the Aα-chain after digestion by thrombin under the same conditions as described above for Scm fibrinogen.

The small amounts of FPB give a direct assessment of Bβ-chain contamination of the Aα-chain preparation. With the thrombin digestion an additional cleavage product was observed to elute at 35 % B that was also present in the Scm fibrinogen digest (Fig.4).Further experiments with purified Aα-chain and different thrombin preparations indicated that the fragment was cleaved from this chain probably by another proteinase present as a contaminant in the enzyme preparation.

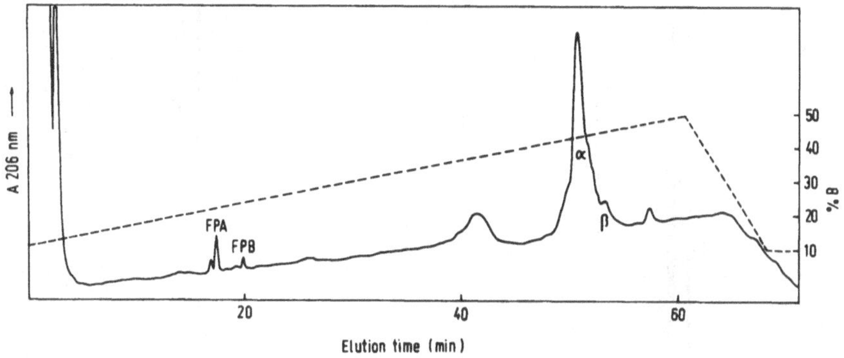

Figure 5: HPLC chromatography of a thrombin digest of purified Aα-chain using a Vydac/TFA system.

Time Course of Release. For the measurement of release kinetics the analytical runs were performed using a short gradient for the elution of fibrinopeptides alone from a sample of the digestion mixture. The chain material that was retained on the column during the injections for a series of time points, could be eluted with the appropriate gradient when the kinetic assays were complete. Examples of two time points are shown in Fig. 6. The integrated peak areas from a series of runs were used to plot the time course of fibrinopeptide release as examplified in Fig.7.

Much higher thrombin activities are required with Scm material than with native fibrinogen to obtain the same cleavage rates. Digests performed under conditions suitable for Scm material but with native fibrinogen as control are difficult to evaluate because of the extreme differences in cleavage rates.

Discussion

The analyses shown for Scm fibrin(ogen) chains and the fibrinopeptides cleaved from them (Fig. 4) may be of aid during the elucidation of variant fibrinogens. The same digests may be used for direct measurements of

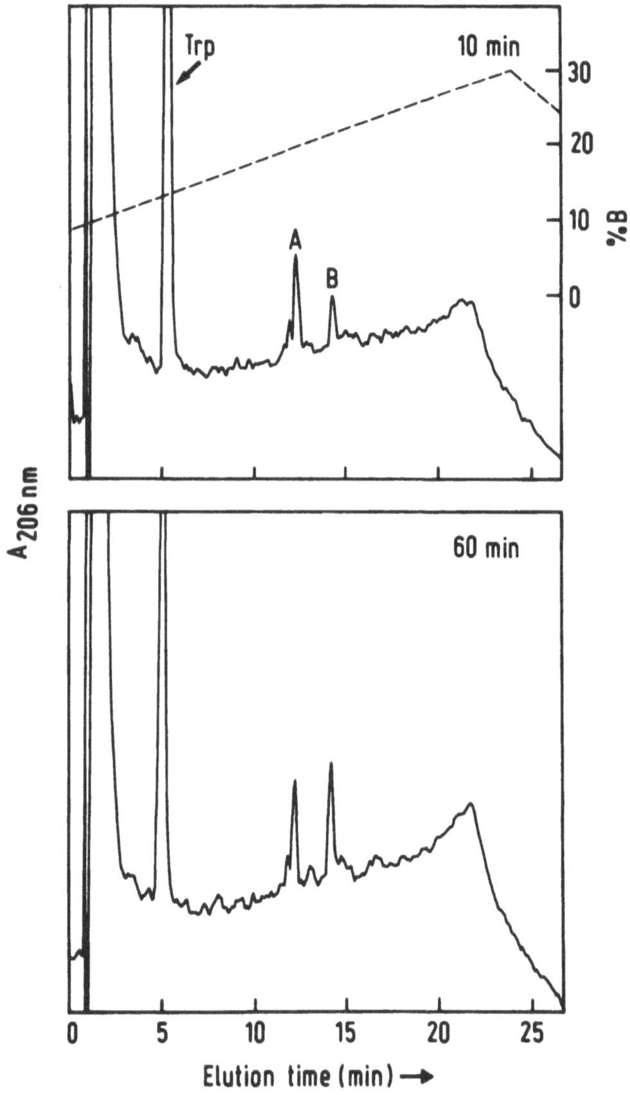

Figure 6: Quantitative fibrinopeptide analysis from a thrombin digest of Scm fibrinogen after 10 and 60 min with a thrombin concentration of 1 u/ml. The Trp peak, the internal standard, was off the chart scale but could still be quantitated by the peak integrator. A Hibar column was used in combination with the 0.1% TFA solvent system.

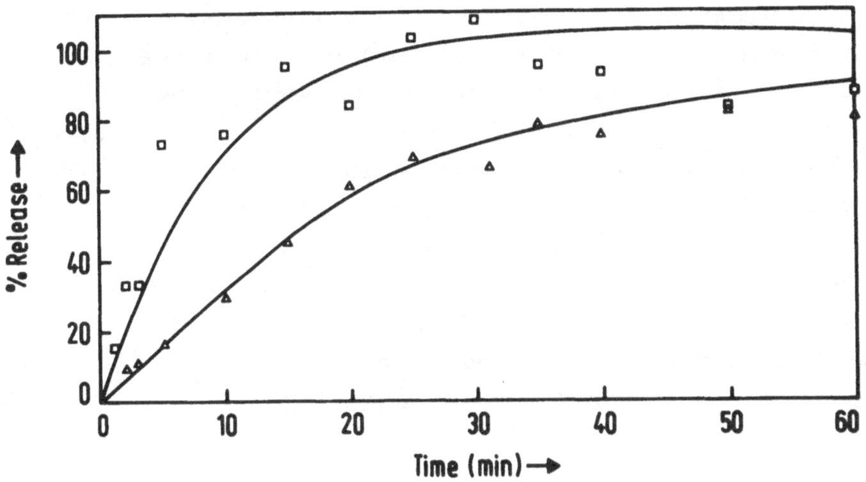

Figure 7: Time course of fibrinopeptides release from Scm fibrinogen under the conditions described in Fig. 6. FPA in indicated by □ and FPB by △.

release stoichiometry by comparison with the internal standard and for collection of fibrin chains for sequence analysis to confirm the presence of non-cleavable chains and of N-terminal deletions or substitutions. The method was found to be particularly useful for examining the purity of isolated chain preparations before further fragmentation for structure analysis.

The result for the time course of fibrinopeptide release (Fig. 7) demonstrates that the direct measurement of cleavage rates of fibrinopeptides from dissociated chain material is possible. The method has the advantage that the enzyme digest can be directly applied to the HPLC column without an additional deproteination step.The low solubility of Scm fibrinogen preparations in the absence of urea makes it, however, necessary to use a pH above 8 in the digest mixture. Although thrombin is able to cleave native fibrinogen in urea solutions (10), the presence of urea might cause unpredictable alterations in the kinetic constants and should therefore be avoided.

Changes in the pattern of fibrinopeptide release caused by the structural defects in molecular variants can be complex and difficult to interpret. Numerous examples have been reported where the observed impairments in FPB release could be an indirect manifestation of slow FPA release and/or defective fibrin polymerisation. In Scm fibrinogen the tertiary structure of the native molecule is to a large extent abolished and therefore any interactions involving polymerisation or long-range effects on the conformation of the cleavage sites are eliminated. A demonstration of slow release of either fibrinopeptide from Scm fibrinogen can thus unequivocably demonstrate that the defect is located within those sequences directly involved in the enzyme cleavage.

References

1. Martinelli, R.A., Scheraga, H.A.: Biochem. 19, 2343-2350 (1980).
2. Kehl, M., Lottspeich, F., Henschen, A.: In: Fibrinogen - Recent biochemical and medical aspects. (A. Henschen, H.Graeff, F.Lottspeich, eds) Walter de Gruyter, Berlin, pp.217-224 (1982).
3. Higgins, D.L., Lewis, S.D., Shafer, J.A.: J. Biol. Chem., 258, 9276-9282 (1983).
4. Southan, C., Lane, D.A., Henschen, A.: In: Fibrinogen - Structure, functional aspects, metabolism. (F. Haverkate, A. Henschen, W. Nieuwenhuizen, P.W. Straub, eds.) Walter de Gruyter, Berlin, pp. 167-182 (1983).
5. Southan, C., Lane, D.A., Henschen, A.: This volume.
6. Kehl, M., Lottspeich, F., Henschen, A.: Hoppe-Seyler's Z. Physiol. Chem. 362, 1661-164 (1981).
7. Kehl, M., Lottspeich, F., Henschen, A.: In:Fibrinogen - Structure, functional aspects, metabolism (F. Haverkate, A. Henschen, W. Nieuwenhuizen, P.W. Straub, eds) Walter de Gruyter, Berlin, pp. 183-190 (1983).
8. Henschen, A., Edman, P.: Biochim. Biophys. Acta, 263, 351-367 (1972).
9. Kehl, M., Lottspeich, F., Henschen, A.: Hoppe-Seyler's Z. Physiol. Chem. 363, 1501-1505 (1982).
10. Brosstad, F., Godal, H.C., Kierulf, P.: Haemostasis, 6, 225-235 (1977).

MODIFICATION OF THE FIBRIN α-CHAIN BY DIPEPTIDYL PEPTIDASE IV

Rolf Mentlein, Gernot Struckhoff, and Eberhard Heymann
Biochemisches Institut der Universität Kiel
Olshausenstrasse 40
D-2300 Kiel, F.R.G.

Introduction

Dipeptidyl peptidase IV (DP IV; EC 3.4.14.5; postproline dipeptidyl aminopeptidase) is a very specific enzyme that cleaves dipeptides from the N-terminal end of peptide chains, provided that the penultimate residue is a proline. Although several small natural peptides with this sequence, e.g. the undecapeptide substance P (Arg-Pro-Lys-Pro-Glu ...) are cleaved by this enzyme (1), the physiological significance is obscure. Since high activities line the capillary endothelium (2), the peptidase may act on a natural substance in the streaming blood.

The last step of the coagulation cascade is the thrombin-catalyzed release of small polar peptides from the N-termini of the fibrinogen α- and β-chains, the resulting fibrin monomers polymerize spontaneously. The fact that (a) the amino acid sequence of the N-terminus of the fibrin α-chain (Gly-Pro-Arg-...) has been conserved in the evolution and (b) small synthetic peptides similar to this sequence (e.g. Gly-Pro-Arg-Pro) competitively inhibit fibrin coagulation (3) indicate that this N-terminus is an important aggregation site. Thus, the clotting capacity of monomeric fibrin should be impaired if dipeptidyl peptidase IV would cleave glycyl-proline from the amino terminus of the fibrin α-chain. In a preliminary communication (4) we reported that dipeptidyl peptidase IV inhibited fibrin polymerization and simultaneously released

Fibrinogen - Structural Variants and Interactions
© 1985 Walter de Gruyter & Co., Berlin · New York - Printed in Germany

glycyl-proline. However, this study was lacking quantitative data on the amount of dipeptide released.

Methods

Human dipeptidyl peptidase IV was prepared from human placenta (5). The enzyme was homogeneous on dodecyl sulfate electrophoresis (M_r 120 000). The activities with chromogenic substrates were estimated at 37°C in 50 mM Tris/HCl-buffer pH 8.0 as described (5).

Glycyl-proline that was cleaved from peptides or proteins was quantitated with a modified amino acid analyser. Samples were acidified (pH ≈ 2) with conc. HCl and applied to a 22 x 0.6 cm thermostated (60°C) column with microbore cation exchanger (Technicon C-3 chromo beds). With 0.1 M citrate buffer pH 3.60 containing 0.4 M Na^+ as eluent Gly-Pro could be separated from all interfering substances (retention time 35 min with buffer pumped at 20 ml/h). The eluate was mixed (1:1) with a buffer containing 0.05% ortho-phthaldialdehyde, 0.2% 2-mercapto--ethanol, and 1 M borate, pH 10.4. The fluorescence (λ_{Ex}=360nm, λ_{Em}=450nm) was continuously monitored after a reaction time of 90s. Minimal amounts of 0.1 nmol glycyl-proline could be estimated.

Varying concentrations of the tetrapeptide Gly-Pro-Arg-Pro (obtained from Bachem, Switzerland) were incubated in 50 mM Tris/HCl pH 8.0 at 37°C with approx. 0.2 µg DP IV for 1 min in a total volume of 0.2 ml, the reaction was terminated by addition of 5 µl conc. HCl. Monomeric desAB-resp. desA-fibrin was obtained from fresh fibrin clots generated with bovine thrombin resp. reptilase (venom from Bothrops atrox, Sigma). The clots were redissolved in 1 M KBr. Samples of 500 µl of these fibrin-solutions were incubated for 60 min at 37°C with 0.5-7 µg purified DP IV in 150 µl 0.2 M phosphate buffer pH 7.5 (final pH 7.4). After acidifying with HCl, the samples were centrifuged and applied to the analyzer. Similar samples containing the same amounts of heat denatured dipeptidyl peptidase served as references. Fibrinogen/thrombin clotting assays were performed using human fibrinogen (Cohn fraction I), purified by gel chromatography and dissolved in 20 mM phosphate buffer pH 7.4 containing 1 mM EDTA,and bovine thrombin (390 NIH-U/mg). Fibrinogen (1-4 mg) was incubated with DP IV (2-40 µg) and thrombin (1-3 µg) at 37°C. Samples with heat denatured DP IV served as reference. After varying times of incubation (max. 120 min), the clots were separated and the fibrin content was estimated with the biuret method. The remaining samples were acidified, centrifuged and analyzed for glycyl-proline. Glycyl-proline released in clotting-assays of this type was also detected on amino-acid-analyzers(Bio--tronik) in the position of norleucine (courtesy of M. Kehl and A. Henschen, Martinsried).

Results and Discussion

Table I summarizes kinetic data for the hydrolysis of Gly-Pro-derivatives with highly purified dipeptidyl peptidase IV. The kinetics found with the tetrapeptide Gly-Pro-Arg-Pro were in the same order as with chromogenic substrates.
This tetrapeptide that has been reported (3) to prevent fibrin aggregation in a fibrinogen/thrombin clotting assay also acts as a competitive inhibitor for the dipeptidyl peptidase (K_i=0.166 mM with Gly-Pro-4-nitranilide). The rapid metabolism of this peptide by endothelial dipeptidyl peptidase IV should be taken into account in in-vivo-studies with this clotting inhibitor. We preincubated the tetrapeptide (0.1 mM) with DP IV (0.35 µg) for 10 min at 37°C in the presence of fibrinogen (1.5 mg) in a total volume of 0.5 ml. During this time the theoretical amount of glycyl-proline was cleaved from the tetrapeptide. After addition of thrombin (2 µg, ca. 0.8 NIH-Units) the same clotting time was observed as in the

Table I Kinetic data for the hydrolysis of Gly-Pro-derivatives with highly purified dipeptidyl peptidase IV at 37°C in 50 mM Tris/HCl-buffer pH 8.0, N_t, maximal turnover number (s^{-1}).

Substrate	K_m (mM)	V (U/mg)	N_t/K_M ($s^{-1}mM^{-1}$)
Gly-Pro-2-naphthylamide	0.200	69	690
Gly-Pro-4-methoxy-2-naphthylamide	0.018	4o[a]	4400
Gly-Pro-4-nitranilide	0.125	168	2690
Gly-Pro-Arg-Pro	0.086	75	1740

(a) Substrate inhibition above 0.1 mM

corresponding assay without tetrapeptide. In contrast, when the dipeptidyl peptidase was added after incubation with thrombin (20 min, 37°C), no clotting could be observed. Apparently, the tetrapeptide cannot be attacked by DP IV once it is bound to monomeric fibrin.

When monomeric desAB- or desA-fibrin (10 µM) was incubated with dipeptidyl peptidase IV in 0.8 M KBr/50 mM phosphate, pH 7.4, at 37°C the amount of glycyl-proline released depended on the dose of added DP IV.

However, the specific activity of the peptidase was low under these conditions (0.18 nmol/(h · µg) = 3 mU/mg). With 10 µM Gly-Pro-4-nitranilide the specific activity was 1500 mU/mg at the same high salt concentration.

In an earlier report (4) we presented evidence for considerably enhanced fibrinogen/thrombin clotting times in the presence of DP IV. We could not confirm these high values with new peptidase preparations. However, with high DP IV activities a dose-dependend release of Gly-Pro could be detected in fibrinogen/thrombin clotting assays (Tab. II). Heat-denatured DP IV or DP IV preincubated (30 min) with the active-site directed inhibitor bis(4-nitrophenyl)phosphate (0.1 mM), which completely abolishes DP IV activity but does not influence thrombin, was without effect. With the highest peptidase activity used 0.17 moles Gly-Pro were obtained per 1 mol fibrin. In all experiments of this type the prolongation of the clotting time as compared to the assays with heat denatured or inhibited enzyme was moderate (Tab. II), but corresponded to values that should be expected in view of the amount of modified fibrin-α-chain.

Table II Action of dipeptidyl peptidase IV (DP IV) in fibrinogen/thrombin clotting assays (pH 7.4, 37°C).

Thrombin	DP IV	Fibrin clotted	Gly-Pro released	clotting time
μg	μg	nmol	nmol	s
1	35	8.2	1.4	71
1	35	4.5	0.6	71
1	14	2.2	0.3	69
1	3	2.7	not dectable	66
1	35 heat denatured	8.2	0.0	64
1	35 inhibitor treated	8.2	0.0	64

1 μg thrombin corresponds to 0.4 NIH-Units or 13 mU with 0.1 mM Tos-Gly-Pro-Arg-4-nitranilide; 1 μg DP IV corresponds to 78 mU with 0.1 mM Gly-Pro-4-nitranilide at pH 7.4, 37°C.

With regard to the high DP IV activities necessary for a significant anticlotting-effect in vitro, a physiological role of the peptidase is questionable. However, the in-vivo situation at the endothelial surface may be different.

References

1. Heymann, E., Mentlein, R.: FEBS Lett. 91, 360-364 (1978).
2. Lojda, Z.: Histochemistry 59, 153-166 (1979).
3. Laudano, A.P., Doolittle, R.F.: Proc. Natl. Acad. Sci. USA 75, 3085-3089 (1978).
4. Mentlein, R., Heymann, E.: Arch. Biochem. Biophys. 217, 748-750 (1982)
5. Püschel, G., Mentlein, R., Heymann, E.: Eur. J. Biochem. 126, 359-369 (1982).

IV. FIBRINOGEN-FIBRIN INTERACTION

ANALYSIS OF COMPOSITION OF SOLUBLE FIBRINOGEN/FIBRIN COMPLEXES BY DIFFERENTIAL ULTRACENTRIFUGATION

John Rieden Shainoff
Research Division, Cleveland Clinic Foundation, Cleveland, OH 44106, USA

Introduction

The solubility of fibrin monomer in blood is determined largely by interaction with fibrinogen (1,2). Soluble complexes formed by the interaction have been shown to have varying dissociability depending on fibrinopeptide content of the dissolved fibrin (3-5). The equilibrium constant for monomer dissociability from complexes formed by α-fibrin lacking fibrinopeptide A has been estimated (3) at 10^{-7} M, while the dissociability of complexes from αβ-fibrin monomers lacking both fibrinopeptides appears to be at least a thousand times less (< 10^{-10} M). Despite the large differences in dissociability, existing data indicate that the solubilities of the two types of fibrin in fibrinogen solution differ by only a factor of three (1,2). The present study addresses the question whether composition of the soluble complexes differ. The relative content of fibrinogen and fibrin in soluble complexes was assessed by the method of differential ultracentrifugation (6), which utilizes resolving power of the Rayleigh interference optical system of the ultracentrifuge for determination of small differences in concentrations of components sedimenting in two aligned chambers (reference and sample).

Materials and Methods

Fibrinogen fraction I-0 (7) was prepared from fresh citrated rabbit plasma, and purified further (8) to remove cryoprecipitable protein and plasminogen. Monomeric α-fibrin was prepared by treating with thrombin at pH 5.3, and then inactivating the thrombin with hirudin as described (9). Monomeric αβ-fibrin was prepared by coagulating fibrinogen with thrombin in 0.3 M TRIS-HCl, dissolving the clot at pH 5.3, and then adding hirudin as described (9). The fibrin monomer preparations were dialyzed at 4°C against a solution consisting of 0.375M TRIS-HCl, 0.015M TRIS-

Fibrinogen - Structural Variants and Interactions
© 1985 Walter de Gruyter & Co., Berlin · New York - Printed in Germany

acetate, and 0.015M benzoyl arginine methyl ester at pH 5.3-5.32, and the fibrinogen preparation was dialyzed against 0.27M NaCl and 0.03M TRIS-HCl at pH 7.4.

Mixtures of fibrinogen and fibrin were prepared by precisely dispensing requisite volumes of the two into separate tubes, and then pouring the fibrinogen and neutralizing buffer into the smaller volume of fibrin solution, and then shuttling the mixture between the tubes several times with care to avoid loss of solution to the outside of the tubes. This method was adopted because of need for rapid admixture of the two solutions and precise dilutions. Reference solutions of fibrinogen, not admixed with fibrin, were prepared by admixing the fibrinogen solution in the same way with dialysate from the fibrin preparation.

Ultracentrifugation was performed with a Spinco Model E equipped with schlieren and Rayleigh optical systems. Alignment of the optical system and filling of the cells were critically assessed to ensure that fringe shifts provided accurate measurements of differences in concentration of sedimenting material in the sample vs. reference chamber. For alignment of the optical system, the mask holder and rotatable cylindrical lens mount were set to obtain nondeviated (straight) fringes across boundaries formed by identical fibrinogen solutions (2 mg/ml) set in both the sample and reference chambers of the cell. A long path length (30 mm) cell equipped with sapphire windows was used for all studies, except for calibration which used a standard 12 mm artificial boundary cell. To ensure identical filling of the two chambers for the alignment, a gasket was placed between the cell and the windows to provide a gap for equilization of fluid levels in the cell. A small volume of liquid fluorocarbon was also added to enable visualization of both the top and bottom menisci of the fibrinogen solutions. With the fibrinogen solution levels thus made identical in the two chambers, there could be no positional differences due to unequal starting points for sedimentation. Such positional differences would lead to transient fringe deviations. The gasket that was used to provide communication between the chambers during alignment was not employed in experimental studies in order to avoid transfer of solution from one chamber to the other. To ensure that the menisci in the sample and reference chamber were the same, the solutions were accurately transferred to the chambers by precision bore, interchangeable syringes fitted to a micrometer (Model SB2 from Micro-Metric Instrument Company, Cleveland, OH). The chambers of the ultracentrifuge cell selected for the study were found to have been perfectly machined, as verified by observation that filling both chambers with identical volumes of fibrinogen solution yielded sharply superimposed menisci and no fringe deviations across the fibrinogen boundaries. The Rayleigh optical system was calibrated by using a 12 mm artifical boundary cell to overlay water above a solution of sucrose of known refractive index, and subsequently determining the vertical deviation in position of the interference fringes across the resultant refractive index gradient. Fringe displacements were measured within 2 microns by a Nikon optical comparator and measuring stage.

Anticipated effects of complex formation on the sedimentation of fibrinogen are illustrated schematically in Figure 1. The upper chart portrays an idealized, rectangular plot of refractive index increment (proportional to concentration) vs. distance in the ultracentrifuge cell during sedimentation. The reference solution consisting of fibrinogen without added fibrin is portrayed as a step function (N_o) representing the boundary of fibrinogen sedimenting at constant concentration (ϕ_o). The concentration of fibrin added to the fibrinogen solution is depicted as f_s, so that the refractive index increment (proportional to total protein) due to f_s represents the difference between the reference and sample solutions. The sample solution containing added fibrin is depicted as a double step function (N_s) corresponding to (left to right) fast sedimenting complexes (P_s') and the slower sedimenting fibrinogen (ϕ_s^{obs}). The fibrinogen boundary in the sample (subscript s) is depicted as sedimenting slower (behind) than the fibrinogen in the reference, because back flow of fluid displaced by the faster sedimenting polymeric complexes causes a reduction in the sedimentation rate which exceeds the reduction imposed by solvent displacement in sedimentation of fibrinogen alone. The concentration of fibrinogen in the sample is depicted as being lower than the reference due to association of some of the fibrinogen with the fast sedimenting complexes. This apparent concentration of fibrinogen will differ somewhat from the actual concentration of fibrinogen (ϕ_s') in equilibrium with the complexes due to a concentrational buildup caused by unequal sedimentation due, in turn, to the viscosity difference across the polymer boundary as described by Johnston and Ogston (10). Likewise, the actual concentration of polymeric complexes (P_s') will be greater than apparent (P_s^{obs}) by an equal increment.

The chart in the center illustrates the difference between refractive indices of the sample vs. the reference solutions by subtraction of the areas of the two. In practice, the step functions are not rectangular, but are sigmoid due to boundary spread by diffusion. A sigmoidal distribution curve is superimposed as a heavy line on the step function plot of refractive index increment. This sigmoidal plot depicts in schematic form the type of pattern anticipated for the Rayleigh interference fringes seen in differential ultracentrifugation, and the data derivable from the fringe deviations are indicated by arrows. The chart at the bottom schematically illustrates the schlieren pattern anticipated for the gradients of refractive index illustrated in the upper chart, but with broadened rather than infinitely sharp peaks to portray spread rather than sharply stepped boundaries.

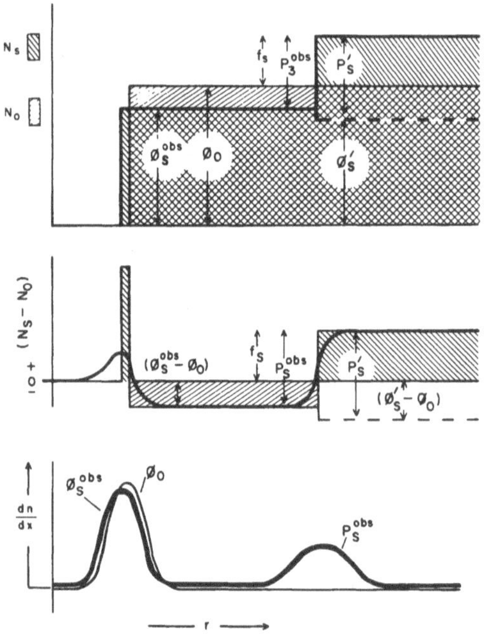

Fig. 1. Schematic description of refractive index differences between solutions in reference and sample chambers containing fibrinogen solution (N) and fibrinogen/fibrin mixture (N_s) for measurement of changes in concentration and sedimentation rates caused by the addition of fibrin. The symbols are explained in the text for ultracentrifuge patterns.

All quantities depicted in the charts are determinable from the ultracentrifuge patterns with exception to the correction ($\phi_s^{obs} - \phi_s'$), which equals the obscured concentration of polymer ($P_s' - P_3^{obs}$), due to the concentration buildup of fibrinogen from Ogston-Johnston effect. The correction can be substantial with large asymmetric proteins, and is calculable from the difference in sedimentation rate (**S**) of the slow component (fibrinogen) and polymeric component (P) in absence (β region) vs. the difference in presence (γ-region) of the polymer, $\phi_s^{obs} / \phi_s' \simeq [s_p^\gamma - s_\phi^\gamma] / [s_p^\gamma - s_\phi^\beta]$.

Viscosities of solutions of fibrinogen and fibrin were determined from flow time (90 to 110 seconds with reproducibility to 0.05 seconds) in Ostwald-type viscometers fitted to a bath regulated at 25 ± 0.02°C. Solutions were centrifuged at 12,000 G for 15 minutes prior to measurement of viscosity. Concentrations of fibrin were determined from the analytical ultracentrifuge patterns.

Results

With the buffer employed (equimolar NaCl and TRIS-HCl at 0.15 ion strength and pH 7.4), α-fibrin required an 8 molar excess of fibrinogen for full dissolution at 25°C, the temperature adopted for centrifugation. This proportion of fibrinogen and fibrin also corresponded to the proportion required for dissolution of the fibrin in saline at physiologic temperature, but with only trace levels of TRIS for buffering, the solubility of fibrin in fibrinogen being somewhat higher in TRIS-HCl than in saline alone. To obtain adequately high concentrations of αβ-fibrin (>0.1 mg/ml) without increasing the relative fibrinogen concentration (<25 molar excess), the ionic strength had to be doubled. The sedimentation characteristics of α- and αβ-fibrin were similar at both 0.15 and 0.3 ionic strength, in that both fibrins formed aggregates which sedimented as a fairly sharp boundary at 24S relative to 8S for fibrinogen at the outset, but gradually broadened while maintaining the modal rate of 24S as centrifugation progressed (Fig. 2, schlieren pattern).

A decrease in the apparent concentration of fibrinogen was discernible from fringe shifts across the fibrinogen boundary in all samples containing added fibrin monomer, due presumably to partial association of the fibrinogen with the added monomer. As anticipated (Fig. 1), there was (Fig. 2) a slight transient upward deviation of the fringe pattern at the very front of the fibrinogen boundary due to a slight difference in position that arose from solvent displaced backwards by the fast sedimenting complexes in the sample. This transient upward deviation was followed by a sharp downward deviation across the fibrinogen boundary, and plateaued at the leading edge of the fibrinogen boundary. The fringe displacement from the plateau level at the meniscus to the plateau at the leading edge of the fibrinogen boundary provided the initial estimate of fibrinogen asssociated with the fast sedimenting complexes. The fringe pattern then deviated upwards across the polymer boundary, and plateaued at the leading edge of the boundary. The plateau region in front of the polymer boundary corresponded to the γ-region with composition corresponding to the original mixtures in containing original concentrations of fibrinogen and complexes but diluted slightly by radial sedimentation. Measurements of concentrations of fibrin dissolved in the solutions, as determined from net displacement of fringes coincided, after slight correction for radial dilution, with those expected from amounts added (cf. table) with exception to a solution formed by adding only a five rather than 8-fold excess of fibrinogen.

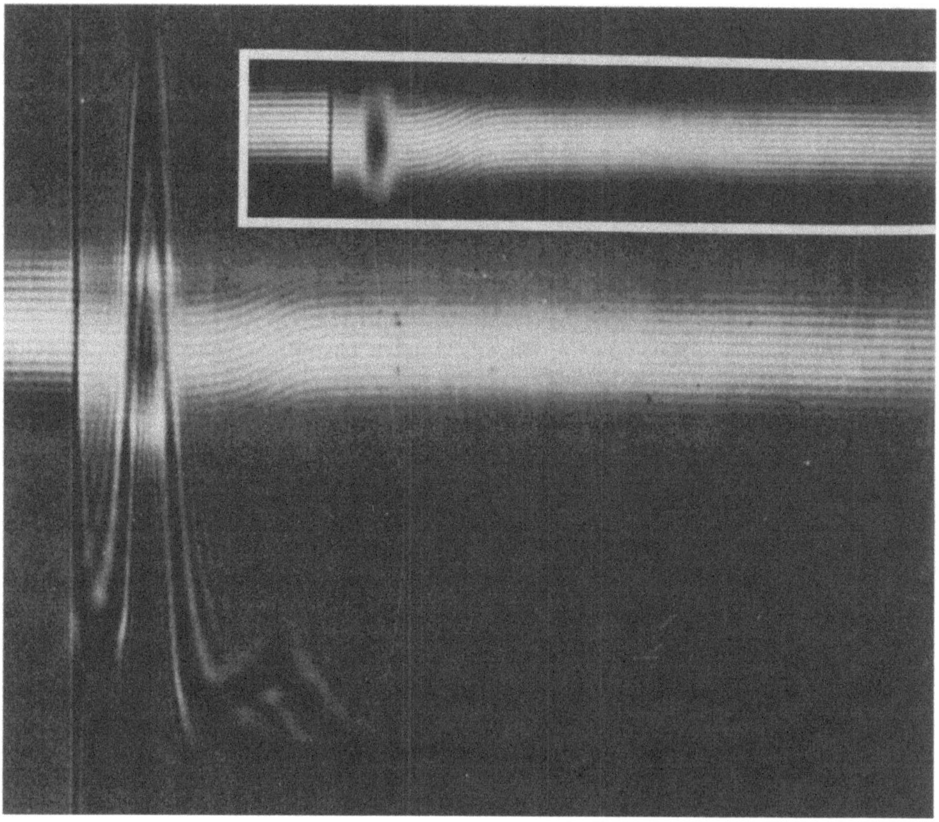

Fig. 2. Schlieren and Rayleigh patterns from ultracentrifugation of fibrinogen (1.73 mg/ml) in reference solution and in sample containing admixed α-fibrin monomer (0.327 mg/ml) at 0.3 ionic strength.

Amounts of fibrinogen appearing to be associated with the fibrin as calculated from observed change in fibrinogen concentration ranged widely between 0.07 and 0.13 mg

fibrin (observed) f_s mg/ml	fibrin (expected) mg/ml	fibrinogen (initial) ϕ_o mg/ml	S_ϕ^β/S_P^γ	S_ϕ^γ/S_P^γ	observed dec. in fibrinogen $(\phi_s^{obs}-\phi_o)$ mg/ml	obscured dec. in fibrinogen $(\phi_s^{obs}-\phi_s')$ mg/ml	apparent fibrinogen in polymers $(\phi_s'-\phi_o)$ mg/ml	$\frac{\text{fibrinogen}}{\text{fibrin}}$ in polymers
0.124	0.126	2.82	0.313	0.302	0.028	0.048	0.076	0.61
.242	.300	1.63	.330	.315	.063	.057	.120	.50
.247	.repeat at 24 h		.310	.287	.054	.050	.104	.42
0.298	.327	1.73	.339	.307	.125	.074	.199	.67
.306	.repeat at 24 h		.307	.282	.080	.069	.149	.49
.124	.130	2.59	.320	.308	.014	.049	.063	.51
.126	.130	1.92	.320	.308	.017	.033	.050	.40
.074	.080	1.74	.336	.330	.028	.015	.043	.58
.238	.271	1.72	.300	.276	.017	.057	.074	.31

Upper rows: αβ-fibrin, 0.30 Γ/2, pH 7.4
Middle rows: α-fibrin, 0.30 Γ/2, pH 7.4
Lower rows: α-fibrin, 0.15 Γ/2, pH 7.4

fibrinogen/mg fibrin depending on the relative concentration of fibrinogen and fibrin in the solutions. This variation was attributable to some of the change in fibrinogen being obscured by the Johnston-Ogston effect. The extent to which viscosity differences contributed to unequal sedimentation of fibrinogen in front of (within) the polymer boundary (γ-region) and behind (outside) the boundary (β-region) was calculated by dividing the observed relative sedimentation rate of fibrinogen/complexes ($S_{\phi}^{\beta}/S_{P}^{\gamma}$) by the relative viscosity of the sample and reference solutions (η sample/η reference), which yielded $S_{\phi}^{\gamma}/S_{P}^{\gamma}$ used for estimation of the enrichment of fibrinogen (Johnston-Ogston effect) behind the fibrinogen boundary. This enrichment was labelled as obscured decrease in fibrinogen in the table. The obscured change in fibrinogen ranged to as high as 3-times the observed decrease. On adding the observed and obscured decrease, consistent values were obtained for the relative amounts of fibrinogen and fibrin in the complexes for both α-fibrin and αβ-fibrin. These values corresponded to approximately 1 molecule of fibrinogen per 2 of fibrin in the polymers, e.g. fibrinogen comprised approximately one-third of the mass of the complex.

The specific viscosity (η/η_0 -1) of the dissolved α- and αβ-fibrin were essentially the same (291±26 SEM and 290±19 per gm/ml). Precision of the viscosity measurements for estimating differential sedimentation across the polymer boundary was verified by observation that hemocyanin, which sedimented well ahead of the complexes, had a 4 percent faster sedimentation rate in fibrinogen than in the fibrinogen-fibrin mixture. Since the mass of the aggregates was approximately three-halves of the fibrin concentrations due to content of fibrinogen, the specific viscosity of the complexes would correspond to about 200 per gm per ml. By comparison to fibrinogen with an intrinsic viscosity of 27 (g/ml^{-1}), we estimate that the complexes had axial ratio on the order of 10 times that of fibrinogen, and corresponded to oligomers similar to the intermediary polymers described by Ferry (11). The modal sedimentation rate of 24S also coincided with that of the fibrin intermediary polymers.

Discussion

Much of the difficulty in measuring association of fibrinogen with fibrin derives from the need to employ large molar excesses of fibrinogen to maintain the fibrin in solution. Complexes precipitated by cold were found to contain essentially equimolar quantities of fibrinogen and fibrin (1). Complexes that were isolated in a soluble form and sedimenting at 24S after stabilizing them by cross-linking with factor XIIIa were found to be composed of approximately 40% fibrinogen, fibrinogen/fibrin ratio = 0.6 - 0.8 (12), essentially the same as the composition determined here by differential ultracentrifugation. However, Sherman et al. (3) observed that some complexes formed by αβ-fibrin could be isolated without stabilization, and these contained large, 2-4:1 proportions of fibrinogen/fibrin. Mahn et al. (14) did not detect reversible complexing in their chromatographic studies on fibrinogen and des AABB fibrin.

The compositions of the complexes indicated by the present study have been taken into account in a preceding study (3) on dissociability of the complexes, and accordingly conform with the proposed structure in which monomers and small oligomers of fibrin are linked by either weak interactions (10^{-7}M K_D) between the E domain of α-fibrin and D-domain of fibrinogen, or by strong ($<10^{-10}$M K_D) interaction for αβ-fibrin.

The enhanced interaction with αβ-fibrin was viewed as a result of coupling of interactions through the A and B-dependant aggregation site, so that the two sites reinforce the staggered overlapping assembly without cross-linking, rather than extend it in separate directions. Despite the large differences in strength of association of fibrinogen with α- and αβ-fibrin, the results of the present study suggest the composition of the complexes formed by the two fibrins do not differ greatly in terms of relative content of fibrinogen and fibrin. However, the two types of fibrin were studied at differing ionic strengths where their solubilities in fibrinogen are similar. Further study will be required to determine the relationship of composition to solubility. By comparison, insoluble fibrils that form when fibrin monomer, either α- or αβ-, are added in excess of solubility contain approximately 14% fibrinogen by weight. Thus, the 30-40% content of fibrinogen observed here for near saturation represents only a factor of two difference.

Summary

Changes in concentration of protein sedimenting in free, nonaggregated form at characteristic rate of 8S upon adding fibrin monomer to fibrinogen solutions were measured by Rayleigh interference optics in the ultracentrifuge. Complexes formed by interaction of the fibrin with the fibrinogen sedimented at a modal rate of 24S, and their content of fibrinogen was determined to correspond to an average of 0.5 mole fibrinogen per mole of monomer from observed change in fibrinogen and correction for differential migration due to viscosity change contributed by the complexes with intrinsic viscosity of 200 $(g/ml)^{-1}$.

Acknowledgement

These studies were supported in part by grants HL-16361 and HL-19767, and the PROPHET computer system of the National Institutes of Health.

References

1. Shainoff, J.R., Page, I.H. J. Exp. Med. 116, 687-707 (1962)
2. Kierulf, P. Thromb. Res. 3, 613-630 (1973)
3. Shainoff, J.R., Dardik, B.N. Ann. N.Y. Acad. Sci. 408, 254-268 (1983)
4. Muller Berghaus, G., Mahn, I., Krell, W. Thromb. Res. 14, 561-572 (1979)
5. Brosstad, F. Thromb. Res. 15, 563-568 (1979)
6. Schachman, H.K. Biochemistry 2, 887-905 (1963)
7. Blomback, B., Blomback, M. Arkiv. for Kemi. 10, 415-443 (1956)
8. Kanaide, H., Shainoff, J.R. J. Lab. Clin. Med. 85, 574-597 (1975)
9. Dardik, B.N., Shainoff, J.R. Thromb. Haemostas. 42, 864-872 (1979)
10. Johnston, J.P., Ogston, A.G. Trans. Faraday Soc. 42, 789-799 (1946)
11. Ferry, J.D. Phys. Rev. 34, 753-760 (1954)
12. Sasaki, T., Page, I.H., Shainoff, J.R. Science 152, 1060-1071 (1966)
13. Sherman, L.A., Harwig, S., Lee, J. J. Lab. Clin. Med. 86, 100-111 (1975)

REVERSIBLE INTERACTIONS OF FIBRIN AND FIBRINOGEN: AN ULTRACENTRIFUGATION STUDY

Klaus T. Preissner, Jürgen Rötker, Eberhard Selmayr,
Hugo Fasold*, and Gert Müller-Berghaus

Clinical Research Unit for Blood Coagulation and Thrombosis, Max-Planck-Gesellschaft, D-6300 Giessen, and *Biochemisches Institut der Universität, D-6000 Frankfurt, Federal Republic of Germany

Introduction

Soluble fibrin-fibrinogen intermediates have been observed in the plasma of patients suffering from disseminated intravascular coagulation (DIC) or incipient thrombosis (1-6). It has been shown that the presence of fibrinogen in plasma suppresses fibrin polymerization to insoluble fibrin strands, and that the interaction of fibrinogen with fibrin enables the nascent fibrin to be transported in a soluble form in blood (7-10). Upon reaction with Factor XIIIa fibrin becomes cross-linked with fibrinogen (11,12) and complexes can be detected in vitro by analytical ultracentrifugation (11). Under in vivo conditions it is questionable whether stable non-crosslinked fibrin-fibrinogen complexes exist in the circulation. Several in vitro studies have been carried out to verify the existence of fibrin-fibrinogen complexes in the absence of Factor XIIIa (3,13-15).

Although fibrin remained in solution provided that a ratio larger than 5:1 of fibrinogen to fibrin is maintained, no stable complexes at 37°C could be detected (16). These data from our laboratory suggest reversible interactions between fibrin and fibrinogen molecules rather than tight complex formation. In an attempt to substantiate the influence of fibrinogen on the polymerization tendency of fibrin, ultracen-

Fibrinogen - Structural Variants and Interactions
© 1985 Walter de Gruyter & Co., Berlin · New York - Printed in Germany

trifugation studies were performed to underline the reversible nature of the fibrin-fibrinogen interaction.

Material and Methods

Human fibrinogen was isolated from freshly drawn ACD-blood by a modified glycine precipitation procedure (17,18). Criteria of purity of the material were assessed by clottability of the protein (95%) and analysis by SDS-polyacrylamide gel electrophoresis. The fibrinogen exhibited three major protein bands and no significant degradation after streptokinase treatment, and no $\gamma - \gamma$ bands after thrombin treatment were detectable.

Labeling of fibrinogen with ^{125}I or ^{131}I was performed using the iodine monochloride method (19) with minor modifications. Radiolabeled fibrinogen contained 0.5-0.8 ^{125}I or ^{131}I atoms per fibrinogen molecule, the clottability of the labeled protein was not decreased in comparison to unlabeled fibrinogen.

Des-AA fibrin and des-AABB fibrin were prepared by incubating labeled fibrinogen with reptilase (0.5 U/mg fibrinogen) and thrombin (2.5 U/mg fibrinogen), respectively, in the absence of urea. After incubation for 5-8 hours the enzymes were quenched with antireptilase or hirudin, respectively, in a five-fold molar excess, the clots were removed from the solution and dissolved in Tris-buffered saline, pH 7.4 (TBS), containing 2.7 M urea.

Fibrin-fibrinogen mixtures were prepared by mixing ^{131}I-fibrinogen and ^{125}I-fibrin (20:1, mole/mole) in urea-containing TBS. After extensive dialysis at room temperature for 2 hours more than 95% of the protein could be recovered. The mixtures were incubated at 20°C or 37°C for further 30 min before ultracentrifugation.

Ultracentrifugation was carried out in a SW 41 rotor (Beckman Instruments) at 20°C or 37°C for 3.5-4 hours at 40,000 rpm in

5-30% linear sucrose density gradients with a cushion of 1 ml 66% (w/w) sucrose. The fibrin-fibrinogen mixtures were centrifuged in triplicates either through a gradient containing equally distributed fibrinogen (2 mg/ml) or a gradient without fibrinogen. All buffers contained 3.5 mM $CaCl_2$ and 5 ATU/ml hirudin and were of physiologic ionic strength and pH 7.4. The gradients were fractionated from the bottom of the tube and the fractions analyzed for radioactivity. In control experiments, des-AA fibrin and des-AABB fibrin dissolved in TBS, containing 2.7 M urea were centrifuged through 5-30% sucrose in TBS without any fibrinogen present.

Results

Des-AA Fibrin, Des-AABB Fibrin

Des-AA fibrin as well as des-AABB fibrin dissolved in an urea-containing buffer were centrifuged without fibrinogen present, neither in the sample nor in the gradient, to mark the positions of both types of fibrin within the gradient profile. Figure 1 depicts the analysis of both runs at 37°C. About 70% of des-AA fibrin and 50% of des-AABB fibrin could be recovered within the 5-30% sucrose density gradient. At 20°C the recovery rates were even less: only half of the applied material could be recovered for both types of fibrin. The major portion of des-AA fibrin and des-AABB fibrin still in solution sedimentated faster than fibrinogen (Fig. 1, arrowheads), which was analyzed in a separate gradient tube in the same centrifugation run. The recovery rate of fibrinogen yielded $98 \pm 5\%$, n=5. In sucrose gradients, dissolved in urea-containing buffer, both types of fibrin sedimented at the same position as fibrinogen.

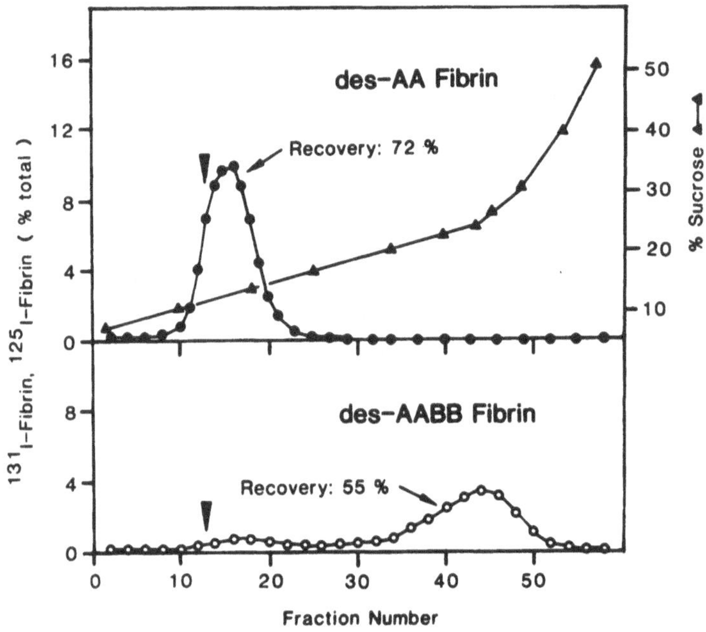

Figure 1: Ultracentrifugation of des-AA fibrin and des-AABB fibrin in 5-30% linear sucrose density gradients in the absence of fibrinogen. Sedimentation direction is to the right. The arrowheads mark the position of fibrinogen, which was analyzed in a separate gradient.

Fibrin-Fibrinogen Mixtures

After centrifugation of des-AA fibrin-fibrinogen (1:20, mole: mole) through the sucrose density gradient at 20°C, separation into des-AA fibrin and fibrinogen occurred. The major portion of polymerized des-AA fibrin migrated well ahead of the fibrinogen peak whereas a minor portion was found in the same position as fibrinogen (Fig. 2 A). The recovery rate of fibrin polymers was about the same as in the control experiment (Fig. 1). When, however, fibrinogen was present within the gradient, the formation of high polymers of des-AA fibrin was suppressed: des-AA fibrin still migrated ahead of sample fibrinogen,

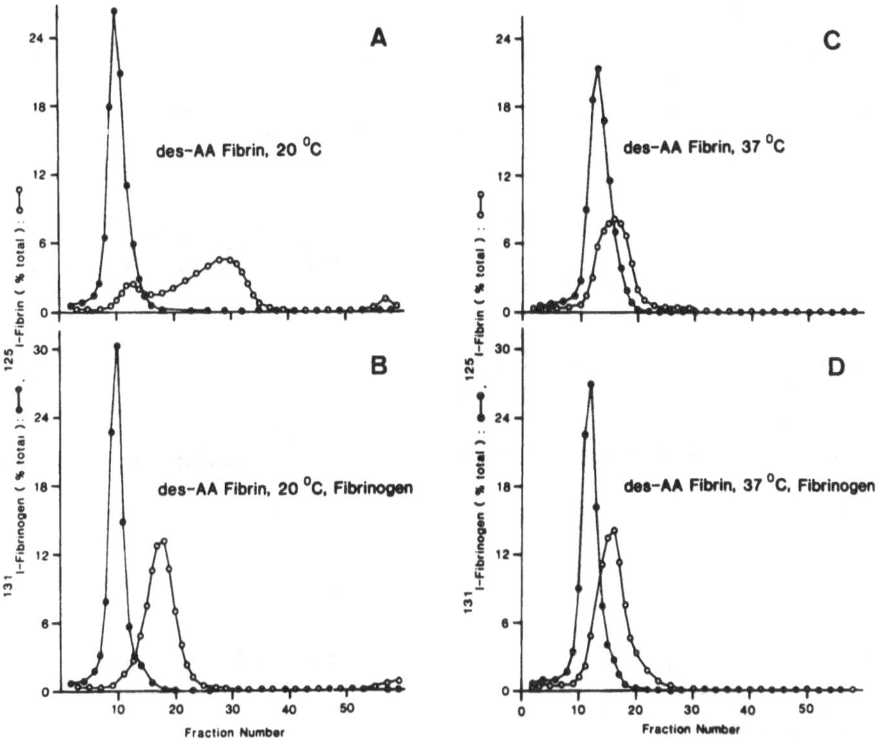

Figure 2: Ultracentrifugation of a mixture of des-AA fibrin and fibrinogen (1:20, mole/mole) at 20°C (A,B) and at 37°C (C,D) through 5-30% sucrose gradients in the presence (B,D) or the absence (A,C) of 2 mg/ml fibrinogen within the gradients. Sedimentation direction is to the right.

but no co-sedimentation was observed (Fig. 2 B). All the fibrin material could be recovered within the gradient.

At 37°C, similar results were obtained, again no loss of fibrin occurred during centrifugation through the fibrinogen-containing gradient (Fig. 2 D). The des-AA fibrin migrated slightly ahead of fibrinogen, yet both protein bands overlapped without an apparent co-sedimentation.

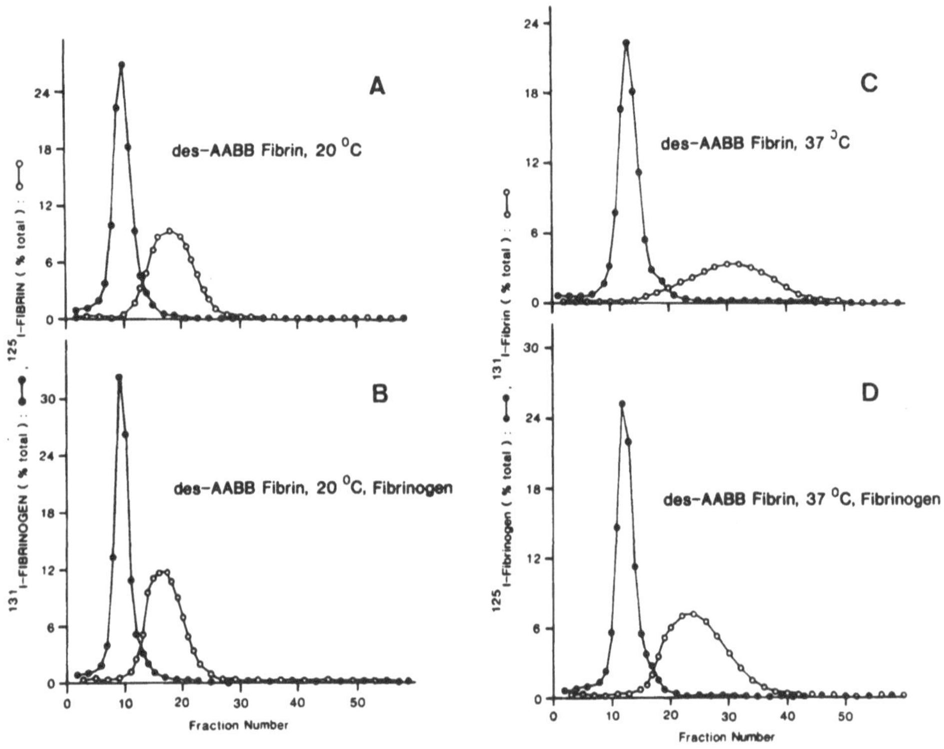

Figure 3: Ultracentrifugation of a mixture of des-AABB fibrin and fibrinogen (1:20, mole/mole) at 20°C (A,B) and at 37°C (C,D) through 5-30% sucrose gradients in the presence (B,D) or the absence (A,C) of 2 mg/ml fibrinogen within the gradients. Sedimentation direction is to the right.

Figure 3 depicts the experiments in which mixtures of des-AABB fibrin and fibrinogen were centrifuged through sucrose density gradients at 20°C and 37°C. As in the former series of experiments the presence of fibrinogen within the sucrose gradient (Fig. 3 B, D) again changed the degree of des-AABB fibrin polymer formation, thereby shifting the fibrin peak closer to the fibrinogen peak. Furthermore, the presence of fibrinogen within the gradient prevented the precipitation of fibrin within the centrifugation tube. Evidently, co-sedimentation of des-AABB fibrin and fibrinogen could not be detected in either case.

Discussion

The following conclusions can be drawn from the present ultracentrifugation experiments performed with fibrin-fibrinogen mixtures at 20°C and 37°C:
The results of the centrifugation runs of fibrin-fibrinogen mixtures through fibrinogen-containing sucrose gradients suggest a solubilizing effect of fibrinogen on des-AA fibrin and des-AABB fibrin at 20°C as well as 37°C. The solubilizing effect was temperature-dependent for both types of fibrin. Co-migration of fibrinogen and fibrin did not occur in all the experiments.
As already noticed in gel filtration experiments (20), des-AA fibrin tended to form high polymers more pronounced at 20°C than at 37°C, when compared to des-AABB fibrin. This could be due to the interaction of only one polymerization site between des-AA fibrin monomers (21). This interaction could partially be disturbed by increased lateral diffusion at higher temperatures. On the other hand, at 37°C des-AABB fibrin always formed heavier polymers than des-AA fibrin (Fig. 3 C, D), since two polymerization sites were involved in this interaction after removal of fibrinopeptide A and B. From kinetic data of the polymerization processes (Schaefer et al., this volume) fibrinogen less effectively competes for the polymerization sites of des-AABB fibrin than for those of des-AA fibrin, thereby expressing a minor solubilizing effect. Although des-AA fibrin migrated slightly faster than fibrinogen at 37°C (Fig. 2 C, D), the overlap of both protein bands seemed to indicate complex formation. From a theoretical point

of view, however, complexes with a 1:1 molar ratio of fibrin to fibrinogen would either lead to a broadening of the radioactive fibrinogen peak or to a decrease in the recovery rate of sample fibrinogen within the gradient, depending on the size of the complexes. In addition, complex formation yielding a stoichiometry greater than 2:1 (fibrinogen to fibrin) would even more affect the shape and position of the fibrinogen peak within the gradient. Thus, the presence of such stable complexes in our experiments seems to be unlikely.

The results of this study rather indicate that fibrinogen is able to keep fibrin polymers in solution by loose protein-protein contact: an equilibrium exchange of sample and gradient fibrinogen with both types of fibrin occurred. During the centrifugation, more and more fibrin polymers became separated from fibrinogen, which lost its capacity to prevent further fibrin-fibrin interaction. Including fibrinogen within the sucrose gradient solution, however, helped to prevent clot formation. Thus, the recovery rates of fibrin in these gradients yielded about 100%. Most likely the fibrin peaks represent a heterogeneous population of polymers. For instance, des-AA fibrin monomers and des-AA fibrin dimers might simultaneously be present in this population, separated from uncomplexed fibrinogen (Fig. 2 B, D).

In conclusion: Although weak interactions between fibrin polymers and fibrinogen account for a solubilizing effect on fibrin, no stable complexes could be detected. Our results indicate that soluble fibrin observed under in vivo conditions might be fibrin polymers interacting with an excess of fibrinogen with rapid exchange, which prevents a further polymerization of these polymers to a fibrin clot.

Acknowledgements

This work was supported by the Stiftung Volkswagenwerk, Hannover, and the Deutsche Forschungsgemeinschaft, Bonn-Bad Godesberg. The authors wish to thank Prof. Dr. E.L. Sattler for his hospitality at the "Zentrale Abteilung des Strahlenzentrums", Universität Giessen.

References

1. Bang, N.U., Chang, M.L.: Sem. Thrombos. Hemostas. 1, 91-128 (1974).
2. Edgar, W., McKillop, C., Howie, P.W., Prentice, C.R.M.: Thromb. Res. 10, 567-574 (1977).
3. Fletcher, A.P., Alkjaersig, N., O'Brien, J., Tulevski, V.G.: Transact. Assoc. Am. Physicians 83, 159-166 (1970).
4. Graeff, H., von Hugo, R., Hafter, R.: Thromb. Res. 3, 465-476 (1973).
5. Niewiarowski, S., Gurewich, V.: J. Lab. Clin. Med. 77, 665-676 (1971).
6. Largo, R., Heller, V., Straub, P.W.: Blood 47, 991-1002 (1976).
7. Shainoff, J.R., Page, I.H.: Circ. Res. 8, 1013-1022 (1960).
8. Kierulf, P.: Thromb. Res. 3, 613-630 (1973).
9. Müller-Berghaus, G., Mahn, I., Köveker, G., Maul, F.-D: Br. J. Haematol. 33, 61-79 (1976).
10. Kröhnke, I., Mahn, I., Krell, W., Müller-Berghaus, G.: Br. J. Haematol. 45, 131-141 (1980).
11. Sasaki, T., Page, I.H., Shainoff, J.H.: Science 152, 1069-1071 (1966).
12. Kanaide, H., Shainoff, J.R.: J. Lab. Clin. Med. 85, 574-579 (1976).
13. Shainoff, J.R., Page, I.H.: J. Exp. Med 116, 687-708 (1962).
14. Brass, E.P., Forman, W.B., Edwards, R.V., Lindan, O.: Thromb. Haemostas. 36, 37-48 (1976).

15. Brosstad, F.: Thromb. Res. 15, 563-568 (1979).

16. Müller-Berghaus, G., Mahn, I., Krell, W.: Thromb. Res. 14, 561-572 (1979).

17. Blombäck, B., Blombäck, M.: Ark. Kemi 10, 415-443 (1956).

18. Mahn, I., Müller-Berghaus, G.: Haemostasis 4, 40-50 (1975).

19. McFarlane, A.S.: Nature 182, 53 (1958).

20. Müller-Berghaus, G., Mahn, I., Krell, W., Bernhard, J.-C.: Fibrinogen: Recent Biochemical and Medical Aspects, Walter de Gruyter, Berlin · New York 1982.

21. Kudryk, B., Collen, D., Woods, K.R.; Blombäck, B.: J. Biol. Chem. 249, 3322-3325 (1974).

V. NORMAL FIBRINOGEN VARIANTS

EVIDENCE THAT THE AMOUNT OF HEPARIN PRECIPITABLE FRACTION IS INFLUENCED BY FIBRINOGEN QUALITY

B. Holm, H.C. Godal
Haematological Research Laboratory, Ullevål Hospital, Oslo, Norway

Introduction

When heparinized plasma is cooled for some hours, a precipitate is formed (1). This heparin precipitable fraction (HPF) consists mainly of fibronectin and fibrinogen (2,3). In plasma from normal subjects, small amounts (0.3-0.6 mg/ml plasma) are found, but in some disease states large precipitates may be seen (5-7 mg/ml plasma). Fibronectin is necessary for precipitation to occur (4), but the plasma concentration of this protein is of minor importance for the amount precipitated (5). Plasma concentration of fibrinogen contributes to some extent to the amount of HPF. However, large variations in HPF may be seen without significant changes in the plasma fibrinogen level (5). Qualitatively changed fibrinogen, precipitable with heparin has been suggested, but direct evidence has not yet been presented. In this context it may be noted that soluble fibrin (fibrin monomer complexes) does not influence the amount of HPF (6).
This presentation includes some observations concerning HPF and qualitative changes of fibrinogen during the course of acute myocardial infarction, and some preliminary data indicating that HPF is closely related to the so-called HMW fibrinogen.

Fibrinogen - Structural Variants and Interactions
© 1985 Walter de Gruyter & Co., Berlin · New York - Printed in Germany

Methods and results

1) Heparin precipitable fraction in acute myocardial infarction. Nine patients suffering from acute transmural myocardial infarction were followed at least 10 days with daily measurements of HPF, plasma fibronectin and plasma fibrinogen.

HPF was prepared as described earlier (6) and quantitated by the Biuret-method (13). Fibrinogen was measured according to Jacobsson (14) and fibronectin by rocket immun-electrophoresis (12).

Results: Typical variations in HPF, fibrinogen and fibronectin are seen in Fig. 1. The variations in plasma fibronectin were small, and the amounts precipitated did not increase with increasing HPF. Accordingly no correlation was found between plasma concentration of fibro-

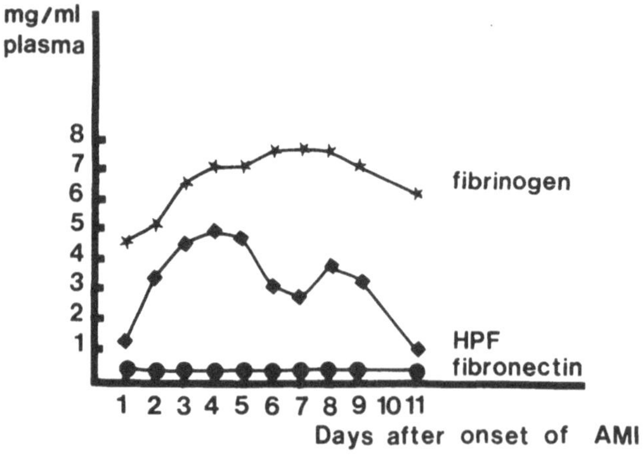

Fig. 1

Variations in the amount of HPF and plasma concentrations of fibronectin and fibrinogen during the course of AMI in a typical patient.

nectin and amount of HPF. On the other hand, when fibronectin was removed from plasma prior to cooling, the precipitation was abolished. We conclude that fibronectin is necessary for precipitation to occur, but that this protein cannot explain variations in amount of HPF.

Increased amounts of HPF were due to increased contents of fibrinogen. The precipitation was to some extent correlated to plasma fibrinogen, but this correlation depended on the stage of the disease. In a typical patient (Fig. 1) the correlation was good during the first days, later on HPF subsided, while plasma fibrinogen remained elevated. A comparison between day 3 and day 10 post infarctum in the nine patients, disclosed a significant reduction ($p < 0.05$) of HPF and HPF/fgn-ratio, and no concomitant reduction of fibrinogen (Fig. 2).

These observations initiated further studies concerning fibrinogen quality.

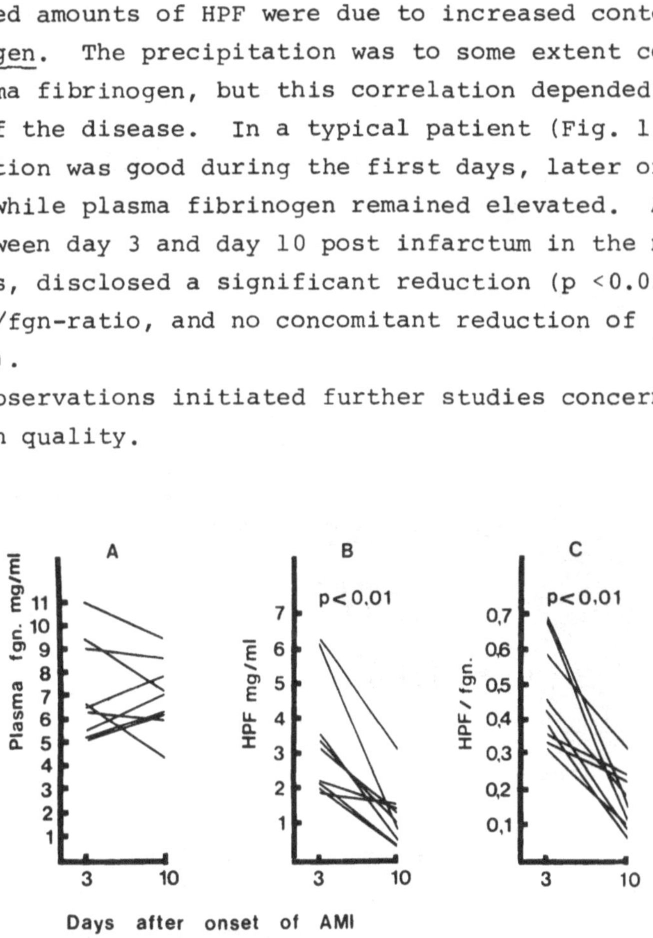

Fig. 2

Variations in the amounts of plasma fibrinogen (A), HPF (B) and HPF/fibrinogen ratio (C) in 9 AMI patients. The values on day 3 after onset of AMI are paired with corresponding values on day 10.

2) Fibrinogen heterogeneity in acute phase.

Fibrinogen in normal human plasma is heterogeneous, containing at least 3 fractions with different molecular weight (7,8,9). Changes in relative amounts of the fractions during acute phase have been described (10).
We have examined heterogeneity of plasma fibrinogen in normal plasma and after AMI and hip joint replacement. Non-crosslinked clots, obtained by clotting in presence of EDTA (9), were dissolved in urea and submitted to SDS-electrophoresis (3% PAG/0.5% agarose) as described by Weinstein and Deykin (8), and quantitated by densitometric scanning using LKB 2202 Ultroscan.

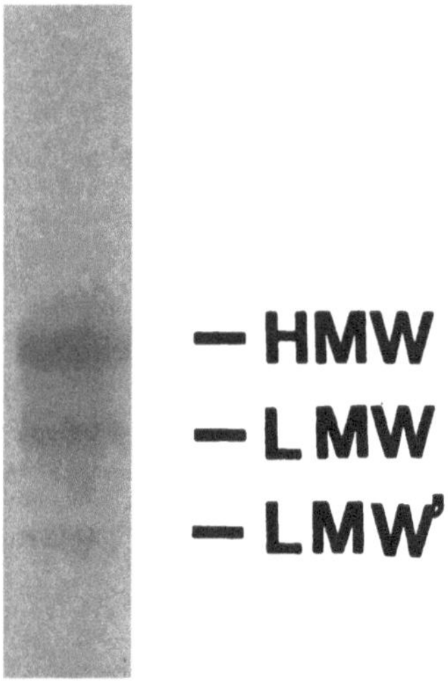

Fig. 3
SDS-electrophoresis (3% PAG) of dissolved, non-crosslinked clot, from a healthy subject. 3 fibrinogens with different molecular weight are seen.

Results. Normal plasma contained regularly 3 fibrinogens of different molecular weight: About 70% high molecular weight fibrinogen (HMW=Fgn I, mw. 340 000), about 26% LMW (=Fgn II, mw. 300 000) and about 4% LMW' (=Fgn III, mw. 270 000) (Fig. 3). In acute phase (Fig. 4) a substantial increase in total fibrinogen was found. This increase was initially due solely to the HMW-increase that started during the first 24 hours after hip joint replacement.

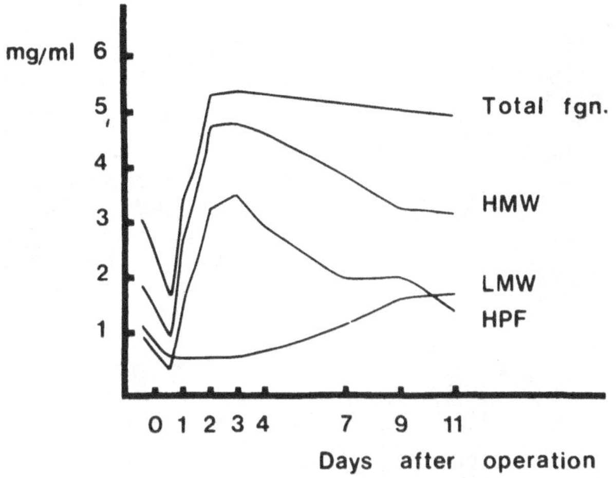

Fig. 4

Variations in HPF, total fibrinogen, HMW and LMW after hip-joint replacement in a typical patient.

Increase of LMW was delayed for about 2 days. After 10-14 days the HMW/LMW ratio was normalized, total fibrinogen, however, was still substantially elevated.

3) Evidence that HPF is influenced by fibrinogen quality.
Correlation between HPF and HMW.

We found a better correlation between HMW and HPF than between total fibrinogen and HPF (Fig. 4). Low molecular weight fibrinogen on the other hand, demonstrated a negative correlation to HPF.

Fibrinogen precipitated in HPF.

SDS-electrophoresis (3% PAG) of dissolved HPF revealed only HMW, documenting that this fraction only contributed to the precipitation (Fig. 5). This was the case both in the small precipitates from normal plasma, and in the greater amounts recovered during acute phase.

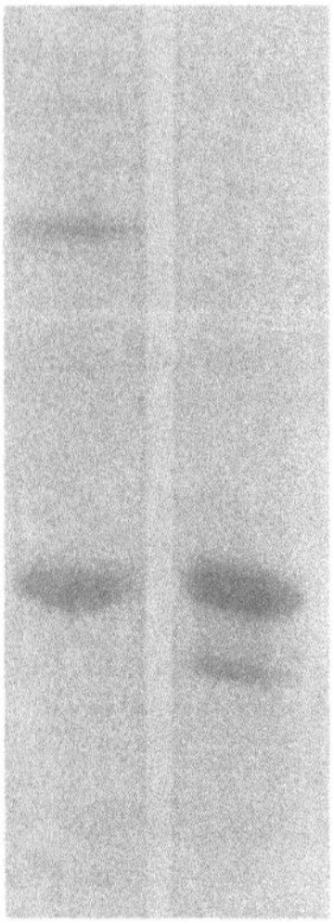

Fig. 5

SDS-electrophoresis (3% PAG) of HPF (left) and dissolved plasma clot (right). Only the HMW-fibrinogen is present in HPF.

Addition of HMW and LMW to normal plasma.

HMW and LMW fractions were purified by fractionated precipitation with saturated $(NH_4)_2SO_4$ (7) of beta-alanine purified fibrinogen (11). HMW and LMW precipitates were dissolved in normal heparinized plasma (about 30 U/ml) dialysed, fibrinogen measured, and HPF prepared and quantitated as described.

From Table 1 is seen that total fibrinogen was increased to the same extent in the samples to be compared, and that this increase mainly was due to HMW in one and LMW in the other.

Addition of HMW led to a substantial increase in HPF. LMW did not influence the amount of precipitation.

Normal plasma, Fibrinogen g/l	2.6	2.6	2.6
Addition of	0	HMW	LMW
Final fibrinogen concentration g/l	2.6	6.0	6.3
Final contents of HMW	65%	78%	38%
HPF mg/ml	0.8	2.5	0.7

Table 1

Addition of HMW and LMW to heparinized plasma from a healthy subject.

Conclusion

Plasma concentration of HMW is fairly well correlated to the amount of HPF, and only this type of fibrinogen is included in the precipitate. Addition of HMW increased HPF, LMW did not. Based on these observations we conclude that the amount of heparin precipitable fraction is influenced by the quality of plasma fibrinogen.

References

1. Thomas, L., Smith, R.T., von Korff, R.: Proc. Soc. Exp. Med. 86, 813-818 (1954).
2. Fyrand, O., Solum, N.O., Kierulf, P.: Thromb. Res. 8, 17-29 (1976).
3. Fyrand, O., Solum, N.O.: Thromb. Res. 8, 659-672 (1976).
4. Stathakis, N.E., Mosesson, M.W.: J.Clin.Invest. 60, 855-865 (1977).
5. Holm, B., Godal, H.C., Kierulf, P.: Thromb. Res. 33, 9-18 (1983).
6. Holm, B., Brosstad, F., Godal, H.C.: Scand. J. Haematol. 30, Suppl. 39, 75-80 (1983).
7. Lipinski, I., Lipinski, B., Gurewich, V.: J. Lab. Clin. Med. 84, 509-516 (1974).

8. Weinstein, M.J., Deykin, D.: Brit. J. Haemat. 40, 617-630 (1978).

9. Tsianos, E.B., Stathakis, N.E.: Thromb. Haemost. 44 (3), 130-134 (1980).

10. Lipinska, I., Lipinski, B., Gurewich, V., Hoffmann, K.D.: Am. J. Clin. Pathol. 66, 958-966 (1976).

11. Jakobsen, E., Kierulf, P.: Thromb. Res. 3, 145-159 (1973).

12. Laurell, C.B.: Scand. J. Clin. Lab. Invest. 29, suppl. 124 (1972).

13. Tietz, N.W.: Fundamentals of clinical chemistry. 188-191. W.B. Saunders Company (1970).

14. Jacobsson, K.: Scand. J. Lab. Clin. Invest. 7, suppl. 14, 1-54 (1955).

THE LOCATION OF A SECOND IN VIVO PHOSPHORYLATION SITE IN THE Aα-CHAIN OF HUMAN FIBRINOGEN

Hans H. Seydewitz, Clemens Kaiser, Irene Witt

Biochemisches Labor der Universitäts-Kinderklinik
D-7800 Freiburg i. Br., Federal Republic of Germany

Introduction

Human fibrinogen contains about one mole phosphorus per mole fibrinogen which is bound exclusively to the Aα-chain (1). 30 - 50% of the phosphorus is bound to Ser 3 of the Aα-chain (2) and is split off with fibrinopeptide A during the fibrinogen-fibrin conversion (3).
The phosphorus content of human fetal fibrinogen is about twice that of the adult form (4). Also in fetal fibrinogen the phosphorus is distributed in nearly equal amounts between the Ser 3 of fibrinopeptide A and one or more binding sites in the rest of the Aα-chain (1).
In the course of our investigations of the fetal Aα-chain we tried to locate these binding site(s) using the difference in phosphorus content of fetal and adult fibrinogen as marker for a rapid tracing of the phosphorylated peptides out of the peptide mixtures of tryptic and chymotryptic digests.

Materials and Methods

Human fibrinogen was prepared from adult blood and from umbilical cord blood as described in (5). The purity of the preparations, in terms of protein material clottable by thrombin, ranged from 95-98%.
Reduction and Carboxymethylation of the fibrinogen disulfide bonds were carried out according to (6). The S-carboxymethylated chains were separated by chromatography on CM-Sepharose CL-6B using a linear gradient between the starting buffer (8 M urea, 0.1 M Tris/Acetate, pH 5.0) and the final buffer

Fibrinogen - Structural Variants and Interactions
© 1985 Walter de Gruyter & Co., Berlin · New York - Printed in Germany

(8 M urea, 0.1 M Tris/Acetate, pH 6.6).
The A α-chains of adult and fetal fibrinogen were cleaved with BrCN and the resulting BrCN peptides separated on Sephadex G-50 superfine according to (7). One of the resulting BrCN pools which still contained 4 peptides was further purified by chromatography on Bio-Gel P-6. The peptide fractions were identified by UV-absorption at 280 nm and by reaction of aliquots with ninhydrin after alkaline hydrolysis.
Trypsin and chymotrypsin cleavages were carried out in 0.1 M NH_4HCO_3 buffer, pH 8.5, with an enzyme to peptide ratio of 1:50. The reaction was stopped after 6 hrs by addition of 3% acetic acid and subsequent lyophilisation.
Sequence determinations were carried out by automated Edman degradation on a Beckman model 890 C sequencer using a modified 0.1 M quadrol program (8). The PTH-amino acids were identified by hplc according to (9).
Peptides were hydrolyzed with 5.7 N HCl at 110°C in evacuated, sealed tubes for 2 hrs (partial hydrolysis for the determination of the phosphorylated amino acids P-Ser and P-Thr) or 24 hrs (total hydrolysis for the determination of amino acid composition). Amino acid analyses were performed on a Liquimat III Amino Acid Analyzer (Kontron Technik, Eching b. München, FRG) using a standard hydrolysate program.
The hplc equipment used for the separation of peptides consisted of 2 Altex model 110A pumps controlled by a microprocessor model 200, a Rheodyne injection valve 7125 and an Uvikon 720 LC spectrophotometer equipped with a 8 µl flow-through cell, all purchased from Kontron Technik. Analytical separations were performed on a Hibar RP-18 column, 4x250 mm, 5 µ (Merck, Darmstadt, FRG) and preparative separations on a Kontrosorb RP-18 column, 10x250 mm, 10 µ (Kontron Technik). Elution was carried out with a linear gradient between aqueous trifluoroacetic acid, pH 2.15 (buffer A) and a 40:60 (v/v) mixture of buffer A and acetonitrile (buffer B). Peptides were detected at 206 nm.

Results

The adult and fetal A α-chains of human fibrinogen were cleaved with BrCN and the resulting peptides separated by chromatography on Sephadex G-50 and Bio-Gel P-6 (Fig. 1).

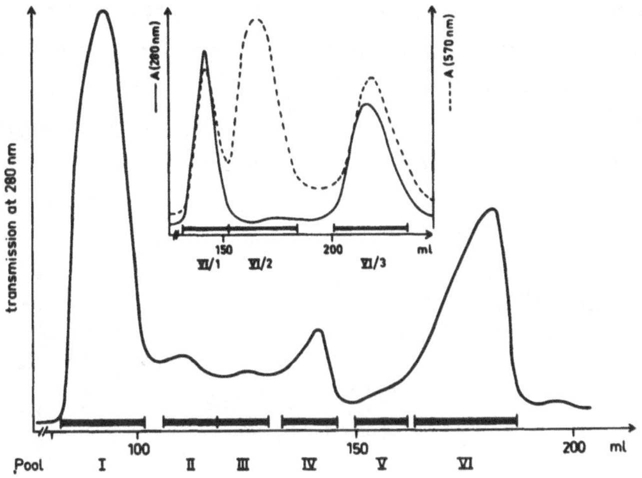

Fig. 1: Separation of the BrCN peptides of fibrinogen Aα-chain on Sephadex G-50 superfine. Insert: Separation of BrCN pool VI on Bio-Gel P-6. Column: 2.6x150 cm; eluent: 1 M formic acid in both cases. There was no difference between fetal and adult Aα-chains.

BrCN-Pool	number of amino acids	position in Aα-chain	N-terminal sequence
CB I	236	241 - 476	E L E R P . . .
CB II	-	-	mixture of uncleaved BrCN fragments
CB III	60	148 - 207	K R L E V . . .
CB IV	67 56	518 - 584 92 - 147	L G E F V . . . E I L R G . . .
CB V	40	52 - 91	K G L I D . . .
CB VI/1	28	208 - 235	K P V P D . . .
CB VI/2	41 26	477 - 517 585 - 610	D L G T L . . . A D E A G . . .
CB VI/3	51	1 - 51	A D S G E . . .

Tab. 1: Contents of the BrCN pools of the fibrinogen Aα-chain.

The BrCN pools obtained consisted of one or at most two peptides (Tab. 1). These peptides were cleaved further with trypsin and the tryptic digests separated by hplc gradient elution.
The resulting chromatograms of fetal and adult tryptic digests were then analyzed for qualitative and quantitative differences.

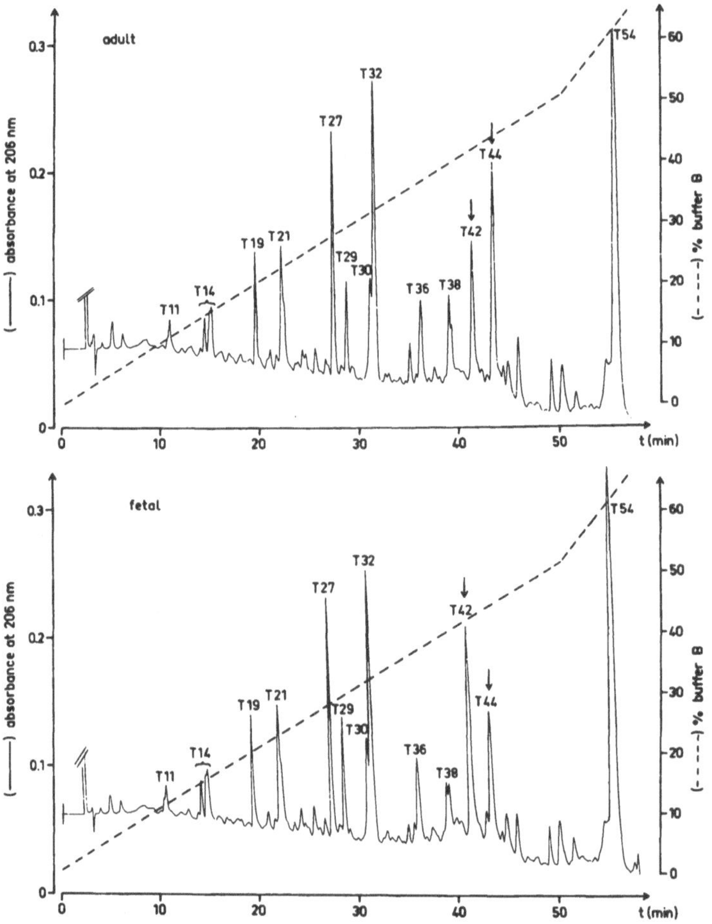

Fig. 2: Hplc gradient elution of the tryptic digests of CB I from adult and fetal fibrinogen Aα-chain.

In the case of the large BrCN fragment CB I the elution patterns shown in Fig. 2 were obtained. As may be seen there is no qualitative but one prominent quantitative difference:

The peak T 42 is distinctly smaller than the peak T 44 in the adult pattern whereas in the fetal pattern the situation is reversed.

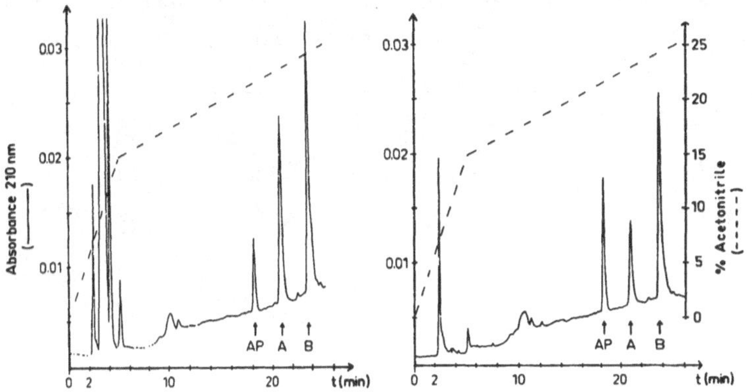

Fig. 3: Hplc gradient elution of the fibrinopeptides from adult (left) and fetal (right) fibrinogen.

This finding is similar to those in the hplc chromatograms of the fibrinopeptides (Fig. 3). The peak corresponding to the phosphorylated fibrinopeptide AP is smaller than the peak of the unphosphorylated fibrinopeptide A in the chromatogram of fibrinopeptides from adult fibrinogen whereas the relation of these two peaks is reversed for fetal fibrinopeptides.

It was therefore suspected that also in CB I the peaks of T 42 and T 44 represent the phosphorylated and unphosphorylated form of the same peptide.

To prove this it had to be shown:

1) that CB I contains phosphorus,
2) if so, that the two peptides have the same N-terminal sequence and amino acid composition, and
3) that one of them contains phosphorus and that the other one does not.

These three implications turned out to be true.

By partial hydrolysis and subsequent amino acid analysis (s. Methods) it was found that CB I contains 0.57 mol P-Ser/mol peptide in fetal and 0.34 mol P-Ser/mol peptide in adult material.

No further BrCN pool of the A α-chain contained significant amounts of P-Ser (i.e. > 0.05 mol P-Ser/mol peptide) with the exception of CB VI/3 which is the N-terminal peptide of the A α-chain (s. Tab. 1) containing the P-Ser of position 3.

By N-terminal sequence determination of T 42 and T 44 up to position 16, it was found that the N-termini of these two peptides are identical and correspond to that of the peptide 290-348 of the known adult sequence (Tab. 2).

^{241}E L E R P G G N E I T R (= T 35)   G G S T S Y G T G S E T E S P R (= T 27)

N P S S A G S W
N S G G S G P G S T G N R = T 32

^{290}N P G S S G T G G T A T W
K P G S S G P G S T G S W
N S G S S G T G S T G N Q* = T 42/44
N P G S P R P G S T G T W
^{342}N P G S S E R^{348}

G S A G H W
T S E S S V S G S T G Q*W
H S E S G S F R P D S P G = T 54
S G N A R P N N P D W G T
F E E V S G N V S P G T R

R   E Y H T E K (in T 21)   L V T S K (in T 21)   G D K   E L R (in T 21)   T G K (in T 11)   E K   V T S G S T T T T R (= T 19)

R S C S K (in T 14, in T 11)   T V T K (in T 14)   T V I G P D G H K (= T 30)   E V T K (in T 14)   E V V T S E D G S D C P E A M(Hse)^{476} (= T 29)

Tab. 2: Sequences of the tryptic peptides of the large BrCN fragment CB I of fibrinogen A α-chain. The sequences shown are those given in (10). The underlined sequences are confirmed on fetal material by the authors. Q* are the acceptor sites for the factor XIIIa-mediated crosslinking of the α-chains (11).

The amino acid composition of the two peptides is also very similar as shown in Tab. 3 and fits in very well with the amino acid composition expected for this peptide. By partial hydrolysis and subsequent amino acid analysis, it was finally shown that T 42 contains about one mol P-Ser/mol peptide and T 44 only trace amounts (Tab. 3).

Thus the hypothesis was proven and the second phosphorylation site of the fibrinogen A α-chain localized to the region 290-348. But this region contains still 14 Ser residues (Tab. 2, 3). In order to localize further the phosphorylation site, T 42 and T 44 were cleaved with α-chymotrypsin and the digests again eluted by hplc. The resulting hplc chromatograms are shown in Fig. 4.

	T 42		T 44		expected for peptide 290 - 348
	adult	fetal	adult	fetal	
P-Ser (mol/mol)	0.92	0.97	0.07	0.10	
Asp	5.42	5.06	5.20	4.96	5
Thr	7.94	7.60	7.14	7.72	8
Ser	12.48	13.22	13.20	12.64	14
Pro	6.96	6.76	6.18	6.28	7
Glu	2.50	2.14	2.84	2.40	2
Gly	17.36	15.78	15.70	15.00	16
Ala	1.62	1.44	1.52	1.48	1
Val	0.18	0.07	0.24	0.26	0
Ile	0	0	0.08	0.12	0
Leu	0	0	0.30	0.30	0
Lys	1.10	0.98	1.08	0.94	1
His	0	0	0.20	0.18	0
Arg set to:	2	2	2	2	2
Arg actually found: (mol/mol)	0.89	1.96	1.55	1.46	

Tab. 3: Amino acid composition of the tryptic peptides T 42 and T 44 relative to their Arg content (24 hrs hydrolysis). The P-Ser content was determined in a separate 2 hrs hydrolysis. These values are corrected by a factor of 1.56 accounting for the experimentally determined destruction of pure P-Ser under the conditions of partial hydrolysis.

There are two major differences between the elution patterns of T 42 and T 44:

1) a difference in elution time of 1.7' in the early eluting peaks C 11 in T 42 and C 13 in T 44; and
2) a prominent peak, C 29, in T 42 which is virtually absent in T 44.

The retention times of the other five main peaks were identical in T 42 and T 44. Quantitative analysis showed that the relative peak fractions were also very similar with the exception of the peak C 33 the relative intensity of which was higher in T 44 by an amount which could account for the lack of C 29.

C 11 and C 13 were sequenced completely; they had the same sequence namely that of the heptapeptide 342-348: Asn - Pro - Gly - Ser - Ser - Glu - Arg. Thus one of them had to be the phosphorylated and the other one the unphosphorylated form of this peptide.

By partial hydrolysis it was found that C 11 contained P-Ser (with a P-Ser to Ser ratio of about one) and C 13 contained no P-Ser as was to be expected already by their hplc retention times.

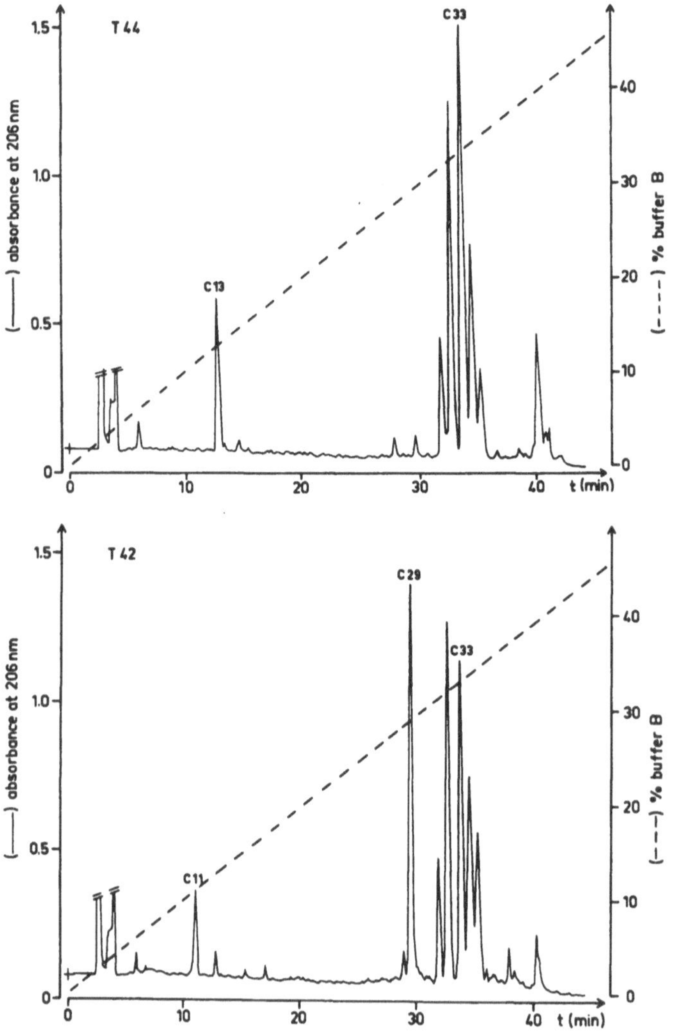

Fig. 4: Hplc gradient elution of the chymotryptic digests of the tryptic peptides T 42 and T 44.

The peptide C 29 was sequenced for 30 steps. It turned out to be the peptide 316-348 not cleaved at Trp 341 by chymotrypsin.

By partial hydrolysis it was found that this peptide contained P-Ser, with a P-Ser to Ser ratio of about 1:7, as could be expected if this peptide were phosphorylated at one of the serines in its terminal segment 342-348.

Thus, the cleavage of Trp 341 had not occured presumably due to the vicinity of the negative charge of P-Ser.

But the possibility remained that the phosphorus in C 29 might be bound to a Ser or Thr in the first part of this peptide. In the first case it should be possible to liberate C 11 from C 29 by prolonged cleavage with chymotrypsin, but not in the second.
In order to test this, C 29 was cleaved with three times the amount of chymotrypsin used in the first cleavage for a three times longer incubation period namely 18 hrs instead of 6 hrs. The hplc elution of this cleavage mixture is shown in Fig. 5.

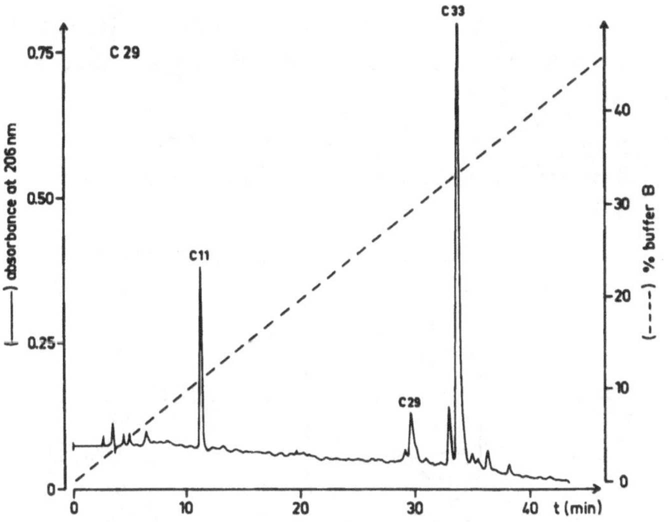

Fig. 5: Hplc gradient elution of the peptide C 29 after prolonged incubation with chymotrypsin.

The peak at 29.3' disappeares almost completely, and instead two new peaks at 10.8' and 33.5' arise. By cochromatography of these peaks with C 11 and C 33 of the first cleavage their identity with these peptides was established. By this experiment it was confirmed that there is only one phosphorylation site in the tryptic peptide T 42, and this is located in the heptapeptide 342-348 of the Aα-chain. Since there are only two serines and no threonine in this peptide, the phosphorylation site is located to these two adjoining serines in position 345 and 346.
Because of lack of material no further experiments to decide between these possibilities were undertaken at the moment.

Discussion

In the preceding section the location of a second in vivo phosphorylation site in the Aα-chain of human fibrinogen at one of the Ser residues 345 or 346 has been demonstrated.

It has to be pointed out that this newly described site is phosphorylated - by an until now unknown protein kinase - to a degree which is equal or even higher than that of the phosphorylation site on fibrinopeptide A at Ser 3 of the Aα-chain and that there is no other site in the fibrinogen Aα-chain which is phosphorylated in vivo to a comparable extent. The other BrCN peptides of the Aα-chain and the tryptic peptides of CB I other than T 42 contained no or only trace amounts of P-Ser or P-Thr.

Though, because of lack of sufficient material we were not able at the moment to decide which of the two Ser residues is the actual phosphorylation site, some arguments favour the position 345 against position 346.

1) If Ser 345 is the phosphorylated one there exists a sequence homology to the known phosphorylation site in position 3 namely that the second amino acid at the carboxyl side of Ser is Glu in both cases:

		3								
A -	D -	S(P)	- G -	E	- G -	D -	F -	L -		
		345								
- P -	G -	S(P)	- S -	E	- R -	G -	S -	A -		
		346								
- G -	S -	S(P)	- E -	R -	G -	S -	A -	H -		

If Ser 346 is the actual phosphorylation site there is no homology at all between these two sites.

2) The observed impedement of the chymotryptic cleavage at Trp 341 is better explained by the presence of the negative charge of P-Ser nearer to this position.

3) From a thermodynamic point of view the introduction of a negative charge would be easier two positions away from an existing one than directly adjacent to it.

KUDRYK et al. (12) have found that in the dog fibrinopeptide A is fully phosphorylated at the time of secretion. If the same holds true for human fibrinogen and also for the second phosphorylation site, the fact that the fibrinogen isolated from whole blood is only partially phosphorylated would imply that it has been dephosphorylated by some phosphatase activity in the circulation. The different degree of phosphorylation would then reflect a different ratio of synthesis and phosphatase activity.

As is the case for the phosphorylation in position 3, the physiological function of the phosphorylation at the site located in this work is completely unclear at the moment.

The phosphorylation site is located on an extremely polar zone with repeats of 13 amino acids with very high sequence homology. We don't know what purpose this remarkable zone serves, but it is possible that the phosphorylation plays a role in this hitherto unknown process.

One possible candidate for this is the interaction of fibrinogen with factor XIII. The fact that the phopshorylation site is located almost exactly in the middle of the two GIN residues 328 and 366 known as acceptor sites for the factor XIIIa-mediated crosslinking of the α-chains (11) might indicate that the phosphorylation at this site is involved in the regulation of α-chain crosslinking. This could occur indirectly in the phase of fibrin polymerisation by disturbing the proper alignment of the α-chains, or, later in the phase of the stabilisation of the fibrin network by direct interference with the action of factor XIIIa.

Furthermore, CREDO et al. (13) have recently reported that fibrinogen is not only a substrate for factor XIIIa but is specificly involved in the generation of active factor XIII from a thrombin-modified but still inactive intermediate. The potential for this specific function of fibrinogen resides in the region of the amino acids 243-424 of the A α-chain, a region which includes the phosphorylation site discussed here.

References

1) Rothweiler, H.: Thesis, Freiburg (1977)
2) Blombäck, B., Blombäck, M., Searle, J.: Biochim. Biophys. Acta 74, 148 - 151 (1963)
3) Blombäck, B., Blombäck, M., Edman, P., Hessel, B.: Biochim. Biophys. Acta 115, 371 - 396 (1966)
4) Witt, I., Mueller, H.: Biochim. Biophys. Acta 221, 402 - 404 (1970)

5) Osbahr, A.J., Gladner, J.A., Laki, K.: Biochim. Biophys. Acta 86, 535 -542 (1964)

6) Henschen, A., Edman, P.: Biochim. Biophys. Acta 263, 351 - 367 (1972)

7) Lottspeich, F., Henschen, A.: Hoppe-Seyler's Z. Physiol. Chem. 359, 1451 - 1455 (1978)

8) Edman, P., Henschen, A. in: Protein Sequence Determination (Needleman, S.B., ed.) 2nd. edn., pp. 232 - 279, Springer Berlin 1975

9) Lottspeich, F.: Hoppe-Seyler's Z. Physiol. Chem. 361, 246 - 249 (1980)

10) Henschen, A., Lottspeich, F., Hessel, B.: Hoppe-Seyler's Z. Physiol. Chem. 360, 1951 - 1956 (1979)

11) Cottrell, B.A., Strong, D.D., Watt, K.W.K., Doolittle, R.F.: Biochemistry 18, 5405 - 5410 (1979)

12) Kudryk, B., Okada, M., Redman, C.M., Blombäck, B.: Eur. J. Biochem. 125, 673 - 684 (1982)

13) Credo, R.B., Curtis, C.G., Lorand, L.: Biochemistry 20, 3770 -3778 (1981)

EVIDENCE THAT THE γ CHAIN POPULATION OF HUMAN PLATELET FIBRINOGEN LACKS THE γ' VARIANT THAT IS PRESENT IN PLASMA FIBRINOGEN

M.W. Mosesson, G.A. Homandberg, D.L. Amrani

The Department of Medicine, Hemostasis Research Laboratory, University of Wisconsin Medical School, Milwaukee Clinical Campus, Mount Sinai Medical Center, Milwaukee, Wisconsin 53233

Introduction

Plasma fibrinogen is a major blood protein that is synthesized and secreted by liver parenchymal cells (1-3). Each molecule consists of three pairs of polypeptide chains (Aα, Bβ and γ, respectively) joined by disulfide bridges and arranged in a symmetrical dimeric structure (4-6). Heterogeneity of each type of chain has been described and has recently been summarized (7); of particular relevance to the studies to be reported here is the existence in normal single donor plasma fibrinogen of two classes of functionally equivalent γ chains (8), here termed γ_A and γ'[1], respectively. The γ' chain is longer than the γ_A chain by virtue of differences in their respective COOH-terminal sequences (10,11). Each type of chain produced by the liver has a unique mRNA (12-15)

---

1. In the first report on this subject (8), the newly recognized minor variant γ chain (~7%) was termed γ' to distinguish it from the major population of γ chains for which the term γ was retained. However, it is somewhat confusing to use the term "γ" for both generic γ chains as well as for a subpopulation of γ chains. Crabtree and Kant (9) circumvented this problem by referring to the two as γ_A and γ_B, respectively. In this paper, we use the term γ_A for the major γ chain variant, but for historical precedent we continue to designate the minor variant chain γ'.

Fibrinogen - Structural Variants and Interactions
© 1985 Walter de Gruyter & Co., Berlin · New York - Printed in Germany

that in rats and humans is derived by differential splicing of the primary mRNA transcript (9,16; A.J. Fornace Jr., G.R. Crabtree, personal communication). Thus, there is a single gene coding for both types of γ chains.

In addition to plasma fibrinogen, there is separate fibrinogen pool in blood platelets (17-24). This fibrinogen pool is contained within the platelets' α-granules (25-27) and evidently is synthesized and packaged in the megakaryocyte (28). Platelet fibrinogen can be secreted by activating platelets with any of several agonists (19,26,29-31); its molecular size is essentially the same as that of nascent plasma fibrinogen (32-36).

Comparative structural studies of platelet and plasma fibrinogen from several subjects with congenital fibrinogen abnormalities have yielded interesting and pertinent information bearing on the genetic mechanism controlling fibrinogen production and storage in platelets. On the one hand, the respective platelet fibrinogens from the Oslo I and Oslo III subjects (37) manifest the same abnormality (Bβ and γ chains, respectively) that is found in their plasma fibrinogens. On the other hand, platelet fibrinogen from the fibrinogen Louisville (D.K. Galanakis, personal communication), Metz (38,39), and Paris I (36) subjects reportedly does not manifest the abnormality characterizing their respective plasma molecules.

The foregoing summarizes relevant background information bearing on the nature of the genetic relationships between the platelet and plasma fibrinogen pools. In an effort to shed new light on this subject, we evaluated whether the plasma

γ_A and γ' chains variants are also present in normal platelet fibrinogen.

Results

Crosslinking of plasma fibrin in the presence of factor XIIIa results in formation of three types of γ-dimers comprised of the various combinations of γ_A and γ' chains (7). As assessed from Coomassie blue-stained Laemmli gels (40), fraction I-2 fibrin, whose γ population consists of 7% γ' chains and 93% γ_A chains, yielded a major γ_A-γ_A dimer band and a minor γ_A-γ' band that accounted for 86% and 13%, respectively, of the total γ dimer population (Fig. 1, lane 1). The γ'-γ' dimers, amounting to <0.5% of the total in this fraction (10), were not visualized under the conditions used in this type of experiment. Peak 1 fibrin (Fig. 1, lane 2), whose γ chain population is essentially 100% γ_A, yielded only one γ-dimer band, γ_A-γ_A. Electrophoresis

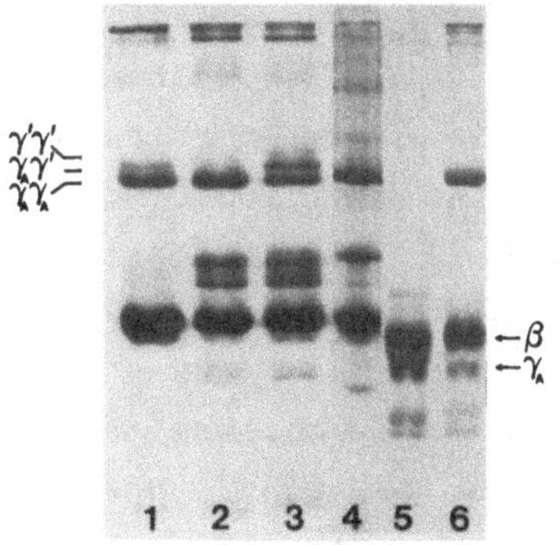

Fig. 1 Slab gel electrophoresis (Laemmli system) of reduced fibrin specimens from three different runs (lanes 1,2,3; lane 4; lanes 5,6) analyzed under the same conditions and stained with Coomassie blue. Lane 1, crosslinked fraction I-2 fibrin; lane 2, crosslinked peak 1 fibrin; lane 3, crosslinked peak 2 fibrin; lane 4, platelet fibrin collected into buffer lacking EDTA; lane 5, platelet fibrin collected into buffer containing EDTA; lane 6, same specimen as in lane 5 after crosslinking. In this lane, in addition to the intense band in the γ_A-γ_A dimer position, a less intense band is also present in the γ_A monomer position.

of mixtures of crosslinked Peak 1 fibrin and fraction I-2 fibrin indicated that under the load and staining conditions we used, γ' chains were readily detectable as γ_A-γ' dimers when they amounted to 1% or more of the total γ chain population.

Electrophoretic analysis of platelet fibrin that had been harvested into a Tris-saline buffer, pH 7.4, revealed that in this case all γ chains had been crosslinked, evidently by virtue of the presence of Ca^{2+} and factor XIII derived from the platelets themselves (Fig. 1, lane 4). An intense band corresponding to the γ_A-γ_A dimer was present as well as prominent bands corresponding to the α and β chain positions. In fibrin specimens that had been harvested into a Tris-saline buffer lacking EDTA, there occasionally were one or more fainter "non-fibrin" bands of uncertain identity, as in the specimen shown in lane 4.

Fibrin collected into EDTA (3mM)-containing buffers yielded few, if any, non-fibrin bands, particularly with respect to those migrating near the position of the three types of γ-dimer (Fig. 1, lane 5). Such material yielded intense bands in the β and γ_A chain positions as well as less intense bands corresponding to α chains and α chain derivatives (7,41). Most notable, however, was the absence of stainable material in the γ-dimer position. When an aliquot of this fibrin was subjected to crosslinking by immersion in a factor XIII, thrombin, Ca^{2+}, cysteine solution, the resulting gel yielded an intense band in the γ_A-γ_A dimer position (lane 6). There was no evidence of dimers containing γ' chains at any load condition.

In order to identify γ chains by virtue of their antigenic

determinants, we prepared transblots (42) of Laemmli gels and reacted them with anti-γ chain antibodies and then with ^{125}I-protein A (Fig. 2). Platelet fibrin harvested in the presence or absence of EDTA (Fig. 2, lanes 3 to 6) yielded a radioactive band in the γ_A chain position at lower loads (lanes 3 and 5) and an additional faint band in the γ_A-γ_A position at higher loads (lanes 4,6).

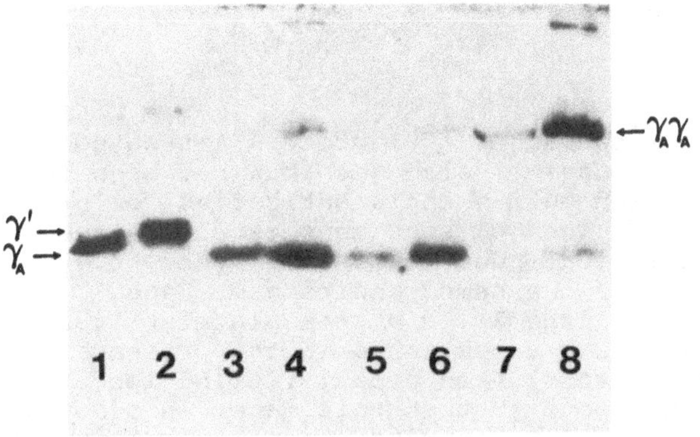

Fig. 2 Radioautogram of the transblot (42) of a Laemmli slab gel that had been reacted with affinity purified anti-γ chain antibodies followed by ^{125}I-protein A. Lane 1, S-CM γ_A chains; lane 2, S-CM γ' chains (S-CM γ_A and γ' chains migrate slightly less anodally than the corresponding uncarboxymethylated chains); lane 3, platelet fibrin harvested into buffer lacking EDTA; lane 4, same specimen, 8x load; lane 5, platelet fibrin harvested into EDTA-containing buffer; lane 6, same specimen, 8x load; lane 7 specimen from the same clot as in lanes 5 and 6, crosslinked in the presence of added factor XIII; lane 8, same specimen, 8x load.

Material collected into EDTA-containing buffers and subsequently crosslinked in the presence of factor XIIIa, yielded a band in the position of γ_A-γ_A dimers and a less intense band, at the higher load, in the monomeric γ_A position (lanes 7,8). In no case under any load condition did we detect the presence of a γ' chain band or a γ'chain-containing dimer band. As indicated previously, had γ' chains amounted to 1% or more of the total γ chain population, they would have been readily detected.

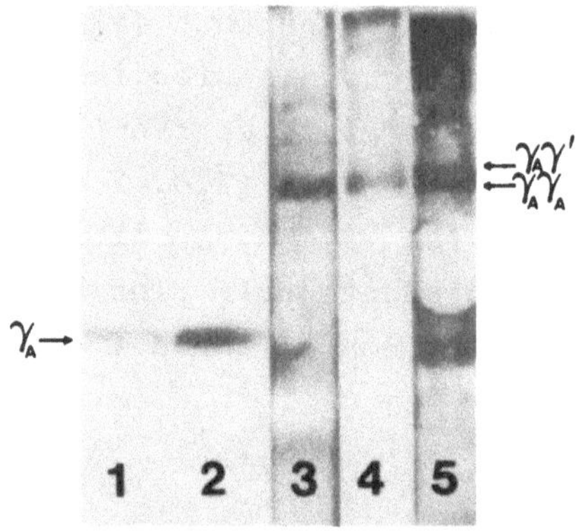

Fig. 3 Radioautogram of ^{125}I protein A-treated transblots of a Laemmli slab gel that had been reacted with affinity purified anti-γ chain antibodies followed by ^{125}I-protein A. The gel specimens were obtained from two different analytical runs (lanes 1,2; lanes 3,4,5) that were carried out under the same conditions. Lane 1, Triton X-100 platelet lysate; lane 2, 10 M urea platelet lysate; lane 3, Triton X-100 lysate crosslinked in the presence of factor XIIIa; lane 4, crosslinked peak 1 fibrin; lane 5, crosslinked peak 2 fibrin - note that in this lane, in addition to the prominent γ_A-γ_A and γ_A-γ' dimer bands, there is a relatively broad immunoreactive band suggestive of a doublet in the monomeric γ chain position reflecting the presence of uncrosslinked γ_A and γ' chains.

We also failed to detect γ' chains in fibrinogen from a Triton X-100 (Fig. 3, lane 1) or 10 M urea (Fig. 3, lane 2) lysate of washed platelets. Furthermore, the Triton lysate, when subjected to crosslinking in the presence of factor XIIIa and Ca^{2+}, yielded a single dimer band in the γ_A-γ_A position (Fig. 3, lane 3).

Discussion

Based upon our analyses of the electrophoretic migration rate of platelet γ chains and crosslinked γ chain dimers, it appears that all, or nearly all (>99%) platelet fibrinogen γ chains are of the γ_A type. This conclusion is based upon the following main observations and assumptions:

1. Electrophoresis of reduced fibrin clots derived from platelets that had been treated with thrombin to first release α-granule fibrinogen and subsequently to convert the released fibrinogen to fibrin, showed the presence of a single class of γ chain or γ chain dimer migrating in the γ_A or γ_A-γ_A position, respectively. The γ chains were identified not only by their migration rates in gels, but also by demonstrating that monomers were capable of forming covalently linked dimers in the presence of factor XIIIa and that both monomers and dimers reacted with antibodies to plasma fibrinogen γ chains.

2. Electrophoresis and transblotting of Triton X-100 or 10 M urea lysates of washed platelets, using γ chain antibodies for their ultimate identification, showed a single class of immunoreactive chain in the γ_A position; the crosslinked, reduced fibrin from the Triton X-100 lysate revealed a single band in the γ_A-γ_A position.

3. We assume that the platelet γ chains are structurally identical to plasma γ_A chains by virtue of their identical electrophoretic migration rate, their immunologic cross-reactivity, their functional behavior (i.e., γ-dimerization), and the fact that studies of the fibrinogen genes indicate that there is a single chromosomal locus for γ, Bβ, and Aα chains (9,13,16). In this regard, the findings in two congenital fibrinogen abnormalities, fibrinogen Oslo I and Oslo III (37), indicate that the respective defects in each molecule are expressed both in their plasma and platelet molecules; this supports the idea of the structural identity

of the platelet and plasma fibrinogen genes and, in particular, that plasma and platelet γ_A chains are derived from identical genes. Our finding that platelet and plasma fibrinogens are functionally equivalent with respect to γ chain crosslinking confirms the results of Doolittle et al. (35), who used an electrophoretic system (43) that does not distinguish γ_A and γ' chains from one another (10).

Our experiments appear to have eliminated the possibility that we were inadvertantly selecting for γ_A chain-enriched fibrinogen molecules by the method we used to isolate and identify the platelet fibrinogen. First, the same results were obtained whether the donor platelets were received from the Blood Center 2 or more hours after collection, or whether they had been prepared from donors who had taken aspirin orally prior to venipuncture and who had been bled directly into a potent platelet inhibitor and anticoagulant solution. Secondly, γ' chains were absent from fibrin obtained after thrombin treatment of washed platelets, as well as from platelet fibrin(ogen) present in Triton X-100 or 10 M urea lysates of washed platelets.

Although it was a popular belief for some time, it now seems highly unlikely that platelet fibrinogen is derived by selective cellular uptake of γ_A chain-containing plasma fibrinogen molecules, expecially since megakaryocytes have recently been shown to synthesize fibrinogen (28). Radioactively labelled plasma fibrinogen does not exchange with the platelet protein (21), nor is there reason to suspect that platelets would be selective with respect to an interaction with γ'- or γ_A-containing plasma fibrinogen molecules (44). Furthermore, it is important to note in the context of "cellular uptake", that plasma molecules bearing the fibrinogen Metz (38,39), fibrinogen Paris I (36), and fibrinogen Louisville (Galanakis, D.K., personal communication) anomalies are not found in the respective fibrinogen from these subjects' platelets.

In rats and humans, γ' chains are produced as a result of differential splicing of the primary mRNA transcript (9,16, A.J. Fornace, G.R. Crabtree, personal communication). Differences in processing of the γ chain gene at this level in the megakaryocyte could result in a gene product lacking γ' chains and seems to be the most probable explanation at this time. Nevertheless, this explanation will have to be reconciled with reports on two congenital plasma fibrinogen abnormalities, evidently resulting from single base mutations [i.e., fibrinogen Metz, Aα Arg 16 Cys (39); fibrinogen Louisville, Aα Arg 16 His, D.K. Galanakis, personal communication], which are not expressed in their respective platelet fibrinogen molecules. A third congenital plasma fibrinogen abnormality, Paris I, characterized by an elongated γ chain (45,46), is also not expressed in the subject's platelet fibrinogen molecules (36). The genetic basis for the Paris I plasma anomaly is not yet known.

Alternative possibilities involving differences in regulation of the γ chain gene should also be mentioned. Failure to translate certain mRNA or to effectively incorporate the translated chains into fibrinogen molecules could also account for the absence of γ' chains from platelet fibrinogen, or for that matter for the failure to observe the Metz, Louisville, and Paris I anomalies in platelet fibrinogen.

Another genetic mechanism that might explain certain observations would be the existence of a second fibrinogen gene that is expressed in megakaryocytes and platelets. This explanation could account for the findings in the Metz, Louisville, and Paris I subjects, but would be inconsistent with the findings in the Oslo I and Oslo III subjects. More importantly, the "separate gene" mechanism seems inconsistent with the findings of Crabtree and coworkers (9,13,16) who found only a single chromosomal locus coding for fibrinogen chains. As pointed out by Kant and Crabtree (47), the existence of a second genetic locus would seem possible, but only if the putative gene retained all the same restriction endonuclease cleavage sites as were found for the gene that has been characterized.

In any case, we feel confident in concluding from our studies that there are significant differences between the normal plasma fibrinogen γ chain population (93% γ_A, 7% γ') and that found in platelets (>99% γ_A). Only further experimental data will enable us to make an unambiguous choice among the several possible mechanisms for this occurrence.

Acknowledgements

This work was supported by NIH program project grant HL-28444. We thank Mr. David Evans, Ms. Mary Falk, and Ms. Martha Bushlow for their excellent technical assistance and Ms. Julie Erickson for her secretarial support.

References

1. Drury, D.R., McMaster, P.D.: J. Exp. Med. 50, 569-578 (1929).
2. Miller, L.L., Bale, W.F.: J. Exp. Med. 99, 125-132 (1954).

3. Barnhardt, M.I., Forman, W.B.: Vox Sang 8, 461-473 (1963).

4. Blombäck, B., Yamashina, I.: Ark. Kemi 12, 299-319 (1958).

5. Mosesson, M.W., Finlayson, J.S.: The search for the structure of fibrinogen, in Spaet, T.H. (ed): Progress in Hemostasis and Thrombosis, Vol 3, New York, Grune and Stratton, p.61-107 (1976).

6. Hoeprich, P.D. Jr., Doolittle, R.F.: Biochemistry 22, 2049-2055 (1983).

7. Mosesson, M.W.: N.Y. Acad. Sci. 408, 97-113 (1983).

8. Mosesson, M.W., Finlayson, J.S., Umfleet, R.A.: J. Biol. Chem. 247, 5223-5227 (1972).

9. Crabtree, G.R., Kant, J.A.: Cell 31, 159-166 (1982).

10. Wolfenstein-Todel, C., Mosesson, M.W.: Proc. Natl. Acad. Sci. (USA) 77, 5069-5073 (1980).

11. Wolfenstein-Todel, C., Mosesson, M.W.: Biochemistry 20, 6146-6149 (1981).

12. Nickerson, J.M., Fuller, G.: Proc. Natl. Acad. Sci. (USA) 78, 303-307 (1981).

13. Crabtree, G.R., Kant, J.A.: J. Biol. Chem. 256, 9718-9723 (1981).

14. Uzan, G., Besmond, C., Kahn, A., Marguerie, G.: Biochem Internat 4:271-277 (1982).

15. Chung, D.W., Rixon, M.W., Que, B.G., Chan, W-Y., Davie, E.W.: Thromb. Haemost. 50(1), 325 abs (1983).

16. Fowlkes, D.M., Kant, J.A., Fornace, A.J., Crabtree, G.R.: Thromb. Haemostas. 50(1), 325 abs (1983).

17. Ware, A.E., Fahey, J.L., Seegers, W.H.: Am. J. Physiol. 154, 140-147 (1948).

18. Salmon, J., Bounameux, Y.: Thromb. Diath. Haemorrh. 2, 93-110 (1958).

19. Grette, K.: Acta Physiol. Scand. (Suppl) 195, 1-93 (1962).

20. Gokcen, M., Yunis, E.: Nature 200, 590-591 (1963).

21. Castaldi, P.A., Caen, J.: J. Clin. Pathol. 18, 579-585 (1965).

22. Nachman, R.L.: Blood 25, 703-711 (1965).

23. Nachman, R.L., Marcus, A.J., Zucker-Franklin, D.: J. Lab. Clin. Med. 69, 651-658 (1967).

24. Nachman, R.L., Marcus, A.J.: Br. J. Haematol. 15, 181-189 (1968).

25. Broekman, M.J., Handin, R.I., Cohen, P.: Br. J. Haematol. 31, 51-55 (1975).

26. Kaplan, K.L., Broekman, M.J., Chernoff, A., Lesznik, G.R., Drillings, M.: Blood 53, 604-618 (1979).

27. Gerard, J.M., Phillips, D.R., Rao, G.H.R., Plow, E.F., Walz, D.A., Ross, R., Harker, L.A., White, J.M.: J. Clin. Invest. 66, 102-109 (1981).

28. Belloc, F., Hourdille, P., Fialon, P., Boisseau, M.R., Soria, J.: Thromb. Haemost. 50(1), 18 abs (1983).

29. Keenan, J.P., Solum, N.O.: Br. J. Haematol. 23, 461-466 (1972).

30. Charo, I.F., Feinman, R.D., Detwiler, T.C.: J. Clin. Invest. 60, 866-873 (1977).

31. Kaplan, K.L., Dauzier, M.J., Rose, S.: Blood 58, 797-802 (1981).

32. Davey, M.D., Luscher, E.F., in Kowalski, E., Niewiarowski, S. (eds): Biochemistry of Blood Platelets. New York, Academic Press, p.9-20 (1967).

33. Solum, N.O., Lopaciuk, S.: Thromb. Diath. Haemorrh. 21, 419-427 (1969).

34. Solum, N.O., Lopaciuk, S.: Thromb. Diath. Haemorrh. 21, 428-440 (1969).

35. Doolittle, R.F., Takagi, T., Cottrell, B.A.: Science 185, 368-370 (1974).

36. Jandrot-Perrus, M., Mosesson, M.W., Denninger, M-H., Ménaché, D.: Blood 54, 1109-1116 (1979).

37. Brosstad, F., Teige, B., Gogstad, G., Olaisen, B., Teisberg, P., Gedde-Dahl, T.: Haemost. Thromb. 50(1), 415 abs (1983).

38. Soria, J., Soria, C., Samama, M., Poirot, E., Kling, C.: Pathol. Biol. (suppl)24, 15-17 (1976).

39. Soria, J., Soria, C., Samama, M., Henschen, A., Southan, C., in Henschen, A., Graeff, H., Lottspeich, F. (eds): Fibrinogen. Recent Biochemical and Medical Aspects. New York, W. de Gruyter, p.129-143 (1982).

40. Laemmli, V.K.: Nature (Lond.) 227, 680-685 (1970).

41. Mosesson, M.W., Galanakis, D.K., Finlayson, J.S.: J. Biol. Chem. 247, 5210-5219 (1972).

42. Burnette, W.H.: Anal. Biochem. 112, 195-203 (1981).

43. Weber, K., Osborn, M.: J. Biol. Chem. 244, 4406-4412 (1969).

44. Zucker, M.B., Mosesson, M.W., Broekman, M.J., Kaplan, K.L.: Blood 54, 8-12 (1979).

45. Budzynski, A.Z., Marder, V.J., Menache, D., Guillin, M.C.: Nature 252, 66-68 (1974).

46. Mosesson, M.W., Amrani, D.L., Menache, D.: J. Clin. Invest. 57, 782-790 (1976).

47. Kant, J.A., Crabtree, G.R.: J. Biol. Chem. 258, 4666-4667 (1983).

DIFFERENCES AND SIMILARITIES BETWEEN HUMAN ADULT AND FETAL FIBRINOGEN FRAGMENTS D_1

Dennis K. Galanakis
Health Science Center, SUNY
Stony Brook, New York

The behavior of human fetal fibrinogen is characterized by delayed fibrin aggregation (1,2), a typically transparent (3,4) and electronmicroscopically distinct(4) fibrin gel, high sialic acid(5,6) and phosphorus(7) content. In addition, fetal fibrin inhibits the aggregation of adult fibrin monomers(1,5); also, these properties are much more marked when samples to be examined are obtained from pre-term (rather than full term) gestation infants(5).

In the present studies we compared fetal fragment D_1 with its adult counterpart. Fetal fibrinogen(fraction I-2) was obtained from full term cord blood(2). Plasmic (Stage 3) digests were prepared (8) in 5 mM $CaCl_2$-containing buffer (50 mM Triethanolamine, TEA, 100 mM NaCl, pH 7.4) and subjected to electrophoresis (Polyacrylamide in the presence of sodium dodecyl sulfate, PAGE-SDS, ref. 9) to ascertain the stage of digestion. Digests for chromatography were dialyzed in 10 mM TEA, 1 mM $CaCl_2$ and applied to a 30 x 1.5 cm (Sigma Chemical Co. St. Louis, Mo.) lysyl-Sepharose column according to the procedure described by Rupp and co-workers (8), briefly described here for convenience. A step-wise elution is achieved by ionic strength increments so that the first protein peak is obtained with the 10 mM (or application) buffer; the second peak is eluted by increasing the TEA concentration to 30 mM, and this (second) protein peak consists of D_1 virtually free from other fragments, as

Fibrinogen - Structural Variants and Interactions
© 1985 Walter de Gruyter & Co., Berlin · New York - Printed in Germany

assessed by PAGE-SDS. In preliminary work our results were in general agreement with those originally described by these investigators. That is, our analyses of pooled samples from the different protein peaks disclosed that the second peak consisted of D_1 and the next two peaks contained smaller D fragments and fragment E, respectively. The latter two peaks are eluted with 50 mM TEA and 20 mM NaCl, and then by increasing NaCl to 100 mM, respectively. Other retained fragments are eluted with a (urea) last step as described (8). Between 40-50 mg of digest protein was applied and chromatographed in this way each time. Elution yield estimates in our experiments were at least 80% of protein applied to the column.

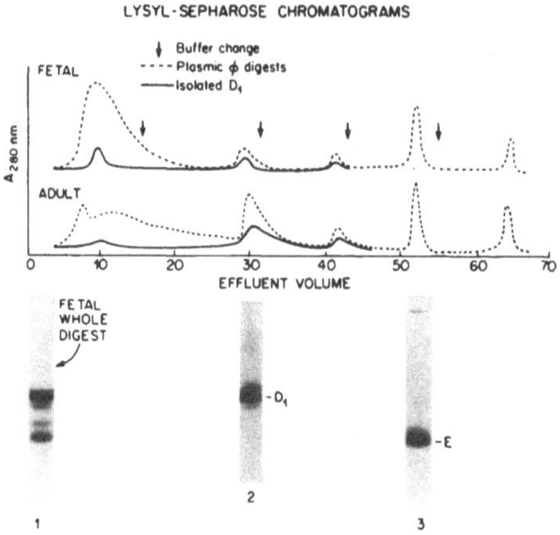

Fig. 1

When fetal and adult fibrinogen digests were compared by PAGE-SDS, a consistently similar electrophoretic pattern was demonstrable, consisting of major bands in the D_1 and E positions and of trace or minor amounts of fragments Y and D_{2-5}. Gels of a typical experiment showing a fetal whole digest and its isolated D_1 and E are illustrated in Fig. 1. Also, the chromatographic peaks appeared similar, Fig. 1, with the important exception of the second elution peak.

This peak appeared at least two-to-three fold greater in all (five) adult than the corresponding peak of (five) fetal samples. Absorbance (280 nm) measurements of pooled and concentrated samples from these peaks as well as electrophoresis and gel scanning densitometry disclosed that 60-80% of fetal D_1 eluted with the first protein peak and the remainder in the second peak. Thus the bulk of fetal D_1 was not retained by the column, in contrast to its adult counterpart which eluted consistently in the peak 2 position, as previously reported (8). Results from analysing isolated single samples of D_1, Fig. 1, were also consistent with this observation in that 60% of fetal D_1 eluted in this (first peak) position, relative to only 24% of adult D_1. In contrast, there were no differences in the elution positions of other fetal and adult Fragments, (i.e. E or D_{3-5}).

Additional electrophoretic analyses disclosed that reduced samples of adult and fetal D_1 displayed identical electrophoretic bands. Moreover, carbohydrate staining of these electrophoretic bands by use of fluorescein-labeled (Sigma Chem. Co.) concanavalin A was carried out. Protein bands in the gels were fixed (in the dark) in a 25% isopropanol-10% acetic acid mixture, washed extensively with distilled water, and incubated in 0.1M acetate buffer, pH 6.5 containing $CaCl_2$ (20 mM), $MgCl_2$ (10 mM) and fluorescein-labeled concanavalin A, (0.3 mg/ml). Following extensive washing with buffer the gels were examined for fluorescence. No differences between fetal and adult D_1 were observed. That is, the larger (β- chain) subunit clearly displayed fluorescence while the second (γ- chain) electrophoretic band lacked fluorescence-staining whether fetal or adult D_1 samples were analyzed (not shown).

Fragment	Sialic Acid (res/mole)
Fetal D_1 (4)	1.3
Adult D_1 (2)	1.4
Fetal E (2)	2.0
Adult E (2)	2.1

Table 1. Sialic acid content of isolated plasmic digest fragments D and E. Values are expressed in residues/mole, assuming molecular weights of 100,000 and 50,000 for D_1 and E respectively (11); variation between similar samples (number in parentheses) was $\leq$10%.

Assays of sialic acid (10) content were also carried out. Single starting fibrinogen preparations, used for this comparison, had the expected sialic acid content, (i.e. 7.6 res/mole and 6.5 res/mole for fetal and adult fibrinogen, respectively); values for fragments D_1 and E, shown in Table 1, indicated significant loss of sialic acid content from both adult and fetal samples. That is to say, the sum of sialic acid residues accounted for 84% of adult and 82% of fetal starting fibrinogen preparations, respectively. These findings may reflect the presence of trace neuraminidases in the buffers or reagents employed. Additional support for this conclusion was provided by repeat measurements of stored fibrinogen (fetal and adult samples). A variable loss of one or more sialic acid residues per mole in all (5 fetal and 4 adult) samples as compared to values obtained when these measurements were made shortly after fibrinogen isolation from plasma. Similar observations on sialic acid measurements have been reported previously (5). Nevertheless, these measurements on plasmic fragments indicated that sialic acid content did not account for the chromatographic behavior of fetal D_1.

Additional studies were carried out in which the inhibition by these fragments of the reaggregation of adult fibrin monomers (5) was used to compare adult and fetal D_1. The two differed in that the inhibition was more marked by fetal D_1, and these results are detailed in a separate report (in preparation).

The studies indicate a structural difference between adult fetal D domains, and raise the possibility that a subterminal oligosaccharide structure(s) may account for these differences. They also suggest the possibility that the carbohydrate on fetal β chains resides on a different asparagine residue than that of the adult β chains. Even though the alternative possibility, that of an amino acid sequence difference in the carboxyl terminal sequence of the fetal γ chains remains to be excluded, available information so far favors the view of a post-transcriptional origin of the distinguishing fetal fibrinogen characteristics. Be that as it may, a structure(s) which accounts at least in part for the distinct fetal fibrinogen behavior seems to reside on its D domain.

References

1. Guillin, M.C., Menache, D.: Thromb. Res. 3, 117-135 (1973).
2. Galanakis, D.K., Mosesson, M.W.: Blood 48, 109-117 (1976).
3. Burstein, M., Lewi, S., Walter, P.: Sang 25, 102-107 (1954).

4. Tesch, R., Trolp, R., Witt, I.: Thromb. Res. 16, 239-243 (1979) and Muller, M., Burchard, W., Witt, I.: Thromb. Res. 24, 339-346 (1981).

5. Galanakis, D.K., Martinez, J., McDevitt, C., Miller, F.: Ann. N.Y. Acad. Sci. 408, 640-643 (1983).

6. Galanakis, D.K., Mosesson, M.W.: VII Intern. Congr. Thromb. Haemost. (Abstr.) p. 79 (1979).

7. Witt, I., Hasler, K.: Biochim, Biophys. Acta 271, 357-362 (1970); and Witt and co-workers(this volume).

8. Rupp, C., Sievi, R., Furlan, M.: Thromb. Res. 27, 117-121 (1982).

9. Weber, K., Osborn, M.: J. Biol. Chem. 244, 4406-4412 (1969).

10. Warren, L.: J. Biol. Chem. 234, 1971-1975 (1959).

11. Marder, V.J., Francis, C.W., Doolittle, R.F.:Fibrinogen Structure and Physiology, in:Hemostasis and Thrombosis-Basic Principles (R.W. Colman et. al., Editors), J.B. Lippincott Co., 1982.

Supported by United States Public Health grant HL27196

VI. ABNORMAL FIBRINOGEN VARIANTS

FUNCTIONAL DEFECTS IN ABNORMAL FIBRINOGENS

Nadia Carrell

Department of Anatomy, Duke University Medical Center, Durham, North Carolina 27710, USA

Jan McDonagh

Department of Pathology, Beth Israel Hospital and Harvard Medical School, Boston, Massachusetts 02215, USA

Defects of fibrinogen activity may exist in all reactions in which fibrinogen plays a role. These most commonly include abnormal fibrinopeptide release and/or a defect in fibrin polymerization. Virtually all reported congenitally abnormal fibrinogens are defective in one or both of these activities (1). Some of the functional activities of fibrinogen are shown in Table I. Various abnormal fibrinogens are defective in one or more of these reactions, and a wide range of combinations of defects should be possible. Abnormalities of fibrin crosslinking, plasminogen binding, plasmin degradation, Ca^{2+} binding, and metabolism have been described (2,3,4,5). One might expect to find abnormal fibrinogens which display defective binding of tissue plasminogen activator, α_2-antiplasmin, and other molecules which interact with fibrinogen and fibrin in physiologically important ways.

Some abnormal fibrinogens are difficult to diagnose, because the defects may not be associated with functions which are assayed in routine coagulation tests. There is as yet no specific relationship between functional defects determined in the laboratory and the clinical symptoms of the pa-

Fibrinogen - Structural Variants and Interactions
© 1985 Walter de Gruyter & Co., Berlin · New York - Printed in Germany

Table I. Functional Activities of Fibrinogen

Fibrin gel formation

thrombin binding
fibrinopeptide A cleavage
fibrinopeptide B cleavage
protofibril formation
lateral association
branchpoint formation

Covalent fibrin crosslinking

factor XIIIa binding
γ- γ dimer formation
α-polymer formation

Fibrin and fibrinogen degradation

plasminogen activator binding
plasminogen binding
plasmin binding and hydrolysis
α_2-plasmin inhibitor binding

Cellular reactions

platelet aggregation
clot retraction
fibroblast migration
macrophage binding - phagocytosis

Other

fibrin-fibronectin interaction
Ca^{2+} binding
solubility
metabolism

tients. Some defects are associated with mild to severe hemorrhage, some with thrombosis, and others with no clinical symptoms. Primary sequence defects have now been described in a number of abnormal fibrinogens (6,7), and as more data become available for more patients, perhaps it will be possible to correlate specific amino acid substitutions with specific functional defects and with clinical diseases.

In addition to presenting interesting clinical problems, abnormal fibrinogens provide a unique opportunity for the study of fibrinogen structure-function relationships. The constellation of specific functional defects which any abnormal molecule exhibits is determined by the specific structural defect and the effect of the structural defect on conformation. Thus, abnormal molecules can provide the way for mapping the functional domains in fibrinogen. Some of the interesting problems in studying these molecules can be illustrated by specific examples.

Fibrinogen Chapel Hill II has decreased fibrinopeptide A release (3). It has recently been determined by Kehl and Henschen that there is a heterozygous substitution at position 16 in the A α chain, from Arg to His (8). This substitution results in an increase in the negative charge of the already highly negatively charged fibrinopeptides. The abnormal peptide is only released in the presence of a large excess of thrombin or reptilase, which releases only the A peptides. With such an excess of enzyme it is possible that the Arg-Val bond at positions 19-20 is cleaved instead of the substituted His-Gly at positions 16-17. This clearly occurs when fragment X from Chapel Hill II is treated with thrombin, since the sequence of the α chain isolated from thrombin-treated fragment X begins at position 20. When the release of fibrinopeptides was quantitated by the reaction of peptide-bound, carboxy-terminal arginine with phenanthrenequinone, complete release was demonstrated. This also implies that Arg_{19} was cleaved in the abnormal protein and that this is the arginine which reacted in the assay.

The fibrin formed by the action of thrombin on this fibrinogen was also studied by laser light scattering (3). A high enzyme to substrate ratio was used in order to study assembly independent of peptide release. The fibrin consisted of short thick fibrils (220 molecules in cross-sectional area as compared to normal fibers which averaged 100 molecules). It is possible that the abnormal clot formation is a secondary effect of the fibrinopeptide defect. This would be expected if the abnormal peptide were not cleaved and the extra negative charges were still present in the E domain. It may also be that the presence of the abnormal peptide in the milieu alters the fiber formation and that if the peptide were removed, assembly would be normal, as has been observed for Fibrinogen Manchester (9).

Crosslinking of Chapel Hill II fibrin was also abnormal. $\gamma-\gamma$ dimers formed at a normal rate, but α chain polymerization was very slow. This could be corrected with a large excess of factor XIII, but not with a normal amount of XIII and excess of thrombin. Plasmin degradation of the fibrinogen was also delayed.

The cluster of defects in this molecule indicates that, not only fibrinopeptide release, but also each step in the formation and degradation of fibrin can be affected by the substitution of a histidine for arginine at position 16 of the Aα chain. This region in the E domain contains a polymerization contact site (10). In addition it may also influence the proper orientation of the carboxy-terminal portion of the α chain in a fibrin clot such that factor XIIIa mediated crosslinking can occur. There is electron microscopic evidence for this part of the α chain in proximity to the E domain (15).

In contrast to the above case, the patient with Fibrinogen Chapel Hill III has a severe thrombotic problem, and the abnormality in fibrinogen function is not detectable in routine coagulation screening (11). All clotting times were normal except those measured with the snake venom enzymes, reptilase and ancrod (Table II). The prolonged clotting time is due to abnormal polymerization, since the release of fibrinopeptides was found to be normal. Polymerization was markedly abnormal with turbidity measurements. When an-

Table II. Plasma Fibrinogen Levels and Clotting Times for the Series of Abnormal Fibrinogens Described from Chapel Hill

Patient	Fibrinogen concentration, mg/dl*			Thrombin time, sec*	Reptilase time, sec*
	A	B	C	D	E
II	116/263	408/235	427/272	29.2/12.4	38.2/14.8
III	258/216	332/235	249/256	12.6/11.5	31.0/16.3
IV	131/219	372/300	352/299	20.3/11.4	90.0/14.0
V	99/225	398/252		24.1/11.4	78.0†/16.3
VI		155/252		14.1/11.5	13.8/16.3
VII	16/250	11/300		34.5/11.6	52.0/14.0

*All values are given for patient/normal.
A: Method of Clauss (12).
B: Radial immunodiffusion (13).
C: Method of Ratnoff and Menzie (14).
D: Determined in a test system containing 0.2 ml plasma, 0.1 ml saline, and 0.1 ml thrombin; final thrombin concentration = 5 u/ml.
E: Determined in the same system as for thrombin with final reptilase dilution of 1:80.
†Indicates appearance of fibers; gelation occurred at ~5 min.

crod was used as the clotting enzyme, there was no increase in turbidity, even though the sample did form a gel. Further studies of fibrin assembly were performed on purified fibrinogen by light scattering. The fibers which formed were very thin, producing a fine, transparent clot. These thin fibers did, however, form a very rigid clot. This abnormal fibrin was highly resistant to plasmin. Once a clot had formed, plasmin was unable to cleave the fibrin molecules, and the constituent polypeptide chains remained intact in the clot. This combination of slow and abnormal assembly of fibrin to produce rigid fibers which are resistant to plasmin provides an explanation for the thrombotic disease in this patient.

Initial characterization of a patient with an abnormal fibrinogen requires several simple laboratory tests. These include: measurement of fibrinogen concentration by a functional (chronometric) test and also measurements of total clottable and immunological levels of fibrinogen in plasma; differentiation between non-clottable fibrinogen and degradation products in serum; measurement of plasma clotting times with low concentrations of thrombin, ancrod, and reptilase; measurement of fibrin polymerization in dilute plasma. Determination of some abnormal function in a relative provides evidence for the congenital nature of the defect. This can be illustrated with several additional patients in which it was possible to provide a clear diagnosis of abnormal fibrinogen. The essential data for these patients are shown in Table II and Figure 1.

Fibrinogen Chapel Hill IV was detected in a woman with inflammatory bowel disease, probably of autoimmune origin. This patient, like Chapel Hill II, had a normal level of fibrinogen when determined immunologically and by the method of Ratnoff and Menzie, which measures total clottable material. However the level determined by a clotting time method was only half as great. The thrombin time was prolonged, and the reptilase clotting time was markedly prolonged. Polymerization was studied with both thrombin and ancrod (Figure 1). With thrombin, the lag phase was normal, but the initial rate of aggregation was decreased. With reptilase, no clot formed within the time span measured. Fibrinogen related material in serum was high (160 μg/ml).

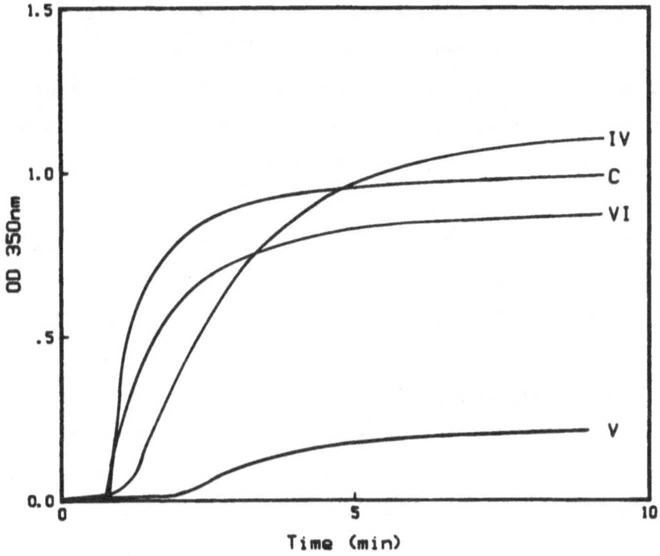

Figure 1A: Fibrin polymerization curves for abnormal fibrinogens IV,V, and VI activated with thrombin. Plasma was diluted 1:2 with Tris-NaCl buffer, and clotting was initiated by addition of 0.1 ml thrombin (10 u/ml). C = normal, control plasma.

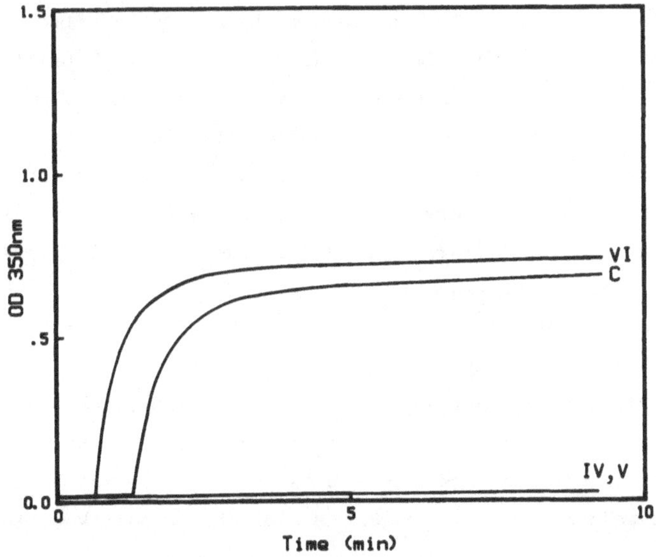

Figure 1B: Fibrin polymerization curves for abnormal fibrinogens IV,V, and VI activated with ancrod. Test system was the same as for thrombin except that ancrod (0.1 ml at 100 ancrod units/ml) was used to clot the fibrinogen.

However in this patient this is a reflection of the poorly clottable abnormal fibrinogen. Actual degradation products of fibrinogen were not detected in the patient's serum. To examine crosslinking, clots were formed in plasma by addition of thrombin, 2.5 units/ml, and calcium, 20 mM. For a qualitative test of crosslinking, clot solubility in an equal volume of 1% monochloroacetic acid was checked. After 24 hours, the clot remained insoluble. A more direct and quantitative assay of crosslinking is to examine the clot on SDS polyacrylamide electrophoresis. For this, the clot is synerised and washed thoroughly with saline before dissolving in SDS-urea-DTT and applying to 5% gels. The subunit pattern obtained, complete dimerization of γ chains and some polymerization of α chains, was identical to the normal control. All of the data concerning this patient indicate that there is a defect in fibrinopeptide A release, and that the delayed rate of fibrin polymerization could be caused by defective peptide release.

Fibrinogen Chapel Hill V was detected in a woman with hypertension and possible previous venous thrombosis; her family history was positive for cerebrovascular disease. The patient had a normal fibrinogen level when measured immunologically and 50% when measured by a clotting time dependent assay. The thrombin clotting time was twice that of normal (Table II) and could be corrected by the addition of normal plasma. Addition of reptilase to the plasma resulted in the slow formation of a few fibers and finally in gelation after five minutes. The polymerization curves are shown in Figure 2. With thrombin there was a pronounced lag time in the patient, probably reflecting delayed fibrinopeptide release, and a slow, small rise in turbidity, indicative of fine fibers as with Chapel Hill III. However with ancrod no clot formed in the time span studied. The same defect was present in the proband, 2 daughters, and a son, so this fibrinogen can be clearly described as congenitally abnormal, with defective fibrin polymerization and probably also defective fibrinopeptide A release. Crosslinking was determined to be normal on the basis of clot insolubility in monochloroacetic acid and SDS-electrophoresis of the crosslinked fibrin.

The plasma fibrinogen concentration of Fibrinogen Chapel Hill VI was 155 mg/

ml when measured immunologically. However, the thrombin clotting time was only slightly prolonged, and the reptilase time was actually shorter than the control. The polymerization curves reflect the results of the clotting times. The rate of polymerization with thrombin was only slightly less than normal. With ancrod the lag time in the patient plasma was much less than in the normal and the aggregation rate was the same; hence gelation occurred more quickly than in the control. The clot was insoluble in monochloroacetic acid. Examination of the subunits structure of the clot by gel electrophoresis revealed that α polymers formed faster in the patient than in normal controls. The observation of increased rate of fibrin polymerization and α chain crosslinking with a decreased fibrinogen concentration is somewhat unusual and probably reflects a primary polymerization anomaly.

Fibrinogen Chapel Hill VII was detected in a young woman with a history of multiple spontaneous abortions in the first trimester. This is clearly a case of hypofibrinogenemia, since the patient's fibrinogen concentration ranged from 11-16 mg/dl by all methods of determination. All clotting times were markedly prolonged due to the low fibrinogen concentration. Mixtures of normal and patient plasma clotted normally. The patient's clots were insoluble in monochloroacetic acid, indicating that covalent crosslinks had formed. Further characterization of this abnormal fibrinogen could not be carried out because of a lack of sufficient material.

These examples illustrate some of the variety of functional defects which may be encountered with abnormal fibrinogens. A principal reason for studying abnormal molecules in detail is that they provide important information concerning normal function. Abnormal molecules have a far greater specificity of amino acid modification than can be obtained by ordinary chemical modification in vitro; they also have the advantage of clinical correlation which would not be available for molecules produced by cloning techniques. As structural information becomes available for more abnormal fibrinogens, it should become possible to use this information to map the functional domains of the fibrinogen molecule. A further challenge will be to correlate specific structural alterations with a specific cluster of functional defects and with the clinical status of the patients.

Acknowledgement

The patients described here were referred to us through the Coagulation Laboratory of the North Carolina Memorial Hospital, University of North Carolina School of Medicine. This research was partly supported by NIH Grant HL 29511.

References

1. Mammen, E.F.: Sem. Thromb. Hemostas. IX, 1-9 (1983).
2. Caen, J.P., Soria, J., Soria, C.: C.R. Acad. Sci. Paris 294, 695-698 (1982).
3. Carrell, N., McDonagh, J.: Brit. J. Haematol. 52, 35-47 (1982).
4. Rupp, C., Kuyas, C., Haberli, A., Furlan, M., Perrot, B.A., Beck, E.A.: In: Fibrinogen - Recent Biochemical and Medical Aspects (Henschen, A., Graeff, H., Lottspeich F., eds.) 167-181 (1982).
5. Gralnick, H.R., Coller, B.S., Fratantoni, C., Martinez, J.: Blood 53, 28-46 (1979).
6. Southan, C., Kehl, M., Henschen, A., Lane, D.A.: Brit. J. Haematol. 54, 143-151 (1983).
7. Southan, C., Henschen, A., Lottspeich, F.: In: Fibrinogen-Recent Biochemical and Medical Aspects (Henschen, A., Graeff, H., Lottspeich, F., eds) 153-166 (1982).
8. Kehl, M., Henschen, A. Personal communication.
9. Lane, D.A., Southan, C., Ireland, H., Thompson, E., Kehl, M., Henschen, A.: Brit. J. Haematol. 53, 587-597 (1983).
10. Laudano, A.P., Doolittle, R.F.: Proc. Natl. Acad. Sci. USA 75, 3085-3089 (1978).
11. Carrell, N., Gabriel, D.A., Blatt, P.M., Carr, M.E., McDonagh, J.: Blood, 62, 439-447 (1983).
12. Clauss, A.: Acta Haematol. 17, 237-246 (1957).
13. Ratnoff, O.D., Menzie, C.: J. Lab. Clin. Med. 37, 316-320 (1951).
14. Mancini, G., Carbonara, A.O., Heremans, J.F.: Immunochemistry 2, 235-254 (1965).
15. Erickson, H.P., Fowler, W.E.: Ann. N.Y. Acad. Sci. 408, 146-163 (1983).

STUDY OF 10 CASES OF CONGENITAL DYSFIBRINOGENEMIA. CLINICAL AND MOLECULAR BIOLOGICAL ASPECTS

Jeannette Soria, Claudine Soria, Meyer Samama, Jacques Caen

Laboratoires des Professeurs Bernadou, Fabiani, Samama, Hôtel-Dieu, 75181 Paris Cédex 04, France.

Laboratoires des Professeurs Caen, et Rousselet, et Unité U.150 INSERM, Hôpital Lariboisière, 75010 Paris, France.

Introduction

Since, the discovery of the first two cases of dysfibrinogenemia by Imperato and Dettori (1) and by Menaché (2), many other dysfibrinogenemia have been found (more than one hundred).

We have analysed several cases of dysfibrinogenemia in our laboratory. The aim of our study is first to analyse the relationship between fibrinogen structure anomaly and abnormal fibrinogen function both in fibrin formation and in platelet aggregation, and second to analyse the relationship between dysfibrinogenemia and clinical disorder, such as haemorrhage, abortion or thrombotic syndrome. In fact, bleeding tendency occurs only in a very few cases of dysfibrinogenemia, but thrombotic syndrome are more often observed. A study was undertaken in order to explain the different mechanisms of thrombosis which may occur in patients presenting with dysfibrinogenemia.

Fibrinogen - Structural Variants and Interactions
© 1985 Walter de Gruyter & Co., Berlin · New York - Printed in Germany

Protocol used for the analysis of dysfibrinogenemia

Dysfibrinogenemia are mainly discovered by chance because of a prolonged prothrombin time and normal level of factors V, VII, X and II.
The anomaly is related to a prolonged thrombin clotting time. Definite diagnosis of dysfibrinogenemia however is made when the abnormal fibrin formation is also present on isolated fibrinogen.
All our abnormal fibrinogens were isolated according to the technique of Blomback (3) which avoids the degradation of fibrinogen because all steps are performed at low temperature. The analysis of isolated fibrinogen concerns :

1) Analysis of the three steps of fibrin formation
   a) Release of fibrinopeptide A (fpA) and fibrinopeptide B (fpB), was performed by various techniques.

   In the oldest one (4), either thrombin or reptilase was added to fibrinogen. The reaction was stopped after various lenghts of time (30 sec., 1 min., 2 min. 10 min., and 30 minutes), by adding trichloracetic acid : fibrin, fibrin monomer, fibrinogen and thrombin were precipitated and the amount of fibrinopeptides released was evaluated in the supernatant by a modified Lowry procedure.

   In our laboratory, we do not use this technique any more because it requires a large amount of fibrinogen and fibrinopeptides structure abnormalities cannot be evidenced. This technique however has allowed us to detect the absence of fpA release in fibrinogen Metz (5).
   For the study of fp release, we now use both :

   . HPLC analysis of fpA and fpB release, according to the procedure described by Kehl and Henschen (6).

By this technique, quantitative determination of fp is possible and also fpA and B structure abnormality can be evidenced (fibrinogen Rouen, fibrinogen Amiens, fibrinogen Clermont-Ferrand).

. Analysis of fpA by immuno-enzymological assay, according to Soria procedure (7). This technique is very sensitive and therefore very helpful for the kinetic study of fpA release mainly in dysfibrinogenemia presenting with defective fpA release.

b) Fibrin monomer polymerization.

Fibrin monomer in 1 % acetic acid solution is prepared according to the procedure of Belitser (8). When diluted in buffer, polymerization of fibrin monomer occurs which is evidenced by recording the O.D. at 350 nm. It must be noted that the rate of polymerization is increased when tris-buffer is used instead of phosphate buffer.

c) Fibrin stabilization

The formation of covalent bonds induced by factor XIIIa leading to gamma-dimers and alpha-polymers is analysed by SDS-PAGE according to the procedure of Schwartz (9).

2) Analysis of fibrinogen structure

a) Analysis of the three chains of fibrinogen

Prior to analysis, the three chains of fibrinogen are dissociated in the presence of urea and mercaptoethanol.

Their apparent molecular weights are determined by SDS-PAGE (9).
Their electrostatic charges are determined either by disc electrophoresis (10) or electrofocusing (11).

b) Analysis of fibrinogen degradation products

Fibrinogen is degraded by plasmin in the presençe or absence of Ca^{++} for various lenghts of time. Then fibrinogen degradation products (unreduced and reduced derivatives) are analysed by SDS-PAGE.
This assay may give some further information about the localization of the fibrinogen abnormality (fibrinogen Pontoise (13)). In addition, this analysis may reveal some abnormalities of fibrinogen which are not detected when studying intact fibrinogen or its dissociated chains (fibrinogen Haïfa (14)).

c) Analysis of fibrinopeptide structure

Fibrinopeptide abnormalities are detected by HPLC analysis, as described previously.

d) Amino-acid analysis

When the fibrinogen anomaly is localized in a well defined part of the molecule, amino-acid sequence of this part of the molecule is performed either by Doctor Henschen or by Doctor O. Bertrand (15).

Our dysfibrinogenemias are divided into two groups, according to this protocol :

Group I : dysfibrinogenemia presenting with abnormal fibrinopeptides release ;

Group II: dysfibrinogenemia presenting with abnormal fibrin monomer polymerization and normal fibrinopeptides release.

Dysfibrinogenemia presenting with abnormal fibrinopeptide release

In this group, we differentiate the cases which present either no release or a decreased fpA release from the cases presenting an abnormal fpA.

1) Dysfibrinogenemia characterized by the absence of fpA release after thrombin addition : Fibrinogen Metz (5, 7)

Fibrinogen Metz is an homozygous dysfibrinogenemia discovered in a female presenting mild bleeding tendency and several abortions.
When performing routine tests, a discrepancy was noted between coagulation of whole blood (20 minutes) and the absence of clottability of citrated plasma in the presence of very high concentrations of thrombin of several origins even when calcium was added to the medium.
The anomaly of fibrinogen Metz was easily detected according to the protocol described previously. The anomaly was located on the A-alpha-chain because when using disc electrophoresis its migration was faster than that of normal A-alpha-chain. Furthermore, no fibrinopeptide A release after thrombin addition was detected by the very sensivite immuno-enzymological assay. Therefore, it was suggested that the anomaly should be located on the A-alpha chain near the site of thrombin action. Indeed the site of mutation was evidenced by the sequential analysis of the N-terminal part of A-alpha-chain carried out by A. Henschen and Coll. who showed that Arg-16 was replaced by cysteine-16.
The process of fibrin formation was studied in fibrinogen Metz because clotting occured without fibrinopeptide A release (17) :
After clotting, the level of fpA was nearly nil in serum (it was checked that all the fibrinogen present in plasma was clotted since fibrinogen related antigen in serum was

below 10 µg/ml). Furthermore, as shown by A. Henschen the fpA sequence remained present in fibrin. In spite of the absence of fpA release, fpB was cleaved by thrombin from fibrinogen Metz.
Indeed the release of fpB by thrombin from fibrinogen Metz was suspected, because according to Lowry procedure half of the amount of fibrinopeptides was evidenced as compared to that released from normal fibrinogen. FpB release in the absence of fpA release was confirmed by A. Henschen and M. Kehl using HPLC analysis.

These results may explain the discrepancy observed between the unclottability of citrated plasma and the coagulation of whole blood. It was suggested that the unclottability of citrated fibrinogen was due to the presence of an excess of negative charges resulting in a repulsion of fibrin monomers. These charges are due to the presence of fpA which is not released by thrombin and the presence of citrate which binds more to fibrinogen Metz than to normal fibrinogen (18). The presence of fpA alone on fibrin monomer is not sufficient to avoid their aggregation.

2) Dysfibrinogenemia characterized by presence of an abnormal fibrinopeptide A.

Three cases of heterozygous dysfibrinogenemia presenting an abnormal fpA were characterized in our laboratory. After thrombin addition, the release of abnormal fpA from fibrinogen was slower than that of normal fpA. The presence of an abnormal fpA was easily detected by HPLC analysis (6). The isolated abnormal fibrinopeptides of each case were then analysed for their amino-acid content. The three cases (manuscript in preparation) are summarized in table 1.

Table 1 : Abnormal fibrinogen presenting an abnormal fpA

	Collaboration with :	Mutation
Fibrinogen Rouen	A. Henschen, M. Kehl, J.Y. Borg, H. Piguet	Gly-12 ⟶ Val
Fibrinogen Amiens	O. Bertrand, J. Maclouf, J. Delobel	Arg-16 ⟶ His
Fibrinogen Clermont-Ferrand	O. Bertrand, J. Maclouf, M.J.Bezou F. Poitevin-Lavenu	Arg-16 ⟶ His

Fibrinogen Rouen is a particular case, because thrombin time is prolonged whereas reptilase clotting time is normal. Therefore, it is suggested that the abnormality does not include the cleavage site but could be located on a site important for thrombin-fibrinogen interaction.

The three patients studied did not present any haemorrhagic nor thrombotic disorder. It is of interest to notice that the mutation Arg-16 ⟶ His , is the most often found in dysfibrinogenemia (many cases were described by A. Henschen).

Congenital dysfibrinogenemia with abnormal fibrin monomer polymerization

This group of dysfibrinogenemia is characterized by :

- prolonged thrombin and reptilase clotting times,
- normal fpA and fpB release
- abnormal fibrin polymerization.

Two different abnormal fibrin monomer aggregation patterns were evidenced when either thrombin or reptilase was added to diluted fibrinogen (19). The increase of turbidity was nonexistent or occurred only at a very low rate as compared to control (Figure 1).

Figure 1 : Fibrin polymerization curve after thrombin addition to fibrinogen solution.

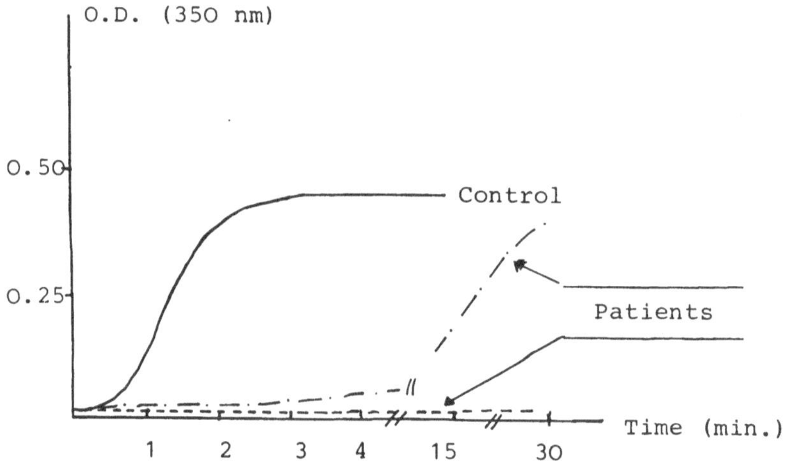

The same patterns were observed when using purified fibrin monomers (8).
In cases of dysfibrinogenemia presenting abnormal fibrin polymerization, it is more difficult to localize the structure abnormality than in cases presenting abnormal fibrinopeptides release. Up to now, the abnormality was localised only in a very few cases.

Fibrinogen Pontoise

This is an heterozygous case of dysfibrinogenemia and its study was performed together with Doctor Haverkate. (12, 13)

The fibrin formation analysis revealed an abnormal clot structure : the polymerization curve obtained with both fibrinogen and fibrin monomers, was not accompanied by an increase in O.D. : a clot was formed, but it was translucent. This anomaly in fibrin formation is related to the presence of an abnormal B-beta-chain, evidenced by SDS-PAGE. As it can be seen in Figure 2, a double B-beta-chain is present in fibrinogen Pontoise, one is normal whereas the other one has a higher molecular weight.

Figure 2

2a : SDS-PAGE of reduced fibrinogen

2b : SDS-PAGE of reduced fragment D

A : patient sample

B : normal sample

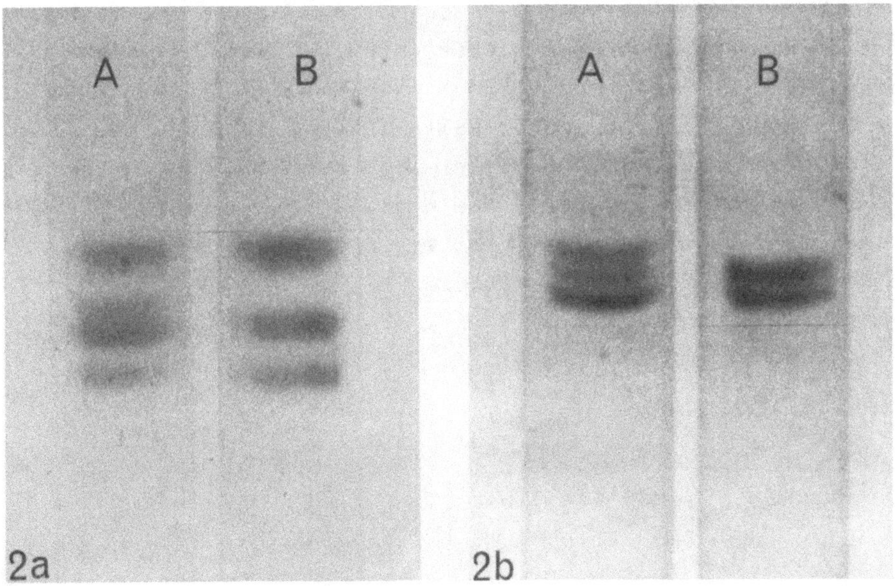

Due to the fact that the increase in the molecular weight (MW) was found also in beta-chain of fragment D (fig.2b) as evidenced by SDS-PAGE (after staining by amino-Schwartz and PAS), it was concluded that the increase of about 2.000 MW of B-beta-chain results in a lenghtening of the C-terminal part of this chain. We therefore postulate the hypothesis of either a mutation of stop codon or a transcription of an intron of DNA. The amino-acid sequence of this part of the molecule (actually studied in Doctor Henschen lab) may give further information.

Fibrinogen Haïfa

This is an heterozygous case of dysfibrinogenemia (14). The work was performed in collaboration with Doctor Rimon, Doctor Tartaski and S. Tavori from Israel (manuscript in preparation). The anomaly is characterized by an abnormal fibrin clot structure (the clot being very weak). The pattern of the SDS-PAGE of the three dissociated chains of fibrinogen Haïfa was the same than that observed with normal fibrinogen. The abnormality has been evidenced by SDS-PAGE analysis of the plasmin induced degradation products. For unreduced samples, when the degradation was performed without calcium, two fragments D were observed with fibrinogen Haïfa (D1 and D3), instead of only one fragment D (D1 or DCate) with normal degraded fibrinogen. On the contrary, when the degradation was carried out in the presence of EGTA, only D3 was obtained from both normal and patient fibrinogens (Table 2).

Table 2 : SDS-PAGE analysis of the reduced sample of normal and Haïfa fibrinogen degradation products obtained in the presence or absence of calcium.

Fibrinogen Tested :	Apparent MW of the chains obtained after fibrinogen degradation by plasmin in the presence of :	
	Calcium	EGTA
Normal	43,000 38,000	43,000 25,000
Haïfa	43,000 38,000 25,000	43,000 25,000

From these results, it appears that the protective effect induced by calcium on plasmin digestion of gamma-chain of fragment D is not observed in the abnormal part of fibrinogen Haïfa. D1 correspond to the normal part of fibrinogen, whereas D3 obtained in the presence of calcium is due to the abnormal part of the fibrinogen (heterozygous case).
Therefore, it is suggested that the mutation in fibrinogen Haïfa is probably located on the calcium-binding site of fibrinogen on fragment D and the localization of the exact mutation could give further information about the fragment D calcium-binding site.

Fibrinogen Ales

It is an homozygous case of dysfibrinogenemia. This work was performed with Doctor Bonnet and Doctor Desvignes (manuscript in preparation).

Thrombin clotting time of isolated fibrinogen is very prolonged (165 sec./Control 14 seconds), and the increase of turbidity after thrombin addition was very delayed.

Figure 3 : Polymerization curve of fibrinogen Ales after thrombin addition.

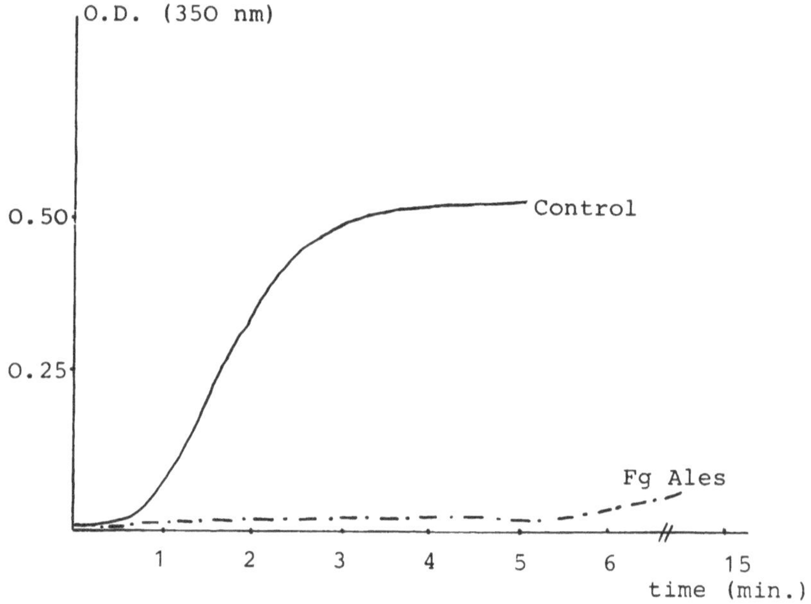

Fibrinogen Ales however remains unclottable with reptilase even at very high concentration of the enzyme. Both fibrinopeptides A and B remain are released normally as shown by HPLC. Therefore, it is concluded that the anomaly of fibrin formation must be related to a great defect in fibrin monomer polymerization. Nevertheless, fibrin stabilization is normal because the clot obtained in the presence of thrombin and factor XIIIa contains alpha-polymers and gamma-dimers as shown by SDS-PAGE. Fibrinogen structure anomaly is evidenced by electrofocusing of the three dissociated chains of fibrinogen and the anomaly is located in the gamma-chain of the molecule.

According to the fact that fibrinogen Ales clots when thrombin is added but remains unclottable in the presence of reptilase and because of the abnormal gamma-chain , it can be suggested that the anomaly is located in the carboxy-terminal part of the gamma-chain which interacts with the complementary NH2-terminal part of the alpha-chain revealed after fpA cleavage. This interaction induces the aggregation of normal fibrin monomers after the addition of reptilase to normal fibrinogen. The study of homozygous dysfibrinogenemia may be of great interest because it may give some informations about polymerization site located on the gamma-chain, which is not identified for the moment.

Table 3 : Hypothesis concerning monomers polymerization process occuring during clotting of fibrinogen Ales.

	Interaction A-a	Interaction B-b	
	- Sites A located at the N-terminal part of $A\alpha$ chain, revealed after FpA release	- Sites B located at the N-terminal part of $B\beta$ chain, revealed after FpB release.	
	- Sites a located at the C terminal part of γ chain	- Sites b located at the C terminal part of $B\beta$ chain	
Thrombin			
Normal fibrinogen	+	+	Coagulation of fibrinogen Ales occurs because of B-b interactions.
Abnormal fibrinogen	-	+	
Reptilase			
Normal fibrinogen	+	-	Coagulation of fibrinogen Ales cannot occurs because of the absence of interaction A-a
Abnormal fibrinogen	-	-	

Clinical aspects of congenital dysfibrinogenemia

The clinical aspects of congenital dysfibrinogenemia are variable . Usually, there are no symptoms and the discovery of dysfibrinogenemia is made by chance (prolonged prothrombin time with normal levels of factors V, VII, X and II).
The patient presenting dysfibrinogenemia Metz has a very mild bleeding tendency and was the only case we observed with haemorrhagic syndrome. Three of our patients however with abnormal fibrinogen charaterized by an abnormal fibrin polymerization suffered from severe thrombotic disease.

In dysfibrinogenemia Nancy, from Professor Streff and Professor Alexandre (21) the patient died at the age of 20 from pulmonary embolism. In dysfibrinogenemia Malmö (22) (work performed with Professors I.M. NILSSON, U. Heldner and Bergqvist) the patient and his brother had suffered from recurrent superficial and deep vein thrombosis.

With Dusard syndrome (23), we described a family presenting dysfibrinogenemia in which four members had severe venous thrombosis with pulmonary embolism and which has been fatal in two cases.

In the cases of dysfibrinogenemia with thrombotic syndrome, we were interested in a possible relationship between the thrombotic disease and the abnormal clot structure and we therefore carried out the following study.

1) The activity of purified fibrinogen in ADP-induced platelet aggregation

   This analysis is carried out by measuring the rate and intensity of ADP-induced aggregation of washed normal platelets in the presence of various concentrations of normal and abnormal fibrinogen. We did never find an abnormal fibrinogen presenting an increase platelet cofactor activity. To our knowledge, only in one case of congenital dysfibrinogenemia recently reported in Oslo (24), the thrombotic disorder was related to an increase of fibrinogen activity in ADP-induced platelet aggregation.

2) The binding of thrombin to fibrin

Thrombin is added to purified normal and patient fibrinogens. After various lenghts of time, the clot is squeezed out and unbound thrombin is measured in the supernatant using chromozym TH (25).
The first case of defective thrombin binding associated with thrombotic disease was reported by Liu and Nossel for fibrinogen New York (26).
In our stydy of dysfibrinogenemia associated with thrombotic disorder, only fibrinogen Malmö presented a defective thrombin binding. Dysfibrinogenemia Malmö is an heterozygous case of hypodysfibrinogenemia. Fibronectin was greatly increased both in plasma (300 % of normal value) and in fibrinogen preparation. It is unlikely that this increase on fibronectin is related to the abnormal fibrinogen because the level of fibronectin was normal in another family member who presented the same abnormal fibrinogen. Furthermore, the fibronectin-free fibrinogen can be obtained using gelatin affinity chromatography.

3) Analysis of fibrin clot lysis.

With Dusard syndrome, we reported the first case in which defective thrombolysis related to an abnormal clot structure may explain the recurrent thrombosis observed in this family.
Defective fibrinolysis may result either from a biochemical modification of the plasmin cleavage sites on the

fibrinogen molecule, or from an abnormal fibrin network more resistant to thrombolysis than normal fibrin clot.

In a first step, we showed that defective thrombolysis occured in Dusard syndrome. In the presence of plasminogen and tissue plasminogen activator released during venous stasis, the patient's fibrin clot was less degraded than control fibrin clot. It seems that the abnormality does not involve the plasmin cleavage sites of the fibrinogen molecule, since the patient's fibrinogen treated with plasmin was degraded similarly to control fibrinogen treated under the same conditions as shown by SDS-PAGE.

In four of the family members studied, the abnormal fibrin polymerization was associated with a decreased activity of Lys-plasminogen bound to the clot (27, 28).
In addition Lijnen and Coll. (29) have demonstrated that generation of plasmin induced by tissue plasminogen activator (t-pA) and various amounts of plasminogen is much lower in the presence of the patient's fibrin than with normal fibrin. These findings may be explained by the fact that plasminogen binding to the patient's fibrin clot is decreased because the binding of plasminogen to fibrin is required for plasminogen activation by t-pA.

Dusard syndrome therefore, represented a new concept in understanding thromboembolic disease, because up to now decreased fibrinolysis has only been related to a decrease of exogenous or endogenous plasminogen activators or to an excess of plasminogen activator inhibitor.

Very recently, N. Carell and Coll. (30) has also reported a case of dysfibrinogenemia in which the thrombotic disease is related to a fibrin highly resistant to lysis by plasmin.

Conclusion

Study of dysfibrinogenemia is very important because it may lead to a better understanding of the mechanism of fibrin formation. Furthermore, abnormal fibrin polymerization may explain some cases of congenital thrombotic disorder.

References

1. Imperato C, Dettori AG : Helv. Paediat. Acta 13, 380-399 (1958).

2. Menache D. : Thromb. Diathes. Haemorrh. 13 (suppl.), 173-183 (1964).

3. Blomback B, Blomback M : Arkiv Kemi 10, 415-443, (1956).

4. Soria J, Soria C, Girard ML : Pathol. Biol. 20, 315-321, (1972).

5. Soria J, Soria C, Samama M, Poirot E, Kling C : Thromb. Diathes. Haemorrh. 27, 619-633 (1972).

6. Kehl M, Lottspeich F, Henschen A. In : Fibrinogen recent biochemical and medical aspects. Editors Henschen A , Graeff H , Lottspeich F , Walter de Gruyter, 217-226 Berlin 1982.

7. Soria J, Soria C, Ryckewaert JJ : Thromb. Res 20, 425-435 (1980).

8. Belitser VN, Varetskaja JV, Malevna EV : Biochim. Biophys. Acta 154, 367-375 (1968).

9. Schwartz ML, Pizzo SV, Hill RL, McKee P : J. Clin. Invest. 50, 1506-1513 (1971).

10. Soria J, Soria C, Boulard C : Biochimie 54, 415-417 (1972).

11. O'Farrel : J. Biol. Chem. 250 (10), 4007-4021 (1975).

12. Haverkate F, Timan G, Soria J, Soria C, Samama M : Thromb. Res. 13, 689-692 (1978).

13. Soria J, Soria C, Guyader AM, Sabatier F : XVII Congress Internat. Soc. Haematol., Paris, Abstract p. 940 (1978).

14. Soria J, Soria C, Tavori S, Samama M, Rimon A, Tatarsky I: VIIIth Internat. Congress Thrombosis Haemostasis, Toronto Abstract 1119 (1981).

15. Henschen A, Lottspeich F, Kehl M, Southan C, Lucas J : In : Fibrinogen, recent biochemical and medical aspects Editors Henschen A, Graeff H, Lottspeich F, Walter de Gruyter, 67-82 , Berlin (1982).

16. Soria J, Soria C, Samama M, Henschen A, Southan C : In : Fibrinogen, recent biochemical and medical aspects Editors Henschen A, Graeff H, Lottspeich F, Walter de Gruyter, 129-143, Berlin (1982).

17. Henschen A, Southan C, Soria J, Soria C, Samama M : VIIIth Internat. Congress Thrombosis Haemostasis,Toronto Abstract 314 (1981).

18. Ladure P, Ricordel Y, Soria J, Soria C, Samama M : In : Fibrinogen, recent biochemical and medical aspects Editors Henschen A, Graeff H, Lottspeich F, Walter de Gruyter 145-151, Berlin (1982).

19. Bouvier CA, Gruedlinger J, Schweizerische Medizische Wochenschrift 93, 1401-1455 (1960).

20. Budzynski AZ, Olexa SA, Pandya BV. : Ann. N.Y. Acad. Sc. 408, 301-313 (1983).

21. Streiff F, Alexandre P, Vigneron C, Soria J, Soria C, Mester L : Thromb. Diathes. Haemorrh. 26, 565-570 (1971).

22. Soria J, Soria C, Ryckewaert JJ, Samama M, Hedner U, Nilsson IM : XVIIIth Congress of the Internat. Soc. Haematol., Montréal (1980).

23. Soria J, Soria C, Caen JP : Brit. J. Haematol 53, 575-586 (1983).

24. Thorsen LI, Stormorken H, Brosstad F, Gogstad G, Solum NO : IX Intern. Congress Thrombosis Haemostasis, Abstract 404, Stockholm (1983).

25. Roka L, Rademacher FG : VIIth Intern. Congress Thrombosis Haemostasis, Abstract 294, London (1979).

26. Liu CY, Nossel H, Kaplan K : J. Biol. Chem. 254, 10421-10425 (1979).

27. Soria J, Soria C, Samama M : Progress in Chemical Fibrinolysis and Thrombolysis, Raven Press, 337-346, New York (1978).

28. Caen JP, Soria J, Soria C : C.R. Acad. Sciences 294, 695-698 (1982).

29. Lijnen R, Soria J, Soria C, Collen D, Caen JP : Thromb. Haemostasis, 51, 108-109 (1984).

30. Carrell N, Gabriel DA, Blatt PM, Carr ME, McDonagh J: Blood 62, 439-447 (1983).

FIBRINOGENS SYDNEY I AND II, A KINETIC STUDY OF {His^{16}}FPA CLEAVAGE AND ITS EFFECT ON FPB CLEAVAGE

Christopher Southan*, Agnes Henschen

Max Planck Institute for Biochemistry, D-8033 Martinsried/ Munich, Federal Republic of Germany

David A. Lane

Department of Haematology, Charing Cross Hospital Medical School, London, W6 8RP, UK

Introduction

Fibrinogens Sydney I and II are two more cases of congenital dysfibrinogenaemia that have recently been identified and they both exhibit a similar clotting defect in plasma and purified solution. Prolonged incubation of the purified fibrinogens with excess thrombin does not cleave all of the expected immunoreactive FPA and yet there is no polymerisation defect associated with the fibrin that is formed (1). Similar results have been obtained in the study of fibrinogen Manchester (2), where it has been shown that half of the total available FPA has an amino acid substitution that delays its cleavage and reduces its immunological crossreactivity. Such cases can, therefore, be functionally distinguishable from those abnormal fibrinogens that have a polymerisation defect associated with and caused by non-cleavage of FPA from the Aα chains (3). In these cases lack of FPA cleavage results in a failure to expose those N-terminal polymerisation domains essential for normal polymerisation.

---

*Present address: Department of Haematology, Charing Cross Hospital Medical School, London, W6 8RP, UK

Fibrinogen - Structural Variants and Interactions
© 1985 Walter de Gruyter & Co., Berlin · New York - Printed in Germany

In this article the amino acid substitution of fibrinogens Sydney I and II is identified as Aα16Arg→His, a kinetic evaluation of the effect of this substitution on the cleavage of fibrinopeptides is reported and a structural interpretation of the binding of histidine rather than arginine in the thrombin active site pocket is considered.

Identification of the structural defect

Fibrinopeptides cleaved from fibrinogens Sydney I and II by thrombin have been analysed by reverse phase HPLC, Figure 1. The elution patterns of fibrinopeptides from both abnormal fibrinogens are similar and only that of fibrinogen Sydney I is illustrated. Comparison of elution patterns obtained with control and variant fibrinogens reveals an additional FPA peak, designated A' in Figure 1, that elutes just before normal FPA. The elution pattern observed for fibrinogen Sydney I is, therefore, essentially identical to that obtained with fibrinogen Manchester (2).

Amino acid analysis of the early eluting FPA peak from fibrinogens Sydney I and II have confirmed an identical substitution to that of fibrinogen Manchester, i.e. the replacement of Aα16Arg by histidine. Comparisons between the chromatograms published for these and other {His^{16}}FPA heterozygotes (5) show a similar phenotypic ratio of 50% (after allowance for a differing molar absorbance for the {His^{16}}FPA) for the proportion of substituted peptide. The consistency of this symmetrical phenotype between geographically widely distributed families strengthens the proposal that the Aα chains of plasma fibrinogen must arise from a single gene locus (6). Independent support for this idea has recently been supplied by an assessment of the number of Aα chain copies present in the human genome (7).

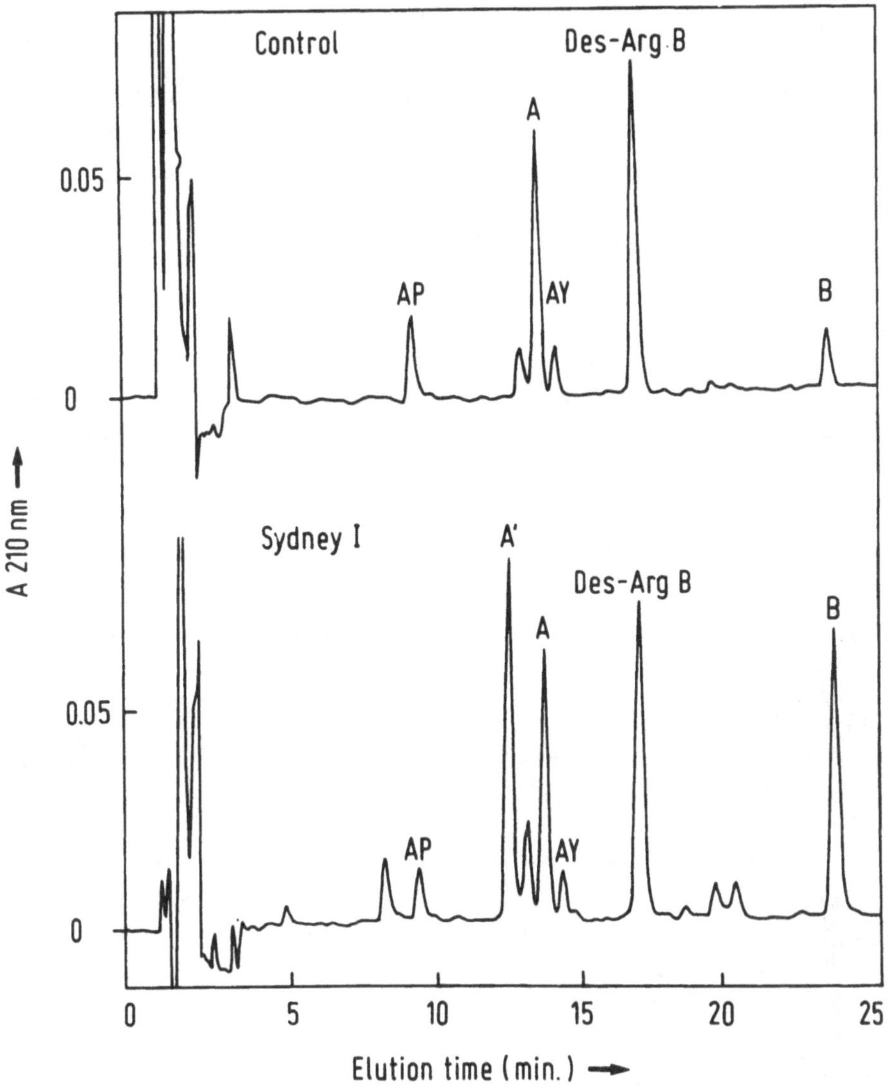

Figure 1. HPLC analysis of the fibrinopeptides released from fibrinogen Sydney I and control fibrinogen after complete cleavage by thrombin (5 u/ml, 16 h, 22°C). Sample preparation elution conditions and peptide designations were as described by Kehl *et al.* (8).

Kinetic analysis of FPA release

A time course of fibrinopeptide release from fibrinogen Sydney I, measured by HPLC, Figure 2, demonstrates a very slow release of {His^{16}}FPA. Additional experiments measuring the earliest phase of release have suggested a 79-fold slower cleavage rate of {His^{16}}FPA rather than FPA.

To establish how the amino acid substitution effects the kinetic parameters Km and kcat, kinetic experiments of fibrinopeptide cleavage have been performed using HPLC, Figure 3. The addition of free tryptophan as an internal standard improved

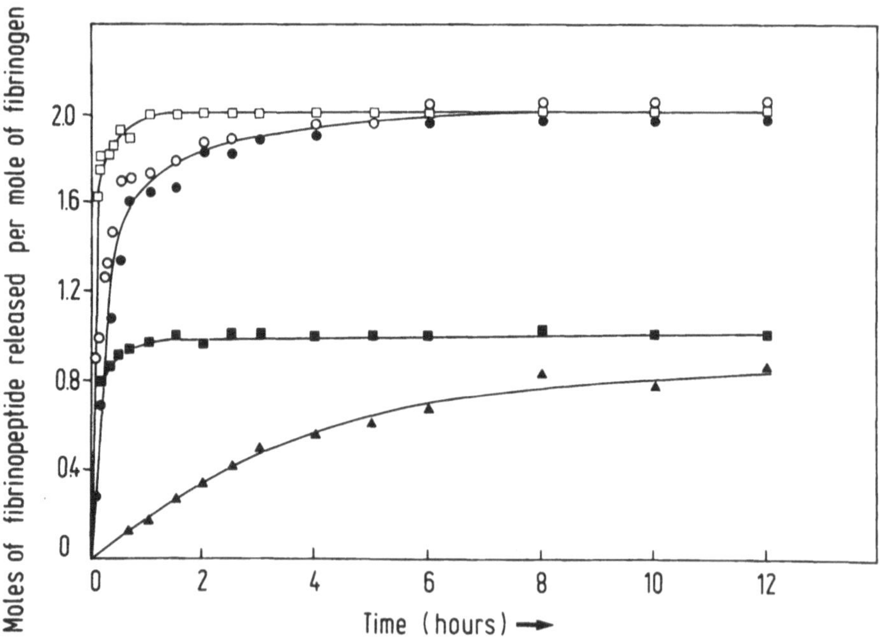

Figure 2. The time course of fibrinopeptide release from fibrinogen Sydney I and control fibrinogen. The thrombin concentration was 0.1 u/ml. Symbols are as follows: FPA control (□); FPB control (○); FPA Sydney I (■); FPB Sydney I (●), {His^{16}}FPA Sydney I (▲).

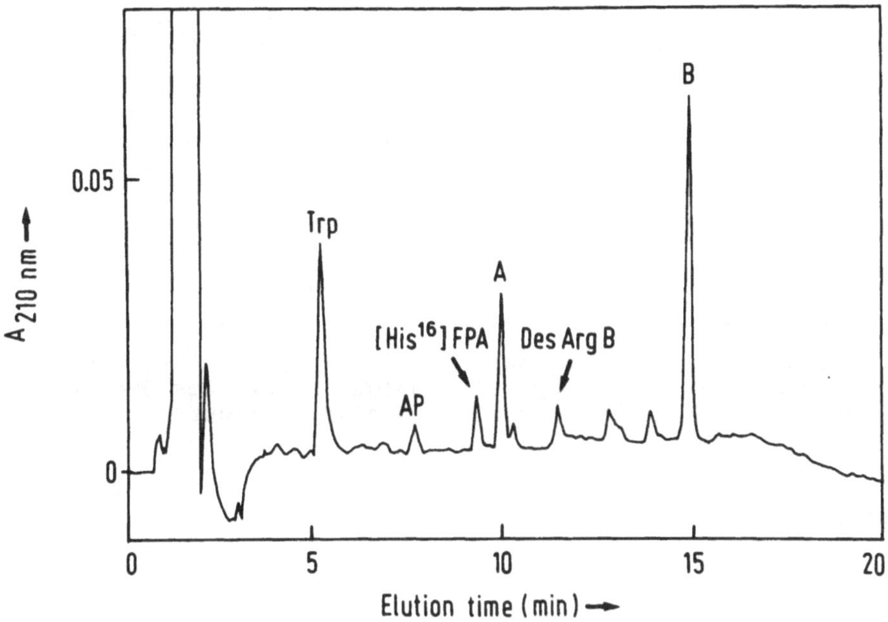

Figure 3. An example of the HPLC analysis used for kinetic determinations. The fibrinopeptides in the sample shown were released from 0.4 mg of fibrinogen. Trp refers to the free amino acid tryptophan added as an internal standard.

the accuracy of measurement and the use of a steeper gradient than that published (8), shortened the analysis time and increased the detection sensitivity. With careful experimentation it proved possible to estimate kcat and Km for cleavage of FPA and {His^{16}}FPA from Lineweaver-Burk plots with only 20 mg of fibrinogen. The results obtained are listed in Table 1.

In circumstances in which competing modes of interaction between enzyme and substrate complicate the determination of the parameters Km and kcat, kcat/Km is a less ambiguous kinetic constant. The results in Table 1 clearly show that kcat/Km is lowered substantially when Aα16Arg is replaced by histidine. A similar result has recently been obtained with another {His^{16}}FPA heterozygote (9).

Table 1. The Michaelis parameters for thrombin cleavage of the Aα chain, calculated from Lineweaver-Burk plots.

	Km μM	kcat s^{-1}	kcat/Km $\mu M^{-1}\ s^{-1}$
FPA control	8.2	30	3.6
FPA Sydney I	18	34	1.9
{His^{16}}FPA Sydney I	0.8	0.094	0.12

Note: It has been assumed that initial cleavage of fibrinopeptides is strictly sequential. Evidence for this assumption is contained in refs (4) and (9) and also in Figure 5.

Substrate binding

The reduction in the specificity constant kcat/Km appears to arise predominantly from a lowering of the rate of cleavage, kcat, rather than the substrate binding step. This is in accordance with what is so far known about the substrate binding of fibrinogen with thrombin because a large proportion of the binding energy is thought to be determined primarily by sequences in the Aα chain extending for some distance from the cleavage site (10).

In an attempt to examine possible differences in substrate binding caused by this amino acid exchange, use has been made of molecular modelling studies based on the known structure of the enzyme/substrate analogue complex between trypsin and the pancreatic trypsin inhibitor (11). This complex is one of the most precisely determined structures of a peptide substrate bound to a serine protease active site. The almost total sequence homology between the active sites of thrombin and trypsin validates the use of the trypsin/PTI complex as an analogue for the thrombin and Aα interactions.

Using a computer graphics display it has been possible to

examine the possible orientation of arginine and histidine within the hydrophobic specificity pocket without necessitating any changes in the structure of the peptide backbone. A two-dimensional representation of the results of this study is shown below, Figure 4.

The preferred specificity of both trypsin and thrombin towards arginine and lysine residues is primarily due to the interaction between the negatively charged side chains and the carboxylate group of Asp 189 at the bottom of the pocket. The result of particular interest from the modelling study is that the shorter histidine side chain can be accommodated within the pocket and that sufficient space is available for the inclusion of an extra water molecule to mediate in the hydrogen bonding interactions between the imidazol ring of the histidine and the carboxylate group of Asp 189.

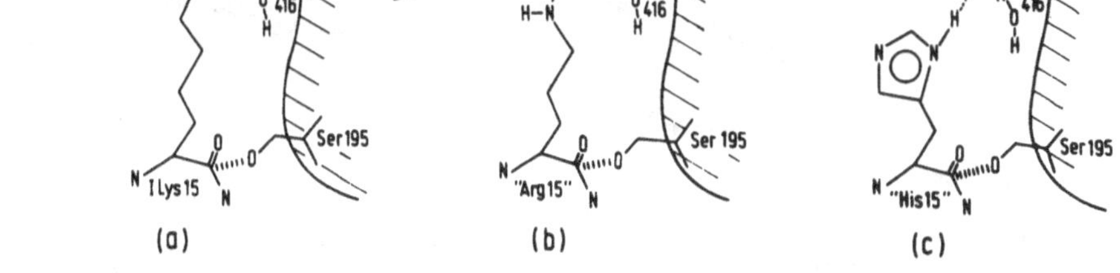

Figure 4. Possible arrangement of P1 side chains in a serine proteinase specificity pocket for lysine (a), arginine (b) and histidine (c). The three residues Gly 219, Asp 189, Ser 195 and the water molecules 414, 416 and X are those directly involved in side chain binding within the pocket. The dotted lines indicate those orientations which should favour the formation of hydrogen bonds.

The purpose of side chain binding in the specificity pocket is to precisely fix the orientation of the scissile bond with respect to the triad of catalytic residues. Although results with modelled structures remain speculative, it is an interesting possibility that a diminished orientation and/or fixation of the scissile bond might cause the increase in activation energy associated with the reduction in kcat.

The release of FPB

In view of the postulated mechanistic link between FPA and FPB release (4) it was of particular interest to examine if FPB

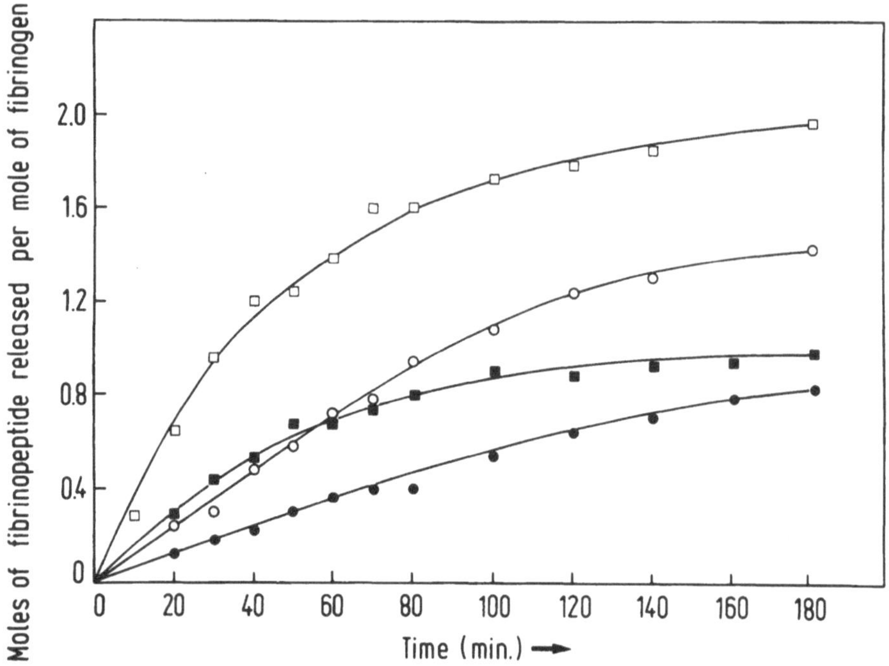

Figure 5. The time course of fibrinopeptide release at a thrombin concentration of 0.04 u/ml from fibrinogen Sydney I and control fibrinogen. The symbols are: FPA control (□); FPB control (○); FPA Sydney I (■); FPB Sydney I (●). The release of {His^{16}}FPA was too low to be measured during the course of this experiment.

release was affected by the slow rate of {His^{16}}FPA cleavage. For the time course of fibrinopeptide release shown in Figure 2 it is clear that FPB release was unaffected by the slow release of total FPA. This does not necessarily contradict the use of a sequential mechanism of fibrinopeptide release in the calculation of the kinetic parameters as non-steady state conditions were employed. When a similar time course to that of Figure 2 was performed at a lower thrombin concentration (0.04 u/ml), Figure 5, the result obtained appeared to support the sequential mechanism of release, i.e. under conditions where only approximately 50% of the available FPA was being cleaved the amount of FPB release was also reduced by approximately 50%.

Table 2. Fibrinopeptide release at different thrombin concentrations from fibrinogen Sydney I and control fibrinogen. After thrombin addition the samples were incubated for 2 h at 22°C. Release is expressed as the % of the total releasable peptide. For fibrinogen Sydney I the 100% release value for normal FPA and {His^{16}}FPA refers to one mole of each peptide per mole of fibrinogen. The molar ratio, A/B for the control fibrinogen is the same as the ratio of % released for the two peptides. For fibrinogen Sydney I ΣA includes the contribution of {His^{16}}FPA to the total FPA released and the molar ratio A/B is calculated using this value. A dash indicates that the particular fibrinopeptide could not be accurately measured.

Thrombin u/ml	Control % Release			Sydney I % Release				
	FPA	FPB	A/B	FPA	{His^{16}}FPA	ΣFPA	FPB	A/B
2.0	100	100	1.0	100	88	94	100	0.96
0.20	99	96	1.0	100	24	62	94	0.65
0.10	98	83	1.1	86	8	51	72	0.72
0.05	89	57	1.6	88	4	46	38	1.2
0.025	61	25	2.4	52	-	28	12	2.3
0.012	38	8.4	4.5	22	-	-	-	-
0.006	19	3.1	6.1	6.6	-	-	-	-

These investigations were extended by performing single point measurements of fibrinopeptide release over a range of thrombin concentrations, Table 2. The results show that a reduction in FPB release from fibrinogen Sydney I becomes appreciable when the thrombin concentration is at or below 0.1 u/ml. Interestingly, the parameter A/B, i.e. moles FPA released/moles FPB released, was similar in spite of the large differences in extents of release. These results confirm a coupling effect between FPA and FPB release in fibrinogen with the Aα16Arg→His substitution (12).

Acknowledgements

The financial support of the Deutsche Forschungsgemeinschaft and The Wellcome Trust are gratefully acknowledged. The authors would like to thank Wolfram Bode for performing the molecule modelling studies.

References

1. Lane, D.A., Ireland, H., Thompson, E., Frost, T.J., Cuddigan, B.: Thrombos. Res. 28, 821-824 (1982).
2. Lane, D.A., Ireland, H., Southan, C., Kehl, M., Henschen, A.: Fibrinogen: Structure, functional aspects, metabolism, Haverkate et al., eds., pp. 155-166, Walter de Gruyter, Berlin (1983).
3. Southan, C., Lane, D.A., Henschen, A.: Fibrinogen, Structure, functional aspects and metabolism. Haverkate et al, eds., pp. 167-182, Walter de Gruyter, Berlin (1983).
4. Blombäck, B., Hessel, B., Hogg, D., Therkildsen, L.: Nature 275, 501-505 (1978).
5. Kehl, M., Lottspeich, F., Henschen, A.: Fibrinogen: Structure, functional aspects, metabolism, Haverkate et al, eds., pp. 183-193, Walter de Gruyter, Berlin (1983).
6. Southan, C., Kehl, M., Henschen, A., Lane, D.A.: Brit. J. Haematol. 54, 143-151 (1983).

7. Kent, J.A., Lord, S.T., Crabtree, G.R.: Proc. Natl. Acad. Sci. USA 80, 3953-3957 (1983).

8. Kehl, M., Lottspeich, F., Henschen, A.: Hoppe-Seylers Z. Physiol. Chem. 362, 1661-1663 (1981).

9. Higgins, D.L., Lewis, S.D., Shafer, J.A.: J. Biol. Chem. 258, 9276-9282 (1983).

10. Scheraga, H.A.: Chemistry and Biology of Thrombin, Lundblad et al., eds., pp. 145-158, Ann Arbor Science, Michigan (1977).

11. Bode, W., Schwager, P.: J. Mol. Biol. 98, 693-717 (1975).

12. Southan, C., Lane, D.A., Bode, W., Henschen, A.: Eur. J. Biochem., 147, 593-600 (1985).

FIBRINOGENS LONDON I - IV, MANCHESTER, SYDNEY I AND II.

Cleavage of fibrinopeptides by thrombin and expression of their polymerisation abnormalities.

David A. Lane, Helen Ireland, Elizabeth Thompson
Department of Haematology, Charing Cross Hospital Medical School, Hammersmith, London W.6.

Christopher Southan and Agnes Henschen
Max Planck Institute for Biochemistry, D-8033 Martinsried/München, Federal Republic of Germany

Introduction

The most commonly used approach to the preliminary investigation of the congenitally abnormal fibrinogens is to examine the three stages involved in fibrin clot formation i.e. fibrinopeptide cleavage by thrombin, polymerisation of fibrin monomers and factor XIIIa catalysed crosslinking. Gross abnormalities associated with the latter two reactions have been relatively easy to demonstrate as the necessary methodology has long been established. The majority of the abnormal fibrinogens have been found to have essential or secondary polymerisation abnormalities, while only a small number have defects in factor XIIIa crosslinking. Until relatively recently it has been difficult to establish an abnormal fibrinopeptide release unless the peptide concerned was uncleavable by thrombin or the patient was homozygous for the defect. The development of radioimmunoassays to the fibrinopeptides (FPA and FPB) (1,2) has enabled the quantative determination of the normal fibrinopeptides and these assays have been used to study variant fibrinogens. Two difficulties have emerged with this use of radioimmunoassays (i) a degradation product of FPB lacking its C-terminus arginine (des Arg FPB) is found in clot supernatants in variable amounts and this does not react strongly to most FPB antisera (3) (ii) amino acid substitutions within the fibrinopeptides also reduce the

Fibrinogen - Structural Variants and Interactions
© 1985 Walter de Gruyter & Co., Berlin · New York - Printed in Germany

crossreactivity of antisera to the peptide (4). Both of these factors complicate the use of radioimmunoassay in the investigation of abnormal fibrinogens.

The development of high performance liquid chromatography (HPLC) for the analysis of fibrinopeptide release has enabled the problems of quantitation to be overcome. Detailed kinetic experiments of peptide cleavage have been performed from which the rate constants of peptide cleavage have also been calculated (5 - 7). We have used one (3) such HPLC method to systematically examine the relationship between fibrinopeptide release and the polymerisation abnormalities of a number of previously described abnormal fibrinogens. The major findings of this study will be summarised here.

Fibrinopeptide cleavage and polymerisation

Before considering these variant proteins it is instructive to briefly consider what is known of some of the molecular events that take place when thrombin initiates fibrin formation, in order that the functional and structural modifications of the variants can be viewed in an appropriate context. The hydrolysis of FPA by thrombin exposes a N-terminal α chain sequence α17-19 (Gly-Pro-Arg) that is part of a N-terminal polymerisation domain (8,9). The domain interacts with an already exposed polymerisation domain causing the formation of protofibrils (10) of various molecular size, depending upon the extent of enzyme action and time after the hydrolysis. Evidence has been presented that the already exposed polymerisation domain is contained within the C-terminal region of fibrinogen (11). This polymerisation domain can be localised to the region of fibrinogen conserved in FgD obtained by plasmin digestion and it has been suggested that it is contained within the γ chain sequence, γ374-411 (12). A further C-terminal polymerisation domain can be demonstrated after factor XIIIa has crosslinked fibrin (13), but this will not be relevant to the ensuing discussion, as the studied interactions have involved monomeric fibrin.

Cleavage of FPA is followed by cleavage of FPB. There are opposing views as to the mechanism involved. One view is that release of FPA results in a conformation change that facilitates FPB cleavage (8), but an alternative possibility is that FPB is cleaved simultaneously to FPA but at a slower rate (5). Regardless of the mechanism, the N-terminal region of the fibrin β chain is exposed by FPB release and this β15-18 (Gly-His-Arg-Pro) sequence is probably part of another N-terminal polymerisation domain. It is possible that the domain exposed by FPA release controls end to end fibrin growth and that the FPB dependent domain is responsible for lateral fibrin association (8).

Abnormal fibrinogens with polymerisation defects secondary to impaired fibrinopeptide release

The results of the investigation of two such variants, fibrinogens London III and Manchester (4,14,15),have been published in detail elsewhere and only a brief resumé of the results will be presented here. Experiments in which the cleavage of fibrinopeptides by thrombin were monitored (by HPLC) as a function of time from fibrinogen London III have demonstrated that only half of the available FPA is cleaved from the Aα chain, even after exhaustive digestion. At low thrombin concentrations this incomplete release of FPA can retard FPB cleavage. The polymerisation abnormality appears to be caused by incomplete exposure of the N-terminal polymerisation domains and is, therefore, secondary to the non-cleavage of a heterozygous abnormal Aα chain. The evidence points to a substitution at position Aα16 in the abnormal Aα chain but the identity of the substituted amino acid has yet to be established.

The most extensively characterised variant to be studied to date using this approach is fibrinogen Manchester. In the original report of the case it was speculated that an observed polymerisation defect was secondary to a demonstrated (by radioimmunoassay) delay in FPA release. This has turned out to be the case. HPLC analysis of the supernatants of a thrombin digest of fibrinogen Manchester has revealed the presence of normal FPA and an additional peptide peak that has been shown to contain a 16AαArg→His substitution. Cleavage by thrombin of this abnormal peptide occurs at a slower rate than the normal FPA component and the polymerisation abnormality is caused by the impaired release of the abnormal peptide delaying the exposure of the necessary N-terminal polymerisation domains. A second consequence of the 16AαArg→His substitution is an altered antigenic expression of the variant peptide. It has been estimated that the crossreactivity of the abnormal fibrinopeptide is five-fold less than native FPA and this complicates the quantitation of the peptide by radioimmunoassay, as mentioned above. The results of the investigation of fibrinopeptide cleavage of fibrinogen Manchester were generally similar to those obtained for fibrinogen Petosky (16) which also contains the 16 AαArg→His substitution and which was reported somewhat earlier.

Subsequent to the finding of a 16AαArg→His substitution in fibrinogen Manchester, two more cases heterozygous for the defect have been identified, fibrinogens Sydney I (17) and Sydney II. Sufficient material has been isolated from fibrinogen Sydney I for the kinetic constants of fibrinopeptide cleavage to be estimated and this will be the subject of a full report elsewhere (7).

Abnormal fibrinogens with essential polymerisation defects

Fibrinogen London II may be such a variant. It is a case of hypodysfibrinogenaemia with abnormal fibrin polymerisation, apparently normal catabolic properties and normal release of FPA (judged by radioimmunoassay) (18). Unfortunately, the limited supplies of plasma available at the time of investigation precluded a study of FPB release or a comprehensive structural investigation. It is, therefore, not possible at this time to be certain of the cause of its polymerisation defect.

The analysis of fibrinongen London IV (14) has revealed a complex manifestation of its molecular defect. A polymerisation abnormality is present even after all the fibrinopeptides have been cleaved. However, careful examination of the rates of cleavage of peptides by the HPLC technique has suggested that the release of FPA and FPB are slightly delayed, but that all of the available peptides are eventually cleaved. Because the isolated FgD from fibrinogen London IV is seemingly normal in its ability to inhibit polymerisation of control fibrin, it has been assumed that the molecular substitution is not at the C-terminus, but near the N-terminus of the molecule. No substitution can be demonstrated by direct sequencing of the N-termini of fibrin α and β chains which indicates that an amino acid substitution is not in the N-terminal polymerisation domains and a normal HPLC elution profile of the fibrinopeptides has suggested them to be normal. On the basis of the available evidence it is suggested that this variant has a substitution that is spatially adjacent to the N-terminal polymerisation domains and the thrombin cleavage sites. An interesting finding with this variant has been the demonstration in one purified sample of a high content of the phosphorylated form of FPA. The significance of this observation is not yet clear.

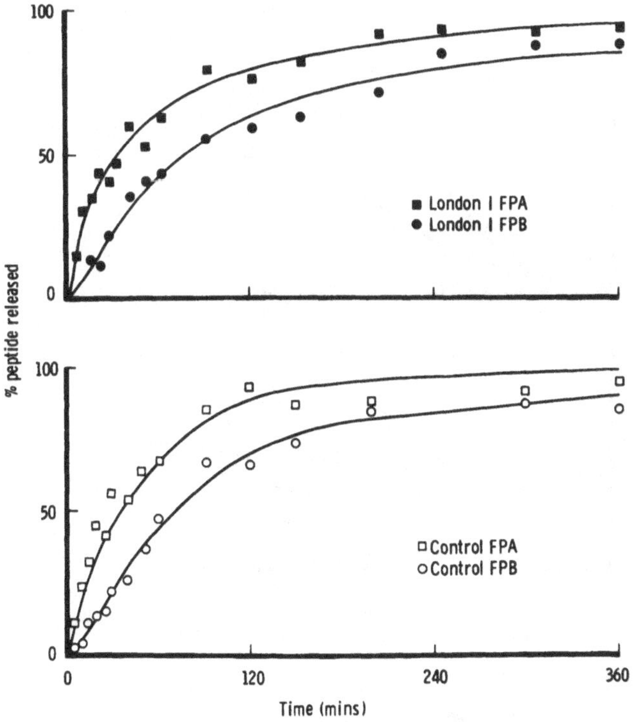

Fig. 1: Release of FPA and FPB by thrombin from fibrinogen London I (upper) and control fibrinogen (lower)

The final case of dysfibrinogenaemia that is to be discussed here is that of fibrinogen London I. In the original report (19) fibrinogen London I was described as having a polymerisation abnormality with normal release of FPA as determined by radioimmunoassay. We have recently re-examined thrombin-induced fibrinopeptide release by HPLC from this variant, Fig. 1, and have confirmed that there is no delay in either FPA or FPB release. The demonstration that a severe polymerisation defect does not retard FPB release even,at low thrombin concentrations, was unexpected.

In order to explore further the nature and location of the polymerisation abnormality, the properties of FgDs isolated by ion exchange chromatography from control and the variant proteins have been compared. Initially, the

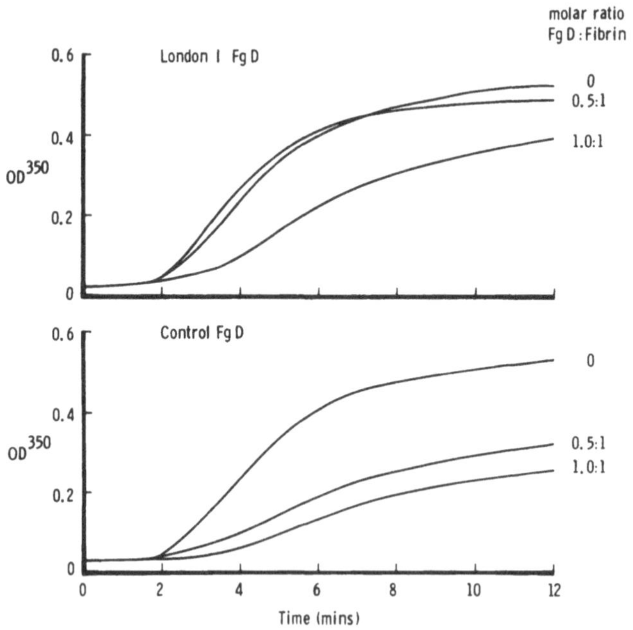

Fig. 2: Inhibition of normal fibrin polymerisation by FgD purified from fibrinogen London I (upper) and from control fibrinogen (lower).

abilities of FgD to inhibit normal fibrin polymerisation were investigated. When the control FgD was added to polymerising fibrin at molar ratios of 0.5:1 and 1.0:1 appreciable inhibition of polymerisation could be demonstrated, Fig. 2. The inhibition observed by similar molar ratios of FgD from fibrinogen London I was consistently (three separate experiments) less.

These observations strongly suggested that the polymerisation abnormality of fibrinogen London I is caused by a molecular abnormality in the C-terminal polymerisation domain. In Fig. 3, terminal plasmin digests of fibrinogen London I and control fibrinogen have been applied to an immobilised fibrin I (Reptilase-treated control fibrinogen) column and the eluted FgD

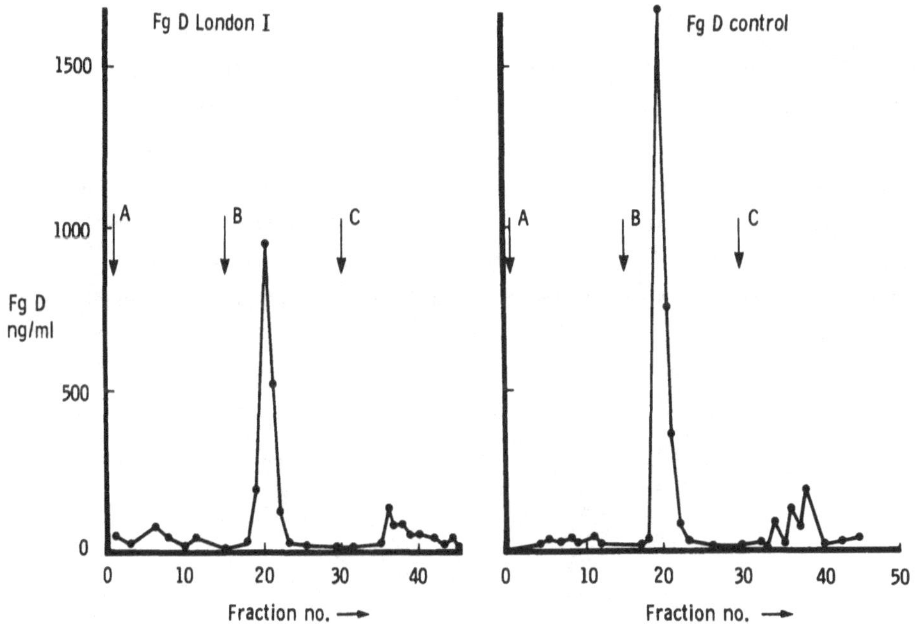

Fig. 3: Binding of FgD from fibrinogen London I and from control fibrinogen to a column of immobilised fibrin I (Reptilase treated control fibrinogen). Exhaustive plasmin digests of fibrinogen London I and control fibrinogens were applied to the column and the amount of FgD eluting with buffers A,B and C (for composition see text)were determined by radioimmunoassay to FgD.

antigen determined by radioimmunoassay. In both digests, no FgD antigen eluted through the column with buffer A(0.05M Tris/0.1 M NaCl, pH 7.5), The vast majority was eluted with buffer B (buffer A+ 1 M NaBr, pH 5.3) and only a small amount could be eluted with buffer C (0.025M acetate/8M urea, pH 5.3). Similar results were obtained when the digests were applied to an immobilised fibrin II (thrombin-treated control fibrinogen) column. These experiments indicated that the gross binding properties of the FgD polymerisation domain of fibrinogen London I are unaltered, rendering affinity chromatography on fibrin an impractical purification procedure for the isolation of the abnormal FgD.

Further experiments have been performed in an attempt to locate the defect more precisely. The C-terminal portion of the chains of control fibrinogen and fibrinogen London I have been isolated essentially by the procedures described by Olexa and Budzynski(12), with some modifications. The heterogeniety of the isolated peptide preparations is at present complicating the search for the substituted amino acid in this variant. It seems that the methodology for the isolation of this C-terminal polymerisation domain may require improvement before it is widely applied to the study of fibrinogen variants.

References

1. Nossel, H.L., Yudelman, I., Canfield, R.E., Butler, V.P., Spanondis, K., Wilner, G.D., Qureshi, G.D.: J. Clin. Invest. 54, 43 - 50 (1974).
2. Bilezikian, S.B., Nossel, H.L., Butler, V.P. and Canfield, R.E.: J. Clin. Invest. 56, 438 - 445 (1975).
3. Kehl, M., Lottspeich, F. and Henschen, A.: Hoppe-Seyler's Z. Physiol. Chem. 362, 1661 - 1664 (1981).
4. Lane, D.A., Southan, C., Ireland, H., Thompson, E., Kehl, M. and Henschen, A.: Brit. J. Haematol. 53, 587-597 (1983).
5. Martinelli, R.A. and Scheraga, H.A.: Biochemistry, 19, 2343-2350 (1980).
6. Higgins, D.L., Lewis, S.D. and Shafer, J.A.: J. Biol. Chem. 258, 9276 - 9282 (1983).
7. Southan, C., Lane, D.A., Bode, W. and Henschen, A.: Eur. J. Biochem. 147, 593-600 (1985).
8. Blombäck, B., Hessel, B., Hogg, D. and Therkildsen, L.: Nature, 275, 501-505 (1978).
9. Laudano, A.P. and Doolittle, R.F.: Proc. Natl. Acad. Sci. USA, 75, 3085-3069 (1978).
10. Hermans, J. and McDonagh, J.: Sem. Thrombos. Hemostas. 8, 11 -24 (1982).
11. Kudryk, B., Collen, D. Woods, K.R. and Blombäck, B.: J. Biol Chem. 249, 3322-3325 (1974).
12. Olexa, S. and Budzynski, A.Z.: J. Biol. Chem. 256, 3544 - 3549 (1981).
13. Olexa, S. and Budzynski, A.Z.: Proc. Natl. Acad. Sci, USA, 77, 1374-1378 (1980).
14. Southan, C., Lane, D.A. and Henschen, A.: In Fibrinogen: Structure, functional aspects, metabolism. Haverkate et al eds., pp 167 - 182, W. de Gruyter, Berlin (1983).
15. Southan, C., Kehl, M., Henschen, A. and Lane, D.A.:Brit. J. Haematol. 54, 143 - 151 (1983).
16. Higgins, D.L. and Shafer, J.A.: J. Biol. Chem., 256, 12013-12017 (1981).

17. Lane, D.A., Ireland, H., Thompson, E., Frost, T.J. and Cuddigan, B.: Thrombos. Res. 28, 821 -824 (1982).

18. Lane, D.A., Allen, A.K., Markwick, J., Thompson, E. and Mackie, I.: In Fibrinogen-Recent Biochemical and Medical Aspects. Henschen et al, eds., pp 183 - 192, W. de Gruyter Berlin,(1982)

19. Lane, D.A., Cuddigan, B., Van Ross, M. and Kakkar, V.V.: Brit. J. Haematol., 44, 483-494 (1980).

ASPECTS OF EVALUATION OF FIBRINOGEN STONY BROOK
A DEFECT RESULTING IN FAILURE TO RELEASE FIBRINOPEPTIDE A

Dennis K. Galanakis
SUNY, Stony Brook, NY

Agnes Henschen
Max-Planck-Institute for Biochemistry
D-8033 Martinsried/Munich, Federal Republic of Germany

Introduction

Of more than 100 reported families with inherited fibrinogen defects the specific amino acid sequence substitution has been determined in less than one third. Most of the defects identified have been found on or near fibrinopeptide A and of these Aα16 Arg substitutions are strikingly common (1). For example, more than 20 apparently unrelated families have been discovered with either an Aα16Arg→His or an Aα16Arg→Cys substitution. The present investigations describe studies on the fibrinogen of a middle-aged black woman who presented with leiomyoma uteri and was found to have an heterozygously expressed Aα chain abnormality of her fibrinogen.

Table. Fibrinogen Stony Brook plasma values for clotting time and fibrinogen level.

Clotting time with thrombin	65 sec	(Normal 18 sec)
" " " reptilase	88 sec	(Normal 21 sec)

Fibrinogen level according to thrombin time method	56 mg/100 ml
" " " " Ratnoff-Menzie method	170 mg/100 ml
" " " " radial immunodiffusion	210 mg/100 ml

Fibrinogen - Structural Variants and Interactions
© 1985 Walter de Gruyter & Co., Berlin · New York - Printed in Germany

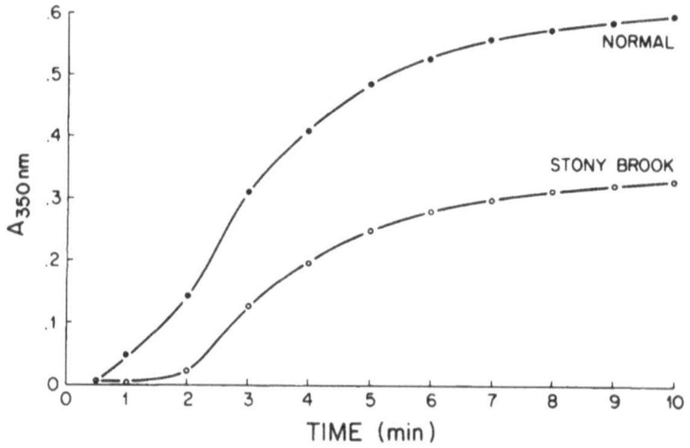

Fig. 1. Thrombin-induced fibrin aggregation curves for normal fibrinogen and fibrinogen Stony Brook.

Results and Discussion

Fibrinogen Stony Brook plasma displayed normal immunoassayable (radial immunodiffusion) and low coagulable (thrombin time) fibrinogen levels, in six plasma samples tested. Measurements of a typical sample are shown in the table. Measurements by the Ratnof- Menzie (2) procedure disclosed plasma levels intermediate to these, suggesting that either some abnormal fibrinogen molecules were coagulable or that they were incorporated in the plasma clot. Isolated fibrinogen (fraction I-2) also displayed prolonged thrombin clotting times. The fibrin monomer aggregation was examined turbidimetrically (Fig. 1) after addition of human thrombin (0.1 U/ml) to the solution of fibrinogen (1 µM). In view of the fact that Stony Brook (I-2) fibrinogen was only partly (~70%) clottable by addition of thrombin, care was taken in these experiments to ascertain that the lower amounts of coagulable protein did not account for any aggregation differences. That is, reducing the concentration of control fibrinogen by 30% (not shown) did not result in delayed onset of clot formation. Thus the delayed clot formation of Stony Brook fibrinogen was not attributable to its lower amount of coagulable protein. This finding suggested that the fibrin clot contained substantial amounts of unclottable or abnormal molecules which

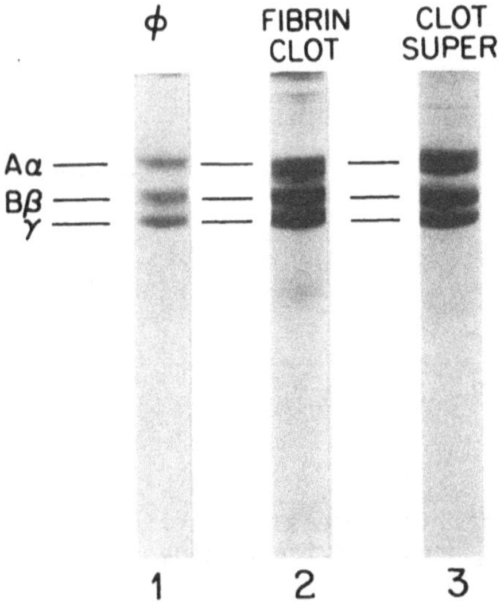

Fig. 2. SDS gel electrophoresis of fibrinogen Stony Brook and its clottable and non-clottable parts.

induced a delay in the aggregation of the normal fibrin monomers. Examination of these fibrinogen samples by sodium dodecyl sulfate polyacrylamide electrophoresis (SDS-PE), Fig. 2, disclosed no abnormality in the size of fibrinogen chains and no appreciable non-fibrinogen contaminants. Thrombin-treated (clottable or unclottable) samples showed, however, formation of an abnormally broad doublet in the α chain position suggesting that a major portion of these chains remained in the Aα chain position, and that not all of A peptide had been released. Additional support for this was provided in other preliminary experiments (not shown) with ancrod-treated samples. Samples incubated with ancrod (3,4) and heat precipitated (56°) to remove certain fragments, were subjected to SDS-PE. In addition to the well known α-chain remnant of 37 kDa (3,4) propositus fibrinogen displayed major amounts of a longer (or abnormal) remnant of 39 kDa,

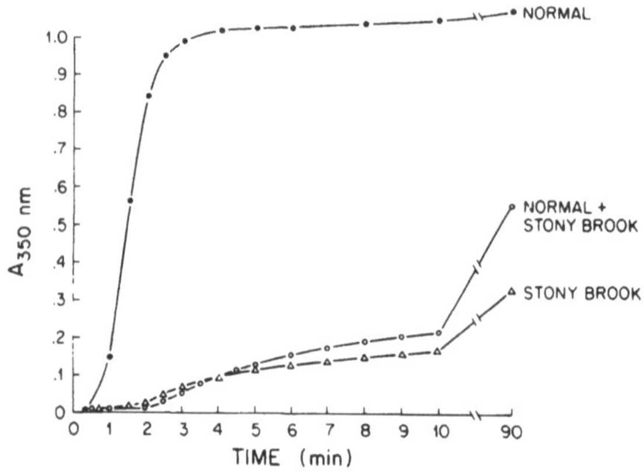

Fig. 3. Ancrod-induced fibrin aggregation curves for normal fibrinogen, fibrinogen Stony Brook and the mixture.

Analyses of thrombin-releasable fibrinopeptides from Stony Brook fraction I-2 were carried out. Assessed by high-performance liquid chromatography (5), approximately half of fibrinopeptide A could not be released from propositus fibrinogen, confirming the electrophoretic studies (above), and indicating an heterozygously expressed abnormality. Releasable fibrinopeptides exhibited an otherwise normal elution profile with respect to retention times. These results and more recent analyses, showing the presence of an Aα16Arg→Cys substitution, form the subject of a separate communication (in preparation).

Examination of the capacity of fibrinogen Stony Brook to inhibit the aggregation (6) of normal fibrin monomers was also carried out. The presence of fibrinogen Stony Brook I-2 exerted a more marked inhibition than that of the normal control. Fig. 3 illustrates this point. Ancrod was added (5 U/ml) to fibrinogen (1.5 µM, pH7) and clot formation monitored turbidimetrically. These experiments established that equimolar mixtures of normal and Stony Brook fibrinogen resulted in a turbidimetric aggregation profile similar to that of Stony Brook alone. This inhibition was even more marked when the abnormal form was isolated from the normal by removing reptilase-clottable material from fraction I-2 (i.e. obtaining a subfraction which possessed electrophoretically intact Aα chains, and was only 20% coagulable with thrombin). Such subfractions displayed a marked inhibitory capacity against fibrin aggregation, relative to the normal controls, both when fibrin monomer reaggregation (6) and when ancrod--induced (Fig. 3) fibrin aggregation was examined. This was consistently shown with all (six) preparations of fibrinogen Stony Brook tested. Thus, this abnormality resulted in a marked increase of the inherent inhibitory capacity of fibrinogen against fibrin monomer aggregation. Similar observations were made with another Aα chain abnormality, Aα16Arg→His (fibrinogen Louisville, ref. 7), and it may therefore be concluded that various substitutions in the Aα16Arg position result in anomalous association of fibrin monomers, and induce an abnormal inhibitory effect on fibrin monomer aggregation.

Additional investigations which disclosed that the capacity of this fibrinogen (Stony Brook) to support ADP-induced aggregation of gel-sieved platelets was diminished, are detailed elsewhere in this volume by Peerschke et al.

This work was supported by United States Public Health grant HL 27196, and by a grant from Deutsche Forschungsgemeinschaft.

References

1. Henschen, A., Kehl, M., Southan, C.: Thromb. Haemost. 50, 416 (Abstr.) (1983).

2. Ratnoff, O.D., Menzie, C.: J. Lab. Clin. Med. 37, 316-320 (1951).

3. Edgar, W., Prentice, C.R.M.: Thromb. Res. 2, 85-96 (1973).

4. Pizzo, S.V., Schwartz, M.L., Hill, R., McKee, P.A.: J. Clin. Invest. 51, 2841-2850 (1972).

5. Kehl, M., Lottspeich, F., Henschen, A.: Hoppe-Seyler's Z.Physiol. Chem. 362, 1661-1664 (1981).

6. Galanakis, D.K., Martinez, J., McDevitt, C., Miller, F.: Ann. N.Y.Acad. Sci. 408, 640-643 (1983).

7. Galanakis, D.K., Henschen, A., Keeling, M., Kehl, M., Dismore, R., Peerschke, E.I.: Ann. N.Y. Acad. Sci. 408, 644-648 (1983).

"FIBRINOGEN TOKYO II": AN ABNORMAL FIBRINOGEN WITH AN IMPAIRED POLYMERIZATION SITE ON THE ALIGNED DD DOMAIN OF FIBRIN MOLECULES

Michio Matsuda, Chizuko Nakamikawa
Institute of Hematology, Jichi Medical School,
Tochigi-Ken, Japan

Masumi Baba and Kohji Morimoto
First Department of Internal Medicine,
Nihon University School of Medicine, Tokyo, Japan

Introduction

Fibrinogen Tokyo II, a recently found hereditary dysfibrinogenemia is characterized by impaired polymerization of fibrin monomers. The other two functions of fibrinogen, i.e., the release of fibrinopeptides by thrombin and crosslinkings by activated factor XIII (XIIIa) appear to be normal (1). In this communication, we are mainly concerned with the impaired polymerization of this abnormal fibrinogen, with special reference to the functional domains involved in the manifestation of abnormal functions.

Results and discussion

1) Clinical aspects of fibrinogen Tokyo II. The proposita is a 39-year-old female who has been operated on for uterine myoma. A routine preoperative examination revealed a significantly decreased level of fibrinogen in plasma that was as low as 27 mg/dl of plasma when determined by the method of Clauss, which is based on the thrombin clotting time (2). The other methods including the turbidimetric and immunochemical techniques yielded normal or nearly normal levels. The thrombin and Reptilase clotting times of plasma and purified fibrinogen

Fibrinogen - Structural Variants and Interactions
© 1985 Walter de Gruyter & Co., Berlin · New York - Printed in Germany.

were markedly prolonged but partially corrected by the addition of calcium ions. These results are consistent with many of the dysfibrinogenemias so far reported (3, 4). Similar abnormalies were found in her father and two sons. None of these affected members have had any manifestations of hemorrhage or thrombosis (Fig. 1).

2) Aggregation of fibrin monomers. Aggregation of fibrin monomers was performed essentially according to Gralnick et al. (5). Aggregation profiles were quite abnormal, especially at higher ionic strengths, and virtually no aggregation occurred when the concentration of NaCl in 60 mM phosphate buffer, pH 6.8, was increased to 0.05 M or higher (Fig. 2). However, the patient's fibrin monomer per se would not interfere with the aggregation of normal fibrin monomers (profiles not shown).

3) Analysis of the postulated polymerization sites. Since the other two functions ascribed to fibrinogen were found to be normal (profiles not shown), we were prompted to investigate whether or not the postulated polymerization sites were functioning normally in this abnormal fibrinogen. Two sets of four polymerization sites have so far been postulated by several investigators on the basis of biochemical observations (6, 7). One is the set of "A" and "a", and the other is composed of "B" and "b". Both "A" and "B" sites must be activated by thrombin. The "a" site is located on the D domain of fibrinogen: therefore, it does not need activation by thrombin. Very interestingly, the "b" site, which is also identified in the D domain of fibrin(ogen), has been claimed to become functional only when two fibrin molecules are linearly aligned to form a DD domain (7). The "b" site may be arranged at or near the crosslink bonds, or these covalent bonds may align and stabilize the bivalent "bb" site as suggested by Olexa and Budzynski (7). The XIIIa-mediated stabilization of the "bb" site may, at least in part, account for an accelerated clotting of fibrinogen by thrombin in the presence of calcium ions. This may also explain why little or no hemorrhagic tendency has been observed in many of the dysfibrinogenemias in spite of a greatly prolonged thrombin clotting time in vitro. Besides these four polymerization sites, three possible modes of polymerization have been proposed by Fowler et al. on the basis of an electron microscopic study (8). In

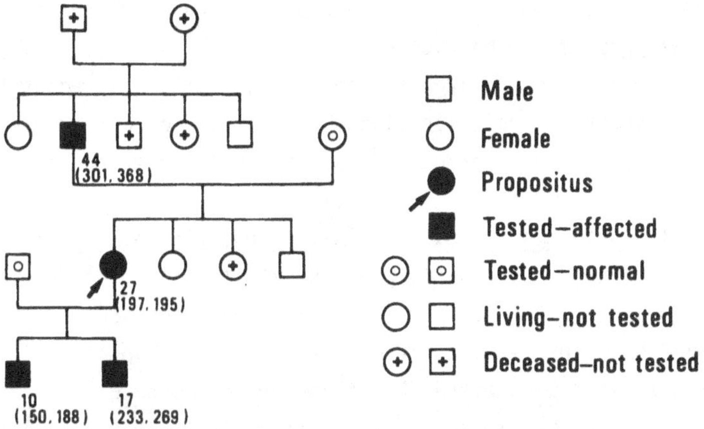

Fig. 1. Pedigree of fibrinogen Tokyo II. Figures represent plasma concentration of fibrinogen determined by the thrombin time method and in parenthesis, those measured by the method of Ratnoff and Menzie and the turbidimetric method, respectively.

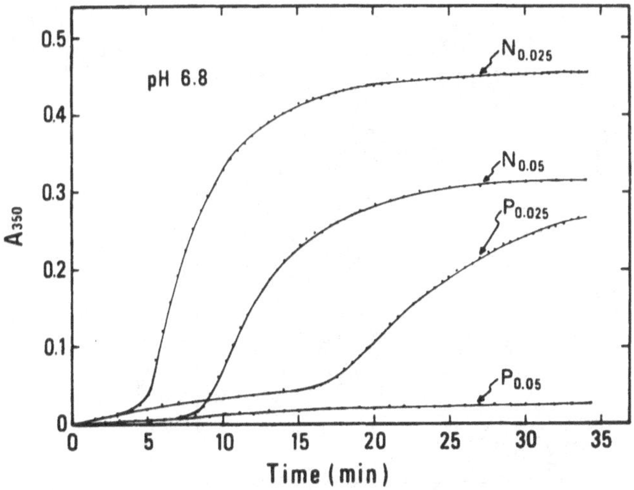

Fig. 2. Aggregation profiles of the patient's (P) and normal (N) fibrin monomers. Figures represent molar concentrations of NaCl in 0.05 M phosphate buffer, pH 6.8, used for the dilution of fibrin monomers.

this communication, however, we are mainly concerned with the polymerization sites proposed by biochemical approaches.
In order to investigate the proposed polymerization sites, we utilized affinity chromatography using insolubilized fibrin monomer or plasmic fragments of crosslinked fibrin derived from normal fibrinogen.

a. Binding of fibrinogen to insolubilized normal fibrin monomer. In order to see if the "a" on the D domain of fibrinogen functions normally, we applied a highly purified fibrinogen fraction (Blombäck I-4 fraction, further purified by DEAE-cellulose chromatography) to insolubilized fibrinogen activated with either Reptilase or thrombin (9). Adsorbed proteins were eluted by employing gradients of NaCl, from 0.1 to 1.0 M and pH, 7.6 to 4.1. Both abnormal and normal fibrinogens were eluted nearly at the same positions from either Reptilase-activated or thrombin-activated fibrinmonomer-Sepharose (Fig. 3). Thus the "a" site should function normally in this abnormal fibrinogen.

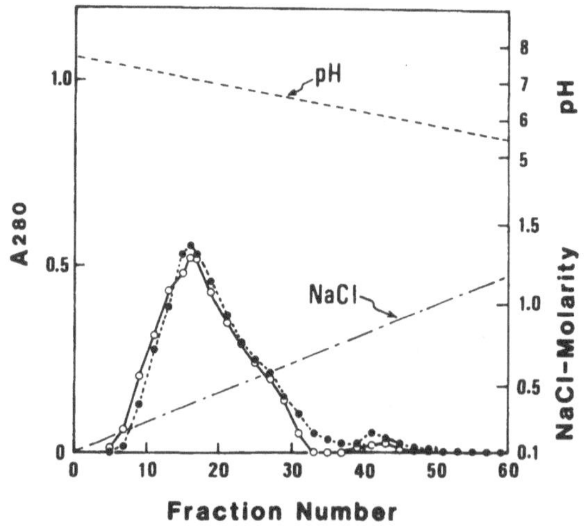

Fig. 3. Affinity chromatography of fibrinogen Tokyo II (o) and normal fibrinogen (•) on fibrinmonomer-Sepharose. (Matsuda, M. et al. (1)).

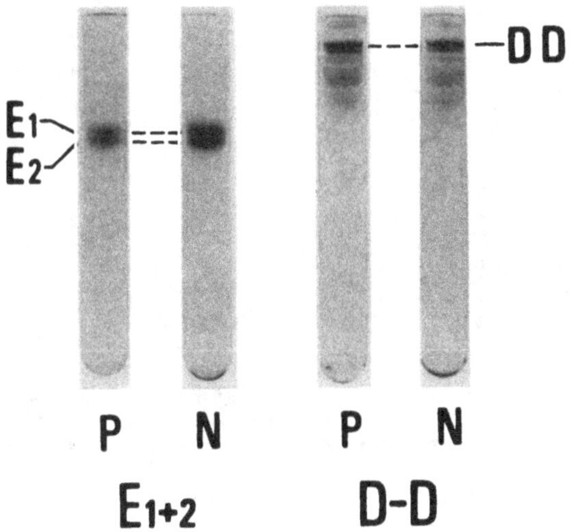

Fig. 4. Sodium dodecyl sulfate polyacrylamide gel electrophoresis of fragments E (E_1+E_2) and DD derived from the patient's (P) or normal (N) fibrinogen.

b. Preparation of fragments E and DD. In order to investigate the "B" - "bb" set, we prepared fragments E, a mixture of E_1 and E_2, and DD according to Olexa and Budzynski (10). These fragments were confirmed to be fairly homogeneous and to hold the proposed molecular weights by sodium dodecyl sulfate polyacrylamide gel electrophoresis (11) (Fig. 4). The respective fragments derived from normal fibrin were conjugated to CNBr-activated Sepharose (12).

c. Binding of fragment E to insolubilized normal fragment DD. Fragment E derived from fibrinogen Tokyo II or that from normal fibrinogen was applied to fragment DD-conjugated Sepharose. The adsorbed proteins were eluted by employing gradients of NaCl, from 0.1 to 1.0 M, pH from 7.6 to 4.1 and urea from 0 to 6 M. As shown in Fig. 5, the adsorbed proteins were eluted in similar fashions at 3.4 M or higher urea concentrations. Thus it appears that the "B" site may function normally to bind to its complementary site "bb" on the DD domain of linearly aligned fibrin molecules.

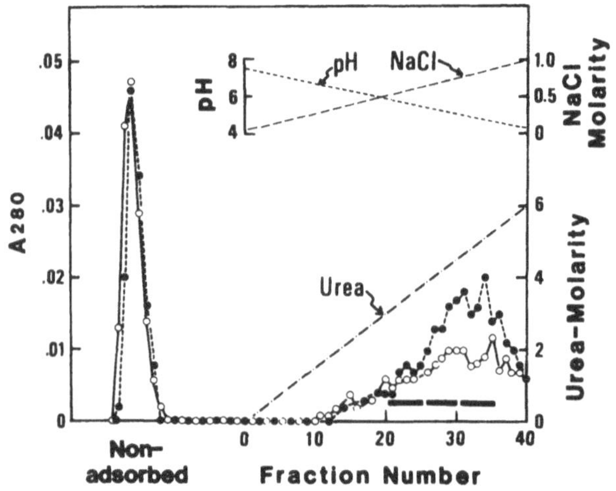

Fig. 5. Affinity chromatography of plasmic fragment E (E_1+E_2) derived from fibrinogen Tokyo II (o) and normal fibrinogen (•) on fragment DD-conjugated Sepharose. (Matsuda, M. et al. (1)).

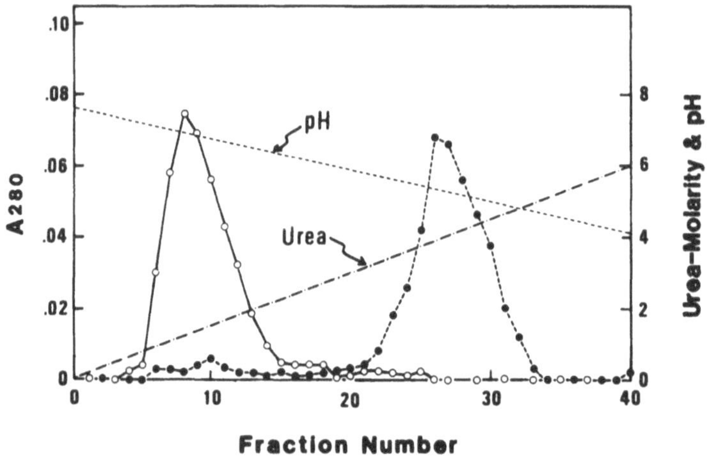

Fig. 6. Affinity chromatography of plasmic fragment DD derived from fibrinogen Tokyo II (o) and normal fibrinogen (•) on fragment E-conjugated Sepharose. (Matsuda, M. et al. (1)).

d. Binding of fragment DD to insolubilized normal fragment E. We then investigated the reverse situation, namely, the affinity of the "bb" site on the DD domain to the "B" site on the E domain. The adsorbed proteins were eluted in the same way as described above. As depicted in Fig. 6, fragment DD derived from fibrinogen Tokyo II was eluted at significantly lower concentrations of urea corresponding to 1.0 to 2.2 M, than the normal one, which was recovered at much higher concentrations of 3.4 to 4.8 M urea. The eluted proteins were confirmed by sodium dodecyl sulfate polyacrylamide gel electrophoresis to hold the original molecular size. Thus it is very unlikely that the reduced affinity of the patient's DD was due to degradation of the molecule during manipulation. The distinctly reduced affinity of the patient's DD to its complementarily functioning fragment E suggests that the "bb" site on the DD domain is malfunctioning in fibrinogen Tokyo II, and that the malfunctioning "bb" site probably gives rise to altered polymerization of fibrin monomers.

4) A model illustrating the impaired polymerization site of fibrinogen Tokyo II. Based on these observations and presumptions, we constructed a model which may illustrate the impaired polymerization site on the DD domain of fibrinogen Tokyo II (Fig. 7) (1). In this model, the defective "bb" site is represented by a loose thin ring connected to each D domain, in comparison with the thick tight one of normal fibrinogen. In the absence of calcium ions or when the two D domains are not crosslinked, this loose ring would not adequately fit the furrow on the E domain which represents the complementary "B" site. This ring may be tightened, however, in the presence of calcium ions and/or upon crosslink formation between the two D domains so as to fit the furrow and to form a more tightly bound (DD)E moiety. The "B" - "bb" set should thus function somewhat properly to promote polymerization. The two "A" sites are symmetrically arranged on a single E domain, to which is bound a short arm which represents the "a" site on a monomeric D domain.

5) Immunoelectrophoresis of fragment DD. The idea that the abnormality reside most probably in the DD domain is supported by more anodal migration of the patient's fragment DD than the normal fragment. Since the mobility of the parent fibrinogen in plasma is indistinguishable from

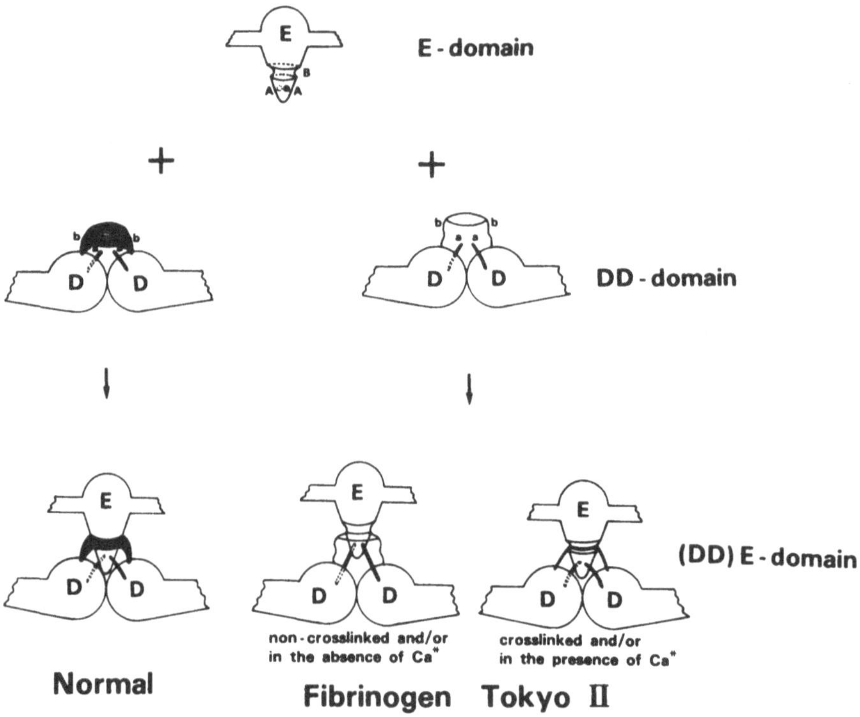

Fig. 7. Schematic presentation of the malfunctioning "bb" site on the DD domain of fibrin molecules derived from fibrinogen Tokyo II. See text for details. (Matsuda, M. et al. (1)).

that of normal, we speculate on the following possibilities: (1) conformational changes may have been induced upon crosslink formation owing to replacement of an amino acid with a less basic or a more acidic one in or near the D domain, or (2) the plasmic cleavage mode may be different due to the mutation, so that the patient's fragment DD may contain a negatively charged segment which should have been cleaved off by plasmin, or conversely, it may have lost a positively charged segment which should have been retained.

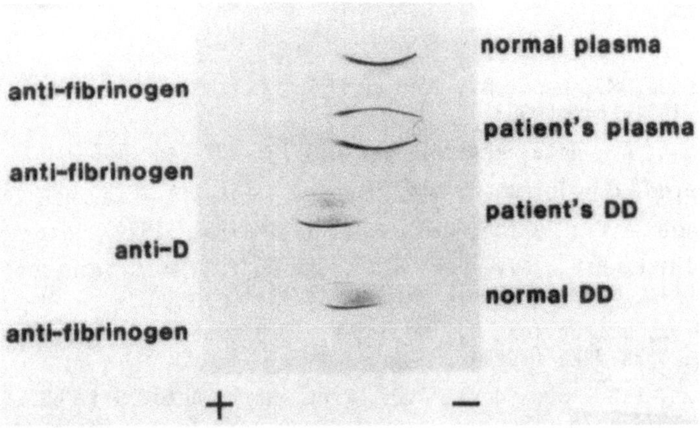

Fig. 8. Immunoelectrophoresis of fibrinogen in plasma and fragment DD derived from fibrinogen Tokyo II and a normal individual.

Concluding remarks

By employing affinity chromatography using insolubilized plasmic fragments of fibrin, we showed that the polymerization site unfolded on a linearly aligned DD domain of fibrin molecules (7) is impaired in the congenital dysfibrinogenemia described as "fibrinogen Tokyo II" (1). This sort of analysis may be helpful to pinpoint the abnormality, on a molecular level, in the functional domains of fibrinogen or fibrin molecules.

Acknowledgements

We are grateful to Drs. S. Iwanaga, University of Kyushu, B. Blombäck, the Karolinska Institute and T. Maeda, the Mitsubishi-Kasei Institute of Life Sciences for helpful discussions. This work was supported in part by a Research Grant from the Public Health Bureau, Ministry of Health and Welfare of Japan.

References

1. Matsuda, M., Baba, M., Morimoto, K., Nakamikawa, C.: J. Clin. Invest. 72, 1034-1041 (1983).
2. Clauss, A.: Acta Haematol. (Basel) 17, 237-246 (1957).
3. Ménaché, D.: Thromb. Diath. Haemorrh. (Stuttg.) 29, 525-535 (1973).
4. Mammen, E.F.: Thromb. Haemostas. 1, 184-201 (1974).
5. Gralnick, H.R., Givelber, H.M., Shainoff, J.R., Finlayson, J.S.: J. Clin. Invest. 50, 1819-1830 (1971).
6. Kudryk, B., Collen, D., Woods, K.R., Blombäck, B.: J. Biol. Chem. 249, 3322-3325 (1974).
7. Olexa, S.A., Budzynski, A.Z.: Proc. Natl. Acad. Sci. USA 77, 1374-1378 (1980).
8. Fowler, W.E., Hantgan, R.R., Hermans, J., Erickson, H.P.: Proc. Natl. Acad. Sci. USA 78, 4872-4876 (1981).
9. Heene, D.L., Matthias, F.R.: Thromb. Res. 2, 137-154 (1973).
10. Olexa, S.A., Budzynski, A.Z.: Biochemistry 18, 991-995 (1979).
11. Weber, K., Osborn, M.: J. Biol. Chem. 244, 4406-4412 (1969).
12. Cuatrecasas, P., Wilchek, M., Anfinsen, B.: Proc. Natl. Acad. Sci. USA 61, 636-643 (1968).

FIBRINOGEN MILAN II: A CONGENITAL DYSFIBRINOGENAEMIA WITH A DEFECTIVE CLOTTING BY THROMBIN, NORMAL CLOTTING BY ARVIN, REPTILASE AND PROTHROMBIN-STAPHYLOCOAGULASE COMPLEX, ASSOCIATED WITH THROMBOTIC EPISODES

F. Haverkate, J. Koopman, C. Kluft

Gaubius Institute, Health Research Division TNO,
Herenstraat 5d, 2313 AD Leiden, The Netherlands

P.M. Mannucci, A. d'Angelo

Hemophilia and Thrombosis Centre A. Bianchi Bonomi, University of Milan, Italy

R.M. Bertina

Thrombosis and Haemostasis Research Unit, University Hospital, Leiden, The Netherlands

Introduction

Up to now, at least 100 cases of inherited dysfibrinogenaemia have been reported. Most of them have been detected by virtue of a delayed thrombin clotting time. The present report describes another congenitally abnormal fibrinogen, with a delayed clotting by thrombin, but surprisingly, with a normal clotting by Arvin®, Reptilase®, and prothrombin-staphylocoagulase complex. Analysis of the fibrinopeptides A and B by High Performance Liquid Chromatography, showed that the peptides have a normal structure. However, their rate of release by thrombin is strongly delayed, whereas the rate of release of fibrinopeptide A by Arvin® appeared to be normal. The fibrin polymerization rate was normal. Evidence is presented that the defective interaction between fibrinogen Milan II and thrombin might be associated with a defective binding of thrombin to the fibrin moiety of the abnormal fibrinogen. This defective thrombin binding might account for the tendency to thrombosis in the

Fibrinogen - Structural Variants and Interactions
© 1985 Walter de Gruyter & Co., Berlin · New York - Printed in Germany

patient with the dysfibrinogenaemia described.

Methods

Venous blood was collected from the patient in tubes containing trisodium citrate (final concentration 0.011 mol/l). Plasma was prepared immediately afterwards, by centrifugation for 30 min at 2300 g and $4^{o}C$.

Fibrinogen was determined in plasma according to Clauss (1), gravimetrically according to Astrup et al. (2), and immunologically according to Mancini (3), using a 1% (w/v) agarose (BDH, Poole, United Kingdom) solution in 0.03 mol/l sodium diethylbarbiturate buffer, pH 8.6 containing 5 mmol/l EDTA.

The dilute blood clot lysis time method was performed with 10% blood (4). The euglobulin clot lysis time method was performed by clotting 0.2 ml of euglobulin fraction (EF) with 0.1 ml bovine thrombin solution (10 NIH/ml) containing 0.075 mol/l NaCl and 0.025 mol/l $CaCl_2$. The activity of a 30 µl drop of the EF prepared at pH 5.9 with a plasma dilution of 1:10 (5) was determined on plasminogen-rich bovine fibrin plates (6) and results expressed in diameters of lysis zones in the plates, after 18 h of incubation at $37^{o}C$. The C1-inactivator resistant activator activity determined (7) represents specifically the vascular or extrinsic activator activity in blood. The total plasminogen activator plus proactivator level in plasma was assayed in the EF by using dextran sulphate (8).

Thrombin, Arvin® and Reptilase® clotting times were determined in plasma (200 µl) after preincubation for 3 min at $37^{o}C$ and subsequent addition of 50 µl of the enzyme solution, with a steel hook which was moved up and down. The enzyme solution used contained bovine thrombin (Leo, Copenhagen, Denmark), 7 NIH/ml, or undiluted Arvin® (Knoll, Ludwigshafen, W.Germany) or undiluted Reptilase® (Boehringer, Mannheim, W.Germany). The thrombin-coagulase time was performed with reagent from Boehringer using the manufacturer's directions. The prothrombin time (PT) was determined (9) with a human brain thromboplastin (Manchester Comparative reagent) provided by Dr L. Poller.

Clotting times of solutions (200 µl) of purified fibrinogen (2 mg/ml) in buffered saline (PBS), pH 7.4 containing 0.15 mol/l NaCl, 1.6 mmol/l KH_2PO_4 and 10 mmol/l Na_2HPO_4 were determined similarly as in plasma. The enzyme solutions contained bovine thrombin, 10 NIH/ml, or purified human thrombin (10), 20 NIH/ml, or undiluted Arvin®, or a 1:1 (v/v) mixture of a 1:10 diluted solution of purified staphylocoagulase (by the courtesy of Dr J. Lindhout, Department of Biochemistry, State University, Maastricht, the Netherlands) and a solution of human prothrombin 4.3 IU/ml (11).

Fibrinogen from patient's plasma and from three normal plasmas was purified essentially according to Van Ruyven-Vermeer and Nieuwenhuizen (12). Plasma was prepared from venous blood containing 0.01% (w/v) disodium EDTA and 40 KIU/ml Trasylol (Bayer, Leverkusen, W.Germany) by centrifugation for 20 min at 3000 g and 4°C. Concentration of the fibrinogen was performed by precipitation in 50% saturated ammonium sulphate solution. The precipitate was dissolved after centrifugation in PBS and dialysed at 4°C against this buffer.

Fibrin monomer solution was prepared from fibrinogen with thrombin as described previously (13). The polymerization time was assessed after addition of 0.15 ml of the fibrin monomer solution (0.6%, w/v) in 0.02 mol/l acetic acid to 0.45 ml of 0.05 mol/l diethylbarbiturate buffer pH 9.0 containing 0.1 mol/l NaCl. The final pH of the solution amounted to 7.4.

Preparation of samples of fibrinopeptides A and B, and subsequent analysis by High Performance Liquid Chromatography (HPLC) was performed essentially according to Kehl et al. (14) with a device manufactured by LKB, Bromma, Sweden. A C-18 reversed phase column, 250 x 4.6 mm (Chrompack, Middelburg, the Netherlands) was used and a gradient solution system was applied using the following solvents: A: 0.025 mol/l sodium phosphate buffer pH 6.0; B: 0.05 mol/l sodium phosphate buffer pH 6.0 mixed with an equal volume of acetonitrile (Uvasol, Merck, Darmstadt, W.Germany). The detection wave-length was 206 nm. The maximum amount of release of fibrinopeptides A and B was determined after incubation overnight at 37°C, of 200 µl fibrinogen solution (4 mg/ml) in PBS with 10 µl of enzyme solution. The enzyme solutions used contained bovine thrombin (20 NIH/ml) or Arvin® (1:5 diluted) or a mixture of staphylocoagulase, 1:10 diluted and human prothrombin (4.3 IU/ml) (11). The rate of release of fibrinopeptides A and B was determined after incubation for different time intervals at 37°C of 200 µl fibrinogen solution (4 mg/ml) in PBS with 10 µl of enzyme solution. The enzyme solutions used contained bovine thrombin (2 NIH/ml) or Arvin® (1:10 diluted) or a 1:1 mixture of staphylocoagulase (1:20 diluted) and human prothrombin (2.2 IU/ml) (11). The reaction was stopped by placing the samples in boiling water for 90 sec. The precipitated fibrinogen and fibrin were removed by centrifugation at 10,000 g for 10 min, after which 50 µl of the supernatant was injected into the HPLC column.

To determine the thrombin binding to fibrin, we added 0.1 ml of fibrinogen solution (2 mg/ml) different volumes of bovine thrombin (5 NIH/ml) varying from 0.1 to 0.3 ml. The total volume was made up with PBS to 0.4 ml, and incubated for 30 min at 37°C. To achieve complete coagulation of the fibrinogen, 0.1 ml of Arvin® was subsequently added, followed by incubation for 30 min at 37°C. The fibrin formed was removed by centrifugation for 10 min at 10,000 g and the thrombin activity measured in the supernatant on the synthetic substrate Chromozyme TH (Boehringer, Mannheim, W.Germany), according to the directions of the manufacturer. The amount of thrombin bound to fibrin was calculated by subtracting the amount of thrombin recovered in the supernatant (equilibrium concentration) from the total amount of thrombin added. The specific activity of the thrombin used was determined by active site titration with p-nitrophenyl p'-guanidinobenzoate HCl (Polysciences, Warrington, U.S.A.) (15). We found that 1 NIH unit of thrombin corresponded with 14 pmol.

Case history

The patient (male, born in 1951) was admitted to our hospital with symptoms of deep venous thrombosis extending into the phemoroiliac axis. Three weeks earlier he had been operated upon successfully because of acute pancreatitis in a peripheral hospital. Laboratory tests performed on admission prior to the institution of heparin therapy showed prolongation of thrombin clotting time, but a normal thrombin-coagulase time. His parents are first cousins and asymptomatic. His brother had a stroke at the age of 21, and also had a prolonged thrombin clotting time and a normal Reptilase® time; the abnormality of his brother was not further characterized. His sister developed deep venous thrombosis at the age of 25, without any predisposing condition. His second brother did not show any signs of thrombosis.

Results

Laboratory assays with the patient's plasma

Table I shows the coagulation parameters of the patient's plasma as compared with normal plasma. As usually found in patients with dysfibrinogenaemia, the prothrombin time was delayed, the thrombin time highly delayed, and the functional fibrinogen assay (Clauss) was lower than the results of the gravimetric and immunological assays. Surprisingly, the clotting by Arvin®

Table I

Coagulation parameters of the patient's plasma as compared with normal plasma

	Normal	Milan II
Prothrombin time (sec)	13.6	21
Thrombin time (sec)	21	> 300
Fibrinogen (mg/ml)		
Clauss		1.9
Gravimetrical		3.2
Immunological		3.4
Arvin® time (sec)	23	23
Reptilase® time (sec)	20	20
Thrombin-coagulase time (sec)	19	22

Table II

Fibrinolysis parameters of the patient's plasma as compared with normal plasma

	Normal	Milan II
Dilute Blood Clot Lysis Time (min), 10% blood	> 119	248
Euglobulin Clot Lysis Time (min)	281	167
Fibrin Plate Assay		
Euglobulin Fraction (EF) (mm)	9- 15	18.1
EF with C1-inactivator (mm)	6- 8	9
EF with Dextran Sulphate (BAU/ml)	85-115	110

Table III

Clotting times (sec) of purified fibrinogen Milan II as compared with normal purified fibrinogen

Clotting enzyme	Normal	Milan II
Bovine thrombin	23	90
Human thrombin	18	190
Reptilase®	65	38
Arvin®	23	16
Prothrombin-Staphylocoagulase complex	28	30

and Reptilase®, which are known to only split off the fibrinopeptides A, is normal. Also the clotting by the staphylocoagulase-prothrombin complex (thrombin-coagulase time) is normal. Table II shows the fibrinolysis parameters which were investigated to find out whether decreased fibrinolytic activity could be an explanation for the episodes of thrombosis in the patient. However, the values of the fibrinolysis assays performed were found to lie within the normal range or were slightly elevated due to a higher level of extrinsic tissue-type plasminogen activator.

Enzymatic clotting of purified Fibrinogen Milan II

The clotting of purified fibrinogen Milan II by both bovine and human thrombin was strongly delayed (Table III), presenting evidence that the fibrinogen is abnormal. The delay in clotting with human thrombin was more pronounced than with the bovine product. The clotting times obtained with Reptilase®, Arvin®, and the staphylocoagulase-prothrombin complex (Table III) were normal, or even shorter, as compared with normal purified fibrinogen. The results agree with the corresponding clotting times in plasma. Calcium ions (10 mmol/l) had no significant influence on the clotting of fibrinogen Milan II by thrombin, indicating that the enzymatic, non-calcium dependent step is the rate-limiting step and consequently associated with the abnormality. This was confirmed by measuring the polymerization time of purified fibrin monomer prepared from fibrinogen Milan II. The polymerization time did not differ significantly from that of normal fibrin monomer: 68 and 77 sec, respectively.
The release of fibrinopeptides A and B from purified fibrinogen Milan II, as measured by HPLC, after prolonged incubation with thrombin, is shown in Figure 1. The profile obtained is similar to that of normal fibrinogen. It indicates that the peptides A and B have no abnormal structure, and secondly, that the amounts of the peptides released after prolonged incubation with thrombin, as reflected by the peak heights, are normal, However, the rate of release of both fibrinopeptides A and B by thrombin, as

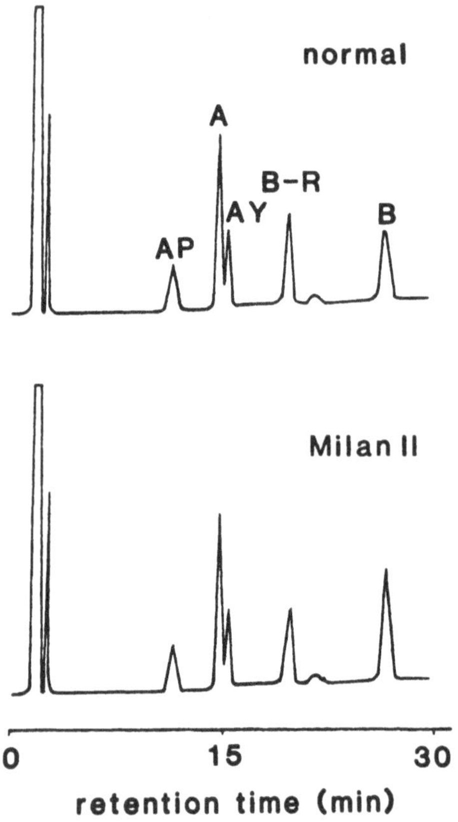

Figure 1. HPLC profiles of fibrinopeptides released after prolonged thrombin action. One molecule fibrinogen released two A and two B peptides. AP: fibrinopeptide A, phosphorylated; A: fibrinopeptide A; AY: Des-arg-fibrinopeptide A; B-R: Des-arg-fibrinopeptide B; B: fibrinopeptide B.

determined by HPLC, after different incubation times, is considerably retarded (Fig. 2), in agreement with the delayed clotting with thrombin. The rate of release of fibrinopeptide A by Arvin® from fibrinogen Milan II appeared to be similar to that from normal fibrinogen (Fig. 3), in agreement with the normal clotting with Arvin® (Table I). The total amount of A and B peptides released by the prothrombin-staphylocoagulase complex

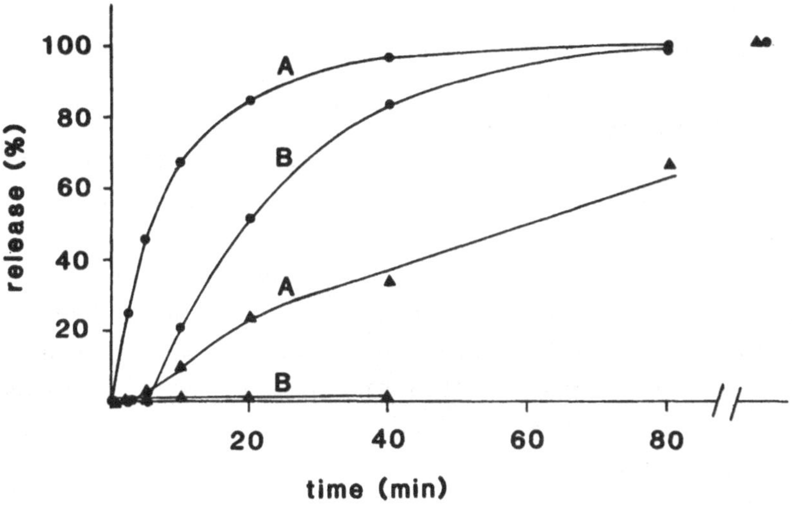

<u>Figure 2.</u> Release of fibrinopeptides A and B by thrombin as a function of time, as determined by HPLC. Normal fibrinogen (●), Fibrinogen Milan II (▲). The points located beyond 80 min have been measured after 16 h.

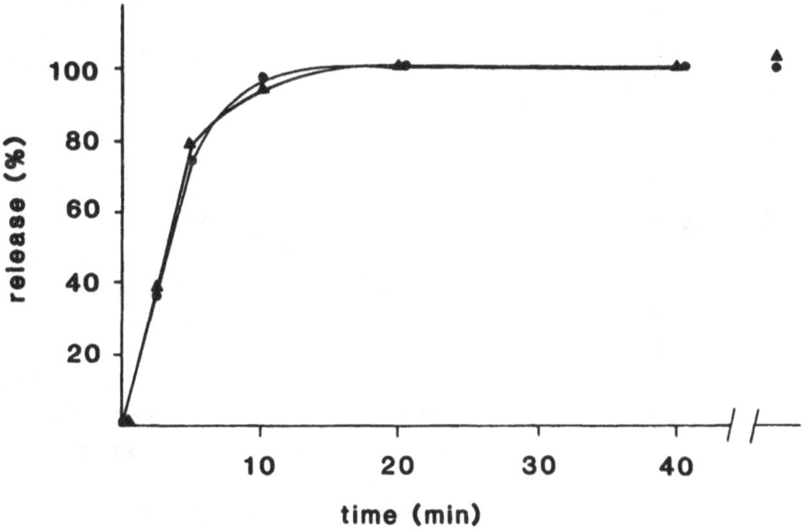

<u>Figure 3</u>. Release of fibrinopeptide A by Arvin® with time, as determined by HPLC. Normal fibrinogen (●), Fibrinogen Milan II (▲). The points located beyond 40 min have been measured after 16 h.

as well as the rate of release was normal, in agreement with the normal thrombin-coagulase time (Table I).

Binding of thrombin to fibrin

From the work of Liu et al. (16), it is known that normal fibrin binds thrombin. As fibrinogen Milan II has a defective interaction with thrombin, we investigated whether it was associated with a defective binding to the fibrin moiety. The binding was determined at a molar ratio fibrinogen/thrombin in the range 30:1 to 170:1. Under these conditions, the amount of thrombin bound to fibrin Milan II expressed at a molar basis, was about 3 times less than the amount bound to normal fibrin (Fig. 4).

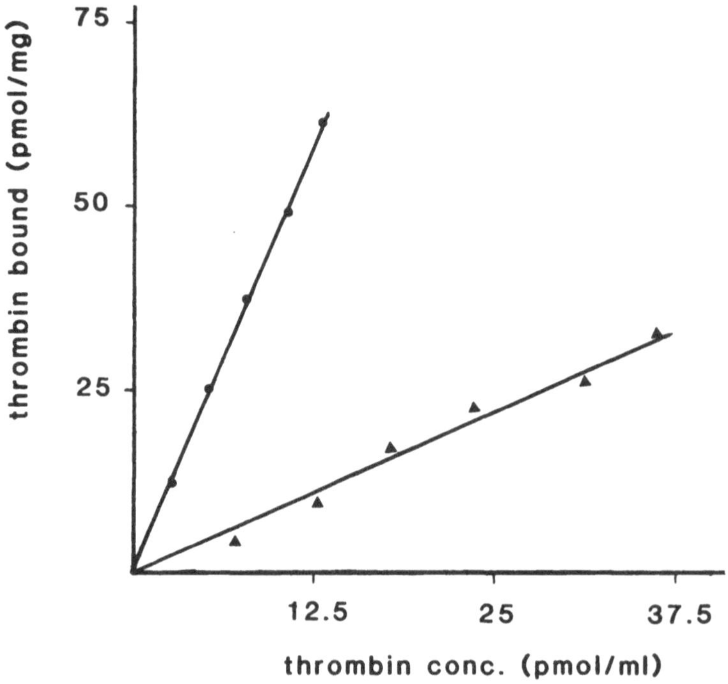

Figure 4. Binding of bovine thrombin to fibrin. Horizontal axis: thrombin equilibrium concentration after binding; vertical axis: thrombin bound to clotted fibrin. Arvin® was added in all experiments to ensure total clotting. Normal fibrinogen (●), fibrinogen Milan II (▲).

It could be argued that thrombin binds only to fibrin formed by thrombin, and not to fibrin formed by Arvin® which was used to clot remaining fibrinogen. Arvin® could thus compete with thrombin in its binding to fibrin. However, high thrombin concentrations and prolonged incubation times resulting in complete clotting of fibrinogen Milan II still showed a defective thrombin binding.

Discussion

The clotting of fibrinogen Milan II by thrombin is considerably retarded, both in plasma and in purified fibrinogen preparations. Surprisingly, the clotting of the abnormal fibrinogen by Arvin® and Reptilase®, known to split off only fibrinopeptide A, is normal, and also that of the prothrombin-staphylocoagulase complex which splits off both A and B fibrinopeptides. Further investigations by HPLC indicate a delayed release of the A and B peptides from fibrinogen Milan II by thrombin, though the structure of the peptides is normal. Prolonged incubation with thrombin results in release of two A and two B peptides as is the case with normal fibrinogen. Consequently, the defect of fibrinogen Milan II manifests itself in the rate of enzymatic release of the peptides. The subsequent fibrin polymerization step proceeds normally as can be deduced from the normal polymerization rate of purified fibrin Milan II.

The delayed release of fibrinopeptides from fibrinogen Milan II by thrombin, the defective thrombin binding, and the normal clotting by Arvin®, Reptilase® and the prothrombin-staphylocoagulase complex indicate the existence of a specific thrombin interaction site present in normal fibrinogen, absent in fibrinogen Milan II. This site is unlikely to be located in the fibrinopeptides A and B as these peptides in fibrinogen Milan II are normal. Location at the N-terminal polymerization sites of the fibrin molecule is also unlikely as the fibrin Milan II polymerizes normally. Our thrombin binding experiments suggest that this thrombin interaction site is located in the fibrin moiety of the normal fibrinogen molecule, and is lacking in the fibrin Milan II. More evidence for the occurrence of such a thrombin binding site in normal fibrinogen may be deduced from the studies of Hogg and Blombäck (17) who found the fibrinogen-thrombin binding to be strengthened by structures within the sequence 34-51 of the fibrin α-chain. It is also conceivable, but less likely, that there is no such thrombin interaction site in normal fibrinogen, but an abnormal structure in fibrinogen Milan II which blocks the interaction with thrombin, not with the other clotting enzymes used.

It is intriguing that fibrinogen Milan II shows a defective interaction with thrombin, and not with the prothrombin-staphylocoagulase complex. Staphylocoagulase activates prothrombin by unmasking the active site possibly through a conformational change, not by limited proteolysis (11). Our study reveals that the active prothrombin complex differs significantly from thrombin in its interaction with fibrinogen.

A possible explanation for thrombotic episodes could be a decreased fibrinolytic activity possibly caused by a defect in the patient's fibrin structure. However, we did not find a decrease in fibrinolytic activity using different clot lysis methods in which the patient's own fibrinogen was present. A decreased binding of fibrinolytic components to fibrin Milan II in vivo cannot be excluded.

Another explanation for the tendency to thrombosis in the patient stems from the studies of Liu et al. (18) who described an abnormal fibrinogen New York I which has some features in common with fibrinogen Milan II in that it shows a defective thrombin binding associated with episodes of thrombosis. Liu et al. (18) associate the tendency to thrombosis of the

New York I patient with the defective thrombin binding which could lead to an increased thrombin level in the bloodstream, and subsequent induction of processes such as platelet aggregation, decrease of antithrombin III level, and protein C activation. At present, no evidence is available for this attractive hypothesis which could also explain the tendency to thrombosis in the patient with fibrinogen Milan II.
It is of interest, in this connection, that fibrinogen New Orleans II (19) in a patient with episodes of thrombosis, also shows a delayed clotting by thrombin, and a normal clotting by the snake venom preparations Ancrod® and Atroxin®. Unfortunately, the thrombin-fibrin interaction has not been reported.

The question whether the defect in fibrinogen Milan II is homozygous or heterozygous remains unanswered. The peptide release curves do not permit further identification. As the parents of the patient are cousins, a homozygous case of dysfibrinogenaemia is fairly probable.

References

1. Clauss, A.: Acta Haematol. (Basel) 17, 237-246 (1957).
2. Astrup T., Brakman, P., Nissen, U.: Sc. J. Clin. Lab. Invest. 17, 57-65 (1965).
3. Mancini, G., Carbonara, A.O., Heremans, J.F.: Immunochem. 2, 235-254 (1965).
4. Chohan, I.S., Vermylen, J., Singh, I., Balakrishnan, K., Verstraete, M.: Thromb. Diath. Haemorrh. (Stuttgart) 33, 226-229 (1975).
5. Kluft, C., Brakman, P., Veldhuyzen-Stolk, E.C.: Progress in Chemical Fibrinolysis and Thrombolysis (Davidson, J.F., Samama, M.M., Desnoyers, P.C. eds.) Vol. 2, pp. 57-65, Raven Press, New York (1976).
6. Haverkate, F., Brakman, P.: Progress in Chemical Fibrinolysis and Thrombolysis (Davidson, J.F., Samama, M.M., Desnoyers, P.C. eds.) Vol. 1, pp. 151-159 (1975).
7. Kluft, C.: Thromb. Res. 13, 135-151 (1978).
8. Kluft, C.: Thrombos. Haemostas. 41, 365-383 (1979).

9. Mannucci, P.M., Frederici, A.B., Sirchia, G.: Vox Sanguinis 42, 113-123 (1982).

10. Bertina, R.M., Broekmans, A.W., Krommenhoek-van Es, T., Wijngaarden, A. van: Thrombos. Haemostas. 51, 1-5 (1984).

11. Hendrix, H., Lindhout, T., Mertens, K., Engels, W., Hemker, H.C.: J. Biol. Chem. 258, 3637-3644 (1983).

12. Ruyven-Vermeer, I.A.M. van, Nieuwenhuizen, W.: Biochem. J. 169, 653-658 (1978).

13. Haverkate, F., Timan, G.: Progress in Chemical Fibrinolysis and Thrombolysis (Davidson, J.F., Samama, M.M., Desnoyers, P.C. eds.) Vol. 2, pp. 67-71 (1976).

14. Kehl, M., Lottspeich, F., Henschen, A.: Hoppe-Seyler's Z. physiol. Chem. 362, 1661-1664 (1981).

15. Chase, T., Shaw, E.: Methods Enzymol. 19, 20-27 (1970).

16. Liu, C.Y., Nossel, H.L., Kaplan, K.L.: J. Biol. Chem. 254, 10421-10425 (1979).

17. Hogg, D.H., Blombäck, B.: Thromb. Res. 12, 953-964 (1978).

18. Liu, C.Y., Nossel, H.L., Kaplan, K.L.: Thrombos. Haemostas. 42, 79 (1979).

19. Andes, W.A.: Blood 60, Suppl. 1, 206a (1982).

PRELIMINARY REPORT CONCERNING TWO NEW CASES OF CONGENITAL DYSFIBRINOGENEMIA (HOMBURG II AND HOMBURG III)

Chieko Miyashita, Jürgen Schwamborn*, Gerhard von Blohn, Ernst Wenzel, Peter Hellstern

Department of Clinical Hemostaseology and Transfusion Medicine,

*Department of Internal Medicine,

Universitätskliniken Homburg, D-6650 Homburg/Saar, Federal Republic of Germany

Introduction

Since Imperato (1) first described a family with dysfibrinogenemia in 1958, more than 88 additional patients have been reported worldwide, of which 5o% were from Europe by 1981 according to Mammen (2). In accordance with recommendations by Beck (3, 4), these abnormalities of fibrinogen molecules were named after the cities in which the particular type of this disorder was first discovered (see Figure 1).

In 1978 we observed a case of inborn dysfibrinogenemia (Homburg I) and reported this family at the former workshop about fibrinogen 1981 (5,6).

In 1982, two further patients, a 61 year old woman and a 22 year old man were observed to have laboratory findings very similar to Homburg I. It must be pointed out that after careful scrutiny no family ties could be found between these families and families within a 15o km radius who have been reported by other centers (Figure 1).

Results: LABORATORY ASSESSMENT

The laboratory analyses of Homburg II and Homburg III proved to be very similar to Homburg I (see Table 1).

Fibrinogen - Structural Variants and Interactions
© 1985 Walter de Gruyter & Co., Berlin · New York - Printed in Germany

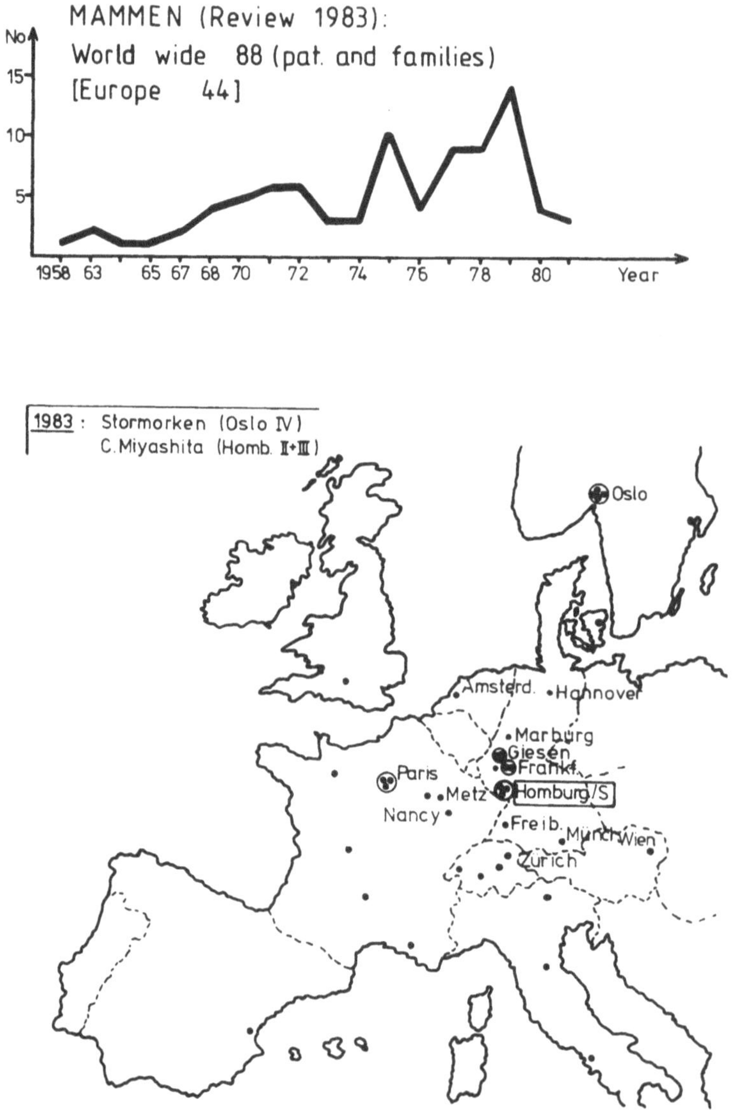

Figure 1. Reported number of patients with dysfibrinogenemia world-wide from 1958 to 1981 according to Mammen (2) (above) and their distribution in Europe (below)

Table 1. Comparison of concentration of fibrinogen to its activity in normals and in afflicted patients' plasmas (Homburg/Saar-I,-II, and -III)

mg / dl	PLASMA - FIBRINOGEN		
	CLAUSS	SCHULTZ	MANCINI
normal values	12o - 27o	18o-36o	19o-45o
H/S - I	58	25o	27o
H/S - II	68	37o	27o
H/S - III	88	51o	43o

There were large discrepancies between fibrinogen levels measured by clotting assay according to Clauss' method (7, 8, see Table 1) in comparison to the immunologically measured concentration by means of Mancini's method (9) or by the method according to Schulz (1o) which determined heat precetable fibrinogen. In the two-dimensional immunoelectrophoresis of fibrinogen, Homburg II and Homburg III were the same as the cross length immunoelectrophoresis from normal plasma or highly purified fibrinogen Kabi Vitrum AB, Stockholm (see Figure 2).

A slightly but hardly significant altered precipitation line was observed with fibrinogen antiserum determinant D as well as E (see Figure 3) indicating a significantly higher serum concentration in Homburg II and III when compared with normal plasma following thrombin treatment. This correlates well with the abovementioned difference between clotting assays in comparison to immunologically measured concentration.

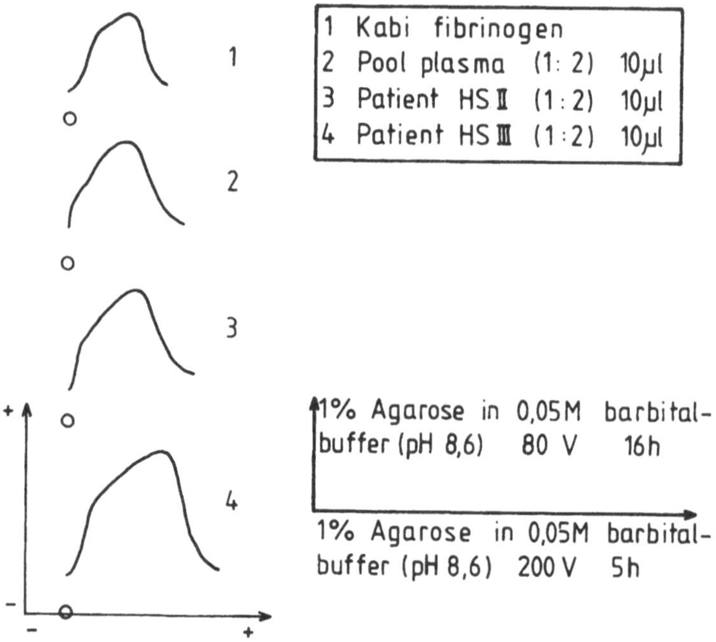

Figure 2. Crossed immunoelectropherograms of purified fibrinogen (7 µg/lo µl) and normal and afflicted patient's plasma against anti-fibrinogen (o.9%) on agarose plate.

Thrombin time (see Table 2), reptilase time, and thrombin coagulase time showed significantly prolonged values in both patients (Homburg II and III), as well as in dysfibrinogenemia Homburg I when compared to normal values. However, when highly purified fibrinogen (Kabi Vitrum AB, Stockholm) was added to the patients' plasmas, the thrombin, thrombin coagulase, and reptilase times returned to normal. The interindividual variations between 5 members in the family with Homburg I and the intraindividual variations of the values in the patient suffering from Homburg II were relatively small (less than 25 % CV).

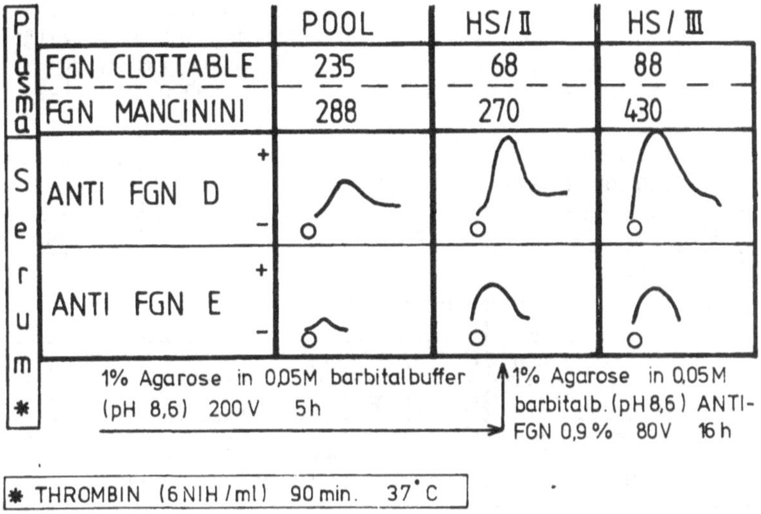

Figure 3. Crossed immunoelectropherograms of normal and afflicted patient's serum against fibrinogen anti-D as well as -E (below) compared with concentration and activity of fibrinogen in each plasma (above).

Table 2. Plasma coagulation times (thrombin-, reptilase- and thrombin-coagulase time) of afflicted patients (Homburg/Saar-I, -II, and -III) compared with normal plasma.

Sec.	THR.T. (2.5 NIH)	REPT.-T. (BOEH.-K.)	THR.C.T. (BOEH.-KIT)
Normal values	< 23	< 18	< 25
H/S I ▵	44*	58*	75*
H/S II ▲	54*	71*	91*
H/S III	48*	59*	81*

* the addition of highly purified FGN (KABI > 7o mg/dl) normalized these values in patient's plasma

▵ Vk interindividual (H/S I, N = 5) < 25%

▲ Vk intraindividual (H/S II, N = 3) < 15%

Table 3. Coagulation status of patients' plasma (above) and values from TEG (thrombelastograph) of blood (below) compared with normal values. (PTZ=prothrombin time test, Quick value; HQ=Hepato Quick value; PTT=activated partial thromboplastin time)

		NORMAL VALUES	HOMBURG I	HOMBURG II	HOMBURG III
PTZ	(%)	> 75	47*	54*	48*
HQ	(%)	>65	75	68	105
PTT	(sec.)	<39	43	40.5	35
TEG	r	< 5.0	4.5	11.0	9.2
	k	< 3.3	5.3	8.0	5.0
	(r+k)	< 8.5	10.2	19.0	14.2
	M_{ε}	>78	104	85	144

* the addition of highly purified fibrinogen (KABI > 70 mg/dl) normalized these values in patients' plasma

The Quick-test (see Table 3) was significantly influenced by the pathological fibrinogen molecule in all three types of dysfibrinogenemia. In comparison to Homburg III, the PTT values of Homburg I and II were prolonged. The coagulation parameters of the thrombelastogram were slightly altered in Homburg I, -II, and -III. All pathological test results of the Quick-test were normalized after adding a minimum of 70 mg/dl fibrinogen to the patients' plasma.

Normal values (see Table 4) were found for factor II through XIII. The fibrin formed in the patients' plasma after the addition of thrombin-calcium was completely stabilized by factor XIII, that means: 45 seconds after adding Thrombin-calcium the formed clot was stable against 1% monochloric acetic acid. The ADP, collagen, and ristocetin-induced aggregation of platelets, the adhesiveness of platelets, platelet count, and platelet morphology and platelet size were found to be normal in repeated trials.

Table 4. Laboratory parameters within normal ranges in Homburg/Saar-I, -II, and -III patients.

NORMAL LABORATORY DATA IN H/S I, II, III
F II - XIII (> 65 %) F VIII-related antigen (70 - 120%)
Fibr.stab.time(< 45 sec in the patient's plasma)
Bleeding time (< 3,3o Duke, < 4.o Ivy)
Platelet count + size / platelet adhesiv. - Hellem II / platelet aggregation: ADP, Collagen, Ristocetin / + Ristocetin cofactor activity / M_ε in TEG.

Summary of Laboratory Assessment and Discussion of its Clinical Relevance

Our preliminary report on these types of dysfibrinogenemia characterizes Homburg I, Homburg II and Homburg III by a substantial amount of functionally inadequate, e.g. non-clottable fibrinogen. Normal platelet function tests and normal bleeding time indicate that the defect does not significantly affect the platelet cofactor property of these abnormal fibrinogen molecule. The normal fibrin stabilization time indicates that the factor XIII susceptibility is also undisturbed. The result of the abnormal screening-test must still be chemically investigated. We have planned to undertake this endeavor in cooperation with Dr. Henschen from Munich.

Table 5. Bleeding tendency and other clinical symptoms of patients' families.

	H/S I*	H/S II	H/S III
FAMILY MEMBERS	18	7	5
EXAMINED	9	2	1
AFFLICTED	6	1	1
BLEEDING TENDENCY	(+)?*	ø	+++
REASON FOR REFERRAL	PT % ↓	PT % ↓	PT % ↓+BL.
INCREASED INCID. OF TE EVENTS	ø	ø	ø
ADD.CLINICAL SYMPTOMS			

* preliminary reports 1979/81: suspected bleeding in 3 unexamined children

Additionally, 9 members of Homburg I, 5 of Homburg II, and 4 of Homburg III families are still unexamined. However, 6 of 9 members of Homburg-I are afflicted (see Table 5). The one patient with dysfibrinogenemia Homburg III suffers from chronic recurrent bleeding complications. This woman had severe bleeding after the second birth, after two aborts, as well as after dental and other surgery. The one patient with dysfibrinogenemia Homburg type II demonstrated few if any episodes of bleeding. This 22 year old patient was referred for consultation because of a pathological prothrombin test (similar to the first patient of family with Homburg I) and with alleged spontaneous petechia. Our laboratory assessment could confirm the pathological prothrombin test. However, no evidence of petechia or of bleeding tendency could be found in the patient. Homburg I patients also characterized by the same laboratory findings showed no bleeding complications in 6 adult members (5). Further assessment of the remaining members of these 3 families must still be undertaken because three unexamined siblings of Homburg I were alleged to suffer from bleeding episodes. This is particularly important because of the sudden occurance of bleeding tendencies in the third generation while no such tendency was ever presented in earlier generations of this Homburg I-family.

In conclusion, we have found that regardless of the familiar history of the laboratory picture in comparison to the clinical manifestation or the asymptomatic form indicating that our three dysfibrinogenemias obviously differ from dysfibrinogenemias described by other surrounding centers. Wiesbaden (11,12) and Marburg (13) for example had hemorrhages as thrombo-embolic diseases. Paris I (14,15) had prolonged wound healing. Severe bleeding was observed in Gießen II (16) and in Metz (17, 18). As always, thorough chemical analysis will unravel the uncertainities or present hypotheses. We wish to thank Dr. A. Henschen for her cooperation in this effort and we hope to receive the understanding compliance of our patients.

References

1. Imperato, Di C., Dettori, A.G.:Helv. Pediatr. Acta 13, 38o - 399 (1958).
2. Mammen, E.F. : Seminars in Thrombosis and Hemostasis 9, 1 - 9 (1983).
3. Beck, E.A. : in: Fibrinogen (Laki, K. ed.) Decker, New York 1968. pp. 269 - 275
4. Beck, E.A. : Blood 24, 853 - 854 (1964).
5. Dumitrescu, E., Taubert, W., Nienhaus, K-H. , Wenzel, E.: Thromb.Haemost. (Abst.) 42, 139 (1979).
6. Miyashita, Ch., Nienhaus, K-H., Wenzel, E., Köhler, M., Hellstern, P.: in: Fibrinogen (Henschen, A., Graeff, H., Lottspeich, F. eds), Walter de Gruyter Berlin, 1982, pp. 193 - 198.
7. Clauss, A. : Acta Haematol. 17, 237 - 246 (1957).
8. Paar, D. : Blut 23, 1 - 6 (1971).
9. Mancini, G., Carbonara, A.O., Heremans, J.F. : Immunochemistry 2, 235 - 254 (1965).
10. Schulz, F.H., Acta Hepatol. 3, 3o6 - 31o (1955).
11. Winckelmann, G. : Thromb. Diath. Haemorrh. Suppl. 55, 345 - 35o (1973).
12. Winckelmann, G. : in: Proceeding of the II. Congress of the International Society of Thrombosis and Hemostasis (Abst.) 1971.
13. Fuchs, G., Egbring, R., Havemann, K. : Blut 34, 1o7 - 118 (1977).
14. Ménaché, D. : in: Proceedings of the IX Congress of European Society of Hematology, Karger Basel, (Abst.), 1963, p. 1255.

15. Ménaché, D. : Thromb. Diath. Haemorrh. Suppl. 13, 173 - 185 (1964).

16. Krause, W.H., Huth, K., Heene, D.L., Lasch, H.G. : Klin. Wochenschr. 53, 781 - 782 (1975).

17. Soria, J., Samama, M., Soria, C., Conrad, J., Kling, C., Bousser, J.: in: Proceed. of the Congress of the International Society of Thromb. and Hemost. (Abst.) 1971.

18. Soria, J., Soria, C., Samama, M., Poirot, E., Kling, C.: Thrombos.Diath.Haemorrh. 27, 619 - 633 (1972).

THE EFFECT OF SODIUM CITRATE ON FIBRIN POLYMERISATION IN PATIENTS WITH LIVER DISEASE

J.L. Francis, N.J. Watson, V.J. Simmonds
University Department of Haematology, Southampton General Hospital, Southampton SO9 4XY, U.K.

D.J. Armstrong
School of Pharmacy, Liverpool Polytechnic, Liverpool L3 3AF, U.K.

Introduction

Acquired dysfibrinogenaemia (DF) is common in patients with liver disease (1,2). This condition is due to defective fibrin monomer polymerisation (FMP), and appears to result from an elevation in the fibrinogen-bound sialic acid content (3,4,5). Plasma from patients with DF fails to produce opaque clots when clotted with thrombin or reptilase, and this property forms the basis of a simple screening procedure for this abnormality (1,2).

During experiments in this laboratory to assess the effects of anticoagulants on plasma FMP, it was noted that sodium citrate had a greater inhibitory effect on the FMP of a patient with cirrhosis of the liver, than on a normal control. As this anticoagulant is commonly used for clotting studies, this observation was investigated more closely.

Materials and Methods

Venous blood was obtained from normal adult controls and patients with a variety of liver diseases. 9 volumes of blood were added to 1 volume of 3.8% trisodium citrate containing 2500 units/ml Trasylol (Bayer).

Plasma FMP was assessed by the FMP ratio method as previously described (2). The normal range for this technique is 2.5 ± 0.25, and abnormal FMP

Fibrinogen - Structural Variants and Interactions
© 1985 Walter de Gruyter & Co., Berlin · New York - Printed in Germany

is indicated by a ratio greater than 3.0. Plasma FMP in the presence of excess citrate was performed in the same way, except that 0.1ml of 3.8% sodium citrate was added to 0.25ml plasma immediately before testing. Fibrinogen was purified by ammonium sulphate precipitation as previously described (5,6), except that Trasylol (200 units/ml) was present throughout the purification procedure.

Whole plasma was treated with neuraminidase (V.cholerae) at a concentration of 0.4 units/ml plasma for 18 hours at 37°C. As a control, the enzyme was replaced by buffer and incubated in parallel. Asialofibrinogen was prepared by treatment with neuraminidase (0.1 units/mg fibrinogen) as previously described (6).

The effect of citrate on the clotting of purified fibrinogen or asialofibrinogen was assessed by adding increasing amounts of 3.8% sodium citrate to 0.2ml volumes of fibrinogen (2.0 mg/ml). Thrombin clotting times were recorded after addition of 0.1ml bovine thrombin (10 units/ml). FMP was determined by diluting the mixtures with 1.0ml of 0.15M NaCl and adding 0.05ml reptilase (20 BU/ml). The absorbance at 350nm was monitored for 10 minutes.

The effect of dialysis on the FMP of plasma was determined as follows: 2.0 ml aliquots of citrated plasma were dialysed overnight at 4°C against 50ml volumes of 0.15M phosphate-buffered saline, pH 7.4 (PBS alone), or a solution containing 5.5 volumes PBS and 1 volume of 3.8% sodium citrate (PBS-citrate). The plasma FMP, with and without addition of excess citrate, was then determined as described above.

Results

Table I shows typical plasma FMP results with and without excess citrate in a normal control and a patient with cirrhosis of the liver. In the normal plasma, the FMP ratio was slightly increased from 2.5 to 2.8. The sample from the cirrhotic patient however, showed a marked increase in the ratio (from 2.4 to 23.7) under the same conditions.

	A^{350nm} after 10 min		Ratio
	Saline	Calcium	(Ca^{++}/sal)
Original plasma			
Normal	0.36	0.90	2.5
Cirrhosis	0.31	0.75	2.4
Excess citrate			
Normal	0.31	0.86	2.8
Cirrhosis	0.03	0.71	23.7

Table I. Typical results for plasma FMP in a normal control and plasma from a patient with cirrhosis, with and without excess sodium citrate.

The results of this procedure applied to 17 normal controls and 37 patients with liver disease are presented in Table II. All patients with cirrhosis, hepatitis and hepatoma developed abnormal plasma FMP in the presence of excess citrate. It must be emphasised that only four of these patients had abnormal FMP ratios when the original plasma was tested. The miscellaneous group was composed of patients with secondary carcinoma of the liver and cases of non-cirrhotic alcoholic liver damage. Only one of these patients developed abnormal FMP in excess citrate. This individual had secondary adenocarcinoma of the liver.

Diagnosis	n	No abnormal in original plasma	No abnormal in excess citrate
Controls	17	0	0
Cirrhosis	19	3	19
Hepatitis	8	1	8
Hepatoma	2	0	2
Miscellaneous	8	0	1

Table II. The effect of excess sodium citrate on plasma FMP in patients with liver disease. The upper limits for the FMP ratio were taken to be 3.0 and 3.5 for tests performed without and with excess sodium citrate respectively.

Table III shows the effect of neuraminidase treatment of plasma on the citrate-induced inhibition of plasma FMP. The results clearly show that this procedure totally abolished the inhibitory effect of excess citrate, compared to the untreated sample.

	A^{350nm} after 10 min		Ratio
	Saline	Calcium	(Ca^{++}/sal)
Control	0.24	0.57	2.37
Control + citrate	0.02	0.44	22.0
Asialo	0.22	0.60	2.73
Asialo + Citrate	0.21	0.40	1.91

Table III. Effect of excess sodium citrate on the plasma FMP of a patient with cirrhosis of the liver, before (control) and after (asialo) treatment of the plasma with neuraminidase.

The effects of excess citrate on the thrombin clotting times and FMP of purified normal and asialo-fibrinogens are shown in Table IV. Increasing citrate concentrations increased the thrombin times of the asialofibrinogen sample, although this may have been partly due to the effects of dilution. The thrombin times of the native fibrinogen however, were more markedly prolonged under the same conditions. Increasing volumes of sodium citrate slightly reduced the extent of FMP of asialofibrinogen, but markedly impaired the FMP of the untreated sample.

The effect of dialysis on the FMP of normal plasma, and plasma from two patients with cirrhosis of the liver, is presented in Table V. Patient 1 had normal FMP when the original plasma was tested, but was abnormal in the presence of excess citrate. The FMP was unaffected by dialysis, but became abnormal again on addition of excess citrate. Patient 2 had defective FMP in the original plasma which was corrected by dialysis against PBS alone, but not by PBS-citrate. The corrected FMP in PBS alone was again impaired by further addition of sodium citrate. The normal control plasma was largely unaffected by dialysis or addition of excess citrate.

Volume of citrate added (ul)	Thrombin time (secs) Native	Thrombin time (secs) Asialo	FMP (A^{350} at 10 min) Native	FMP (A^{350} at 10 min) Asialo
0	12.1	7.6	0.251	0.493
25	25.0	11.5	0.117	0.367
50	36.1	15.0	0.054	0.332
75	105.5	35.5	0.034	0.324
100	187.5	63.0	0.015	0.316

Table IV. Effect of increasing volumes of sodium citrate on the thrombin clotting time and FMP of 0.2ml volumes of fibrinogen before (native) and after (asialo) neuraminidase treatment.

PRE-DIALYSIS:	Original plasma	Excess citrate	
Control	1.8	2.0	
Patient 1	2.9	11.7	
Patient 2	48.0	55.5	
POST-DIALYSIS:	Against PBS alone	Against PBS-citrate	Excess citrate
Control	1.4	1.75	2.0
Patient 1	2.7	2.0	8.0
Patient 2	2.0	10.0	15.5

Table V. Effect of dialysis against PBS or PBS-citrate on the FMP ratios of plasma from a normal control and two patients with cirrhosis.

Discussion

Fibrin polymerisation is commonly impaired in patients with chronic liver disease, although estimates of the frequency of the defect have varied (1,2,7). In most cases, acquired dysfibrinogenaemia (DF) appears to be due to elevation of the fibrinogen-bound sialic acid content (3,5). However, failure to reproduce the abnormality in purified fibrinogen suggests that the presence of plasma inhibitors of fibrin polymerisation may play a role in the pathogenesis of this condition (4,8).

The present work has shown that the addition of excess amounts of sodium citrate inhibits the FMP of plasma from patients with cirrhosis, hepatitis and hepatoma to a much greater extent than that of normal plasma. Most of the patients described had normal plasma FMP when the original plasma was tested. This simple procedure therefore resulted in complete discrimination between normal controls and patients with biopsy-proven cirrhosis, hepatitis and hepatoma.

Treatment of plasma with neuraminidase totally abolished the inhibitory effects of sodium citrate. In addition, the effect of citrate on the clotting of purified asialofibrinogen was markedly less than that on native fibrinogen. These data suggest that sialic acid residues are somehow involved in citrate-induced inhibition of fibrin polymerisation. However, the mechanism of such an interaction is far from clear, particularly as both molecules are negatively charged, making an electrostatic association unlikely.

It is possible that the fibrinogen produced in some patients with liver disease binds more citrate than normal fibrinogen. The congenital dysfibrinogenaemia, fibrinogen Metz, is completely incoagulable in a citrated medium (9), and has been shown to bind four times as much citrate as normal fibrinogen (10). A significant proportion of the circulating fibrinogen in normal individuals has undergone limited proteolytic attack, resulting in the appearance of a lower molecular weight Aα-chain fragment termed Aα/4 (11). This fragment has increased anionic binding, and as patients with liver disease commonly have evidence of increased Aα-chain degradation (8), this may result in increased binding of citrate ions.

It is postulated that increased citrate binding results in a corresponding increase in the number of negative charges on the fibrinogen molecule, thereby impairing fibrin polymerisation. Removal of sialic acid residues would reduce the total number of negative charges and hence improve polymerisation, as demonstrated in the present study. It therefore follows that less citrate would be required to impair the FMP of fibrinogen with a high sialic acid content. This would explain why samples from patients with liver disease, in which elevations of fibrinogen-bound sialic acid are common (5), are more sensitive to the effects of excess citrate.

Even if the amount of citrate bound by the abnormal fibrinogen of liver disease was not increased, the extra negative charges contributed would be expected to render the molecule acutely more sensitive to slight increases in the sialic acid content. These observations may explain why abnormal fibrin polymerisation in plasma is not always mirrored in purified fibrinogen or fibrin monomer preparations (4). In the present study, the defective FMP of a patient with cirrhosis of the liver was corrected by removal of citrate ions by dialysis. The involvement of citrate in the pathogenesis of defective plasma FMP was further suggested by the fact that no improvement was observed after dialysis of plasma against buffer containing a comparable citrate concentration. Fibrinogen has been shown to bind citrate ions with a relatively low (Kd = 1.4×10^{-4}M) affinity (12), and it is therefore possible that these ions could be removed by dialysis. Preliminary experiments in this laboratory (unpublished) have shown that storage of plasma at -20°C results in deterioration of plasma FMP. This effect is more marked in samples from patients with liver cirrhosis and can also be reversed by dialysis. This suggests that fibrinogen can bind further citrate ions during storage, even at low temperatures. The excess ions are readily removed by dialysis, and possibly represent binding to the second class of very low affinity (Kd > 10^{-3}M) binding sites suggested by the computer analysis of Nieuwenhuizen and co-workers (12).

It is concluded, that whatever the mechanism of citrate binding to fibrinogen, the presence of citrate in plasma renders the FMP considerably more sensitive to small increases in the fibrinogen-bound sialic acid content. If this is so, the incidence of acquired DF may be far higher than previously reported, and the demonstration of defective plasma FMP in the presence of excess citrate may be a useful aid in the differential diagnosis of liver disease. Further studies are required to substantiate this suggestion, but preliminary studies (unpublished) with an experimental animal model of cirrhosis, have shown that the prolongation of the thrombin time in the presence of excess citrate correlated well with the development and severity of the cirrhosis.

References

1. Green, G., Thompson, J.M., Dymock, I.W., Poller, L. Brit. J. Haematol. 34, 427-439 (1977).

2. Francis, J.L., Armstrong, D.J. J. Clin. Pathol. 35, 667-672 (1982).

3. Martinez, J., Palascak, J.E., Kwasniak, D. J. Clin. Invest. 61, 535-528 (1978).

4. Soria, J., Soria, C., Ryckwaert, J.J., Samama, M., Thompson, J.M., Poller, L. Thromb. Res. 19, 29-41 (1980).

5. Francis, J.L., Armstrong, D.J. Haemostasis 11, 215-222 (1982).

6. Francis, J.L., Armstrong, D.J. Haemostasis 11, 223-228 (1982).

7. Barr, R.D., Allardyce, M., Brunt, P.W., McPhie, J.L. J. Clin. Pathol. 31, 89-92 (1978).

8. Weinstein, M.J., Deykin, D. Brit. J. Haematol. 40, 617-630 (1978).

9. Samama, M., Soria, J., Soria, C. In: Recent Advances in Blood Coagulation (Ed: Poller, L.), Churchill Livingstone, pp 313-336 (1977).

10. Ladure, P., Ricordel, Y., Soria, J., Soria, C., Samama, M. In: Fibrinogen - Recent Biochemical and Medical Aspects (Eds: Henschen, A., Graeff, H., Lottspeich, F.), W. de Gruyter, pp 145-152 (1982).

11. Galanakis, D.K., Mosesson, M.W., Stathakis, N.E. J. Lab. Clin. Med. 92, 376-386 (1978).

12. Nieuwenhuizen, W., Vermond, A., Hermans, J. Thromb. Res. 22, 659-663 (1981).

VII. FIBRINOGEN DEGRADATION PRODUCTS

STUDIES OF THE PROTEOLYTIC FRAGMENTS OF THE C-TERMINAL PORTION OF THE α-CHAIN OF HUMAN FIBRINOGEN

Elemer Mihalyi

Hematology Service, Clinical Pathology Department, National Institutes of Health, Bethesda, Maryland 20205

Introduction

A large portion of the α-chain of fibrinogen is removed by proteolytic enzymes in a fast reaction preceding the fragmentation into the large protein-like Y, D and E fragments. This is an appreciable segment, amounting to approximately 23% of the molecule. For various reasons it was assumed that it is a flexible, structure-less appendage, hanging loose on either end of the molecule (1). It contains cross-linking sites, but no other function has been assigned to this portion of the molecule. It would be rather surprising if such a large segment had no fixed steric structure or more important function.

Studies of the mechanism of formation and properties of the fragments derived by proteolytic digestion were successful in answering questions about structure and function with a large number of proteins including the main portion of the fibrinogen molecule (2). Therefore, application of these methods to resolve the status of the C-terminal segment of the α-chain of fibrinogen seems reasonable. The release of the large fragments of this portion does not appear to be the result of random cleavages, but follows a certain pattern. Several bands that originate from this region have been reported in the SDS-polyacrylamide gel-electrophoresis patterns of digests (3, 4, 5). However, only one of them, variously denoted as fragment A (6, 7), H (8), or P21 (5), has been characterized to some extent. Isolation and further characterization of

Fibrinogen - Structural Variants and Interactions
© 1985 Walter de Gruyter & Co., Berlin · New York - Printed in Germany

these fragments following either trypsin or plasmin digestion seemed to be a necessary step in these studies and some attempts in this direction will be described in this paper.

Materials and Methods

Human fibrinogen, grade L, lot No. 31513, 59801, 82700 81, purchased from AB Kabi, Stockholm, Sweden, was utilized without further purification, but it was dialyzed against 0.3 M NaCl. A standard plasmin preparation (9) was obtained from the Michigan Department of Health and was used for the preparation of the fragments, while the experiment recorded on Fig. 1 was performed with plasmin obtained by activation of Glu-plasminogen by urokinase. There was no difference in the behavior of either the fibrinogen lots or the two plasmin preparations. Trypsin, TPCK-treated, was purchased from Millipore Biochemical Co., Freehold, N.J., lot No. 39A 686, 31N 840. Both trypsin and plasmin were standardized by TAME-assay in the pH stat. Soy bean trypsin inhibitor was also obtained from Millipore Corp. Sephadex G-75 superfine, lot No. 13594 was a product of Pharmacia AB, Uppsala, Sweden.

Digestions of fibrinogen by either trypsin or plasmin were performed at pH 8.00 and 25° in CO_2-free atmosphere, monitored by the pH stat. At the desired level of base consumption the reaction was stopped by adding an excess of soy bean trypsin inhibitor. The arrested digestion mixtures were brought to pH 7.00 by the pH stat. Aliquots of these were used for heat precipitation, trichloroacetic acid (TCA) precipitation, measurement of the total optical density and SDS-polyacrylamide gel-electrophoresis (SDS-PAGE). For heat precipitation, the samples were kept at 95° for 8 minutes, then centrifuged. TCA precipitation was performed by mixing ice cold equal volumes of samples and 1.836 M (30%) TCA, allowing the mixtures to stand in ice bath for 2 hours, then centrifuging. The supernatants were read at 280 and 320 nm in the Zeiss PMQ II spectrophotometer. The 7.5% gels for SDS-PAGE were formed in quartz tubes and after the runs were scanned at 280 nm _in situ_ in the Zeiss Chromatogram Scanner. The Sephadex G-75 chromatography was performed in Pharmacia K 16/100 columns, with a bed height of 95 cm. Effluent was monitored in an LKB Uvicord S monitor at 206 or 276 nm, then collected in a fraction collector. Amino acid analyses were performed with the Beckman System 6300 amino acid analyzer. Specific optical densities were calculated from the readings in the PMQ II spectrophotometer, or Cary Model 14 spectrophotometer, on the basis of total nitrogen estimates, or the peptide concentration obtained from the amino acid estimates in the hydrolysates. Most of the details of the above mentioned methods were described in a previous publication (10).

Results

For a general orientation in the course of the reaction the experiment depicted in Fig. 1 was performed. This shows the increase with digestion time of the heat stable and of the TCA-soluble fractions. The former includes fragment E and all fragments smaller than this, the latter all the peptides smaller than approximately 10,000 MW, the cut-off point of precipitation at the TCA concentration employed. Thus, the difference between the two represents the peptide segments of approximately 10,000 MW and larger. These can originate only from the C-terminal two-thirds of the α-chain, because there are no chain segments outside of the D and E domains long enough for their formation. On the other hand, the small peptides may come from any of the chain extremities originally present or exposed after the initial separation of the D and E domains.

Next, the contribution of fragment E to the total heat stable fraction was estimated either from SDS-PAGE scans of whole digests, or of the heated supernatants. Fragment E appears early in plasmic digests and contributes significantly to the total optical density even during the fragmentation of the α-chain, as shown for example in Fig. 5. With trypsin however, it appears late, after the α-chain fragmentation is practically completed. The curves with fragment E substracted, as seen in Fig 1, level off at approximately 20% of the total optical density

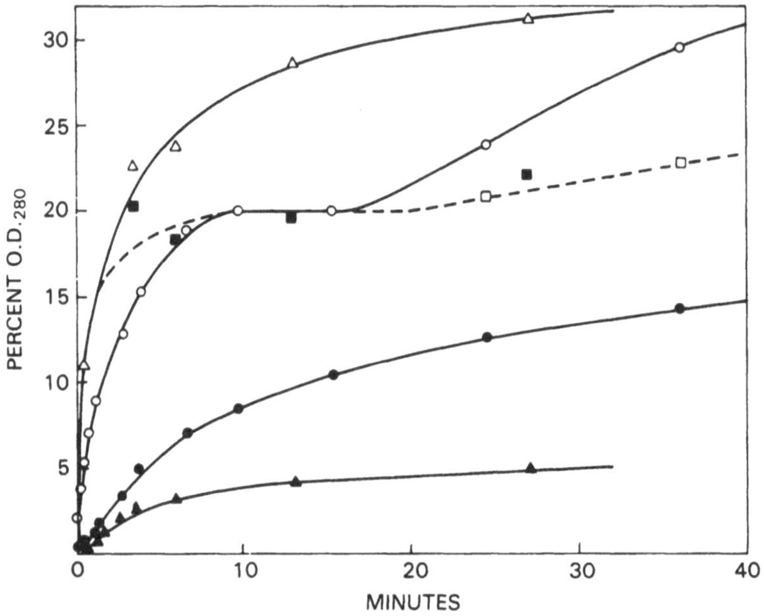

Fig. 1. Relative O.D.$_{280}$ of heat-stable and TCA-soluble fractions of fibrinogen digests plotted against reaction time. Broken line: heat-stable fraction minus fragment E. Reaction conditions: 0.3 M NaCl, pH 8.00, 25°; 19.6 mg fibrinogen per ml, 0.0448 mg trypsin or 4.79 CTA units plasmin per ml reaction mixture. Symbols: △ heat-stable fraction, ▲ TCA-soluble fraction of plasmin digests; ○ heat-stable fraction, ● TCA-soluble fraction of trypsin digests; □ heat-stable fraction minus fragment E of trypsin; ■ of plasmin digests.

of fibrinogen both with trypsin and plamsmin digests. The plateau region suggests that the release of the α-chain C-terminal portion is completed early in the digestion. This was evident already from the very first studies of the chain segments that arose during the fragmentation of fibrinogen (3). The magnitude of the fraction of optical density indicates that the whole segment from at least Trp 229 to the end of the chain is involved in this process, since the 8 tryptophan and 2 tyrosine residues present there contribute 19.3% to the total optical density of fibrinogen. A few more remarks are in order: Although the molar concentration of trypsin in the reaction mixture is slightly higher than that of plasmin, 1.87×10^{-6} M vs. 1.61×10^{-6} M, still the reaction with plasmin appears to be much faster. Further, as mentioned already, fragment E appears early with this enzyme, whereas with trypsin only after a considerable lag period. The TCA-soluble fraction, on the other hand, is more than twice higher with trypsin. The latter phenomenon has its explanation, as it will be evident latter, in the secondary cleavage of the main fragment into two pieces, one of which is not precipitated by TCA. The slow continuing buildup of TCA-soluble material after the above reactions is a reflection of the progressive unwinding and cleavage of small peptides from the large protein-like D and E as well as of the α-chain fragments. It should be emphasized that the magnitude of the optical densities in these fragments at 280 nm depends on their specific optical density, i.e. their aromatic amino acid content. Therefore, the relative amounts of optical density in these fractions does not reflect the mass distribution in them. In fact, the TCA-soluble peptides are poor in aromatic amino acid residues, consequently their contribution to the total mass may be 4 - 5 times higher than what would appear to be from Fig. 1.

Large α-chain fragments with either trypsin or plasmin were prepared based on the data mentioned in the preceding paragraph. Two sets of digests were fractionated: one at approximately half-way along the curve of formation of the α-chain fragmennts and the other at completion of this reaction. For these preparations the digestions were run at about 1/2 of the trypsin and 1/10 of the plasmin concentration of the experiment in Fig. 1. The supernatants of the heat coagulated digests were fractionated on the Sephadex G-75 superfine column. Figure 2 shows the elution curves of the two trypsin digests superimposed, recorded at 206 nm. The early digest shows 4 main components and 3 shoulders. Of these the heavier disappear on further digestion and two lighter ones become the predominate component. In the same time the low molecular weight peak becomes asymmetric and increases in size. In these runs the absorption was monitored at 206 nm, which has the advantage of detecting also peptides that lack aromatic residues and thus are missed

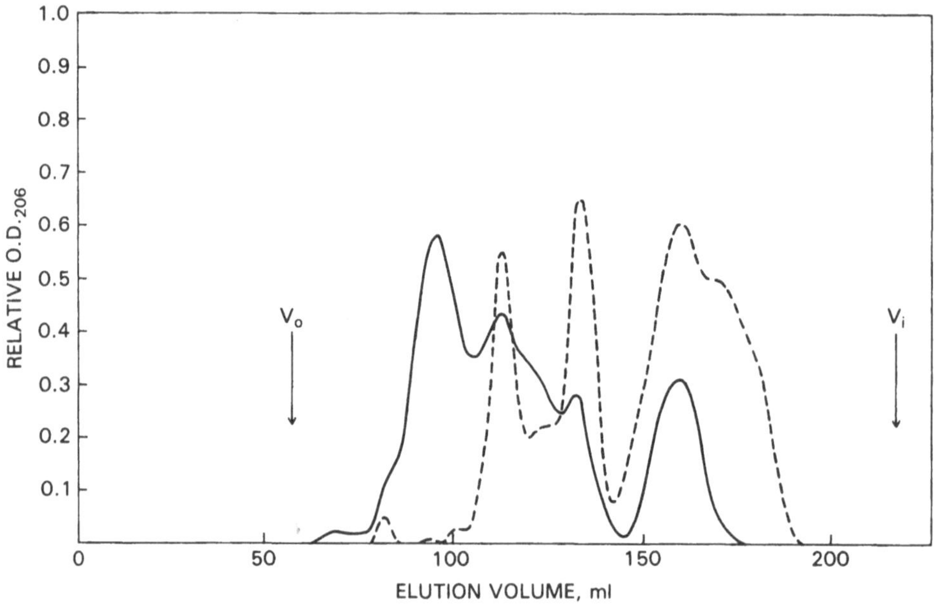

Fig. 2. Gel filtration chromatography on Sephadex G-75 superfine of the heat-stable fraction of fibrinogen digested by trypsin. Elution patterns recorded at 206 nm. Solid line: sample digested to the extent of 13.6 moles NaOH per 340,000 g fibrinogen consumed in the pH stat. Broken line: digestion to 43.0 moles NaOH per 340,000 g fibrinogen. Above figures multiplied by 1.389 give number of bonds cleaved per mole, 18.9 and 59.7 respectively (see Ref. 10). Vo and Vi: elution volumes of blue dextran and phenol red respectively.

at 280 nm. Further, absorption even in the presence of aromatic chromophores is 10 - 20 times higher at 206 than at 280 nm. To show this effect, elution diagrams of the more advanced digest were recorded at the above two wave length and are superimposed in Fig. 3. To make the curves comparable, the ordinate of the 280 nm curve is 8 fold expanded with respect to the 206 nm curve by applying 4 times larger sample and operating the monitor at a 2 times higher sensitivity. It is evident on the recordings that a large amount of small peptides

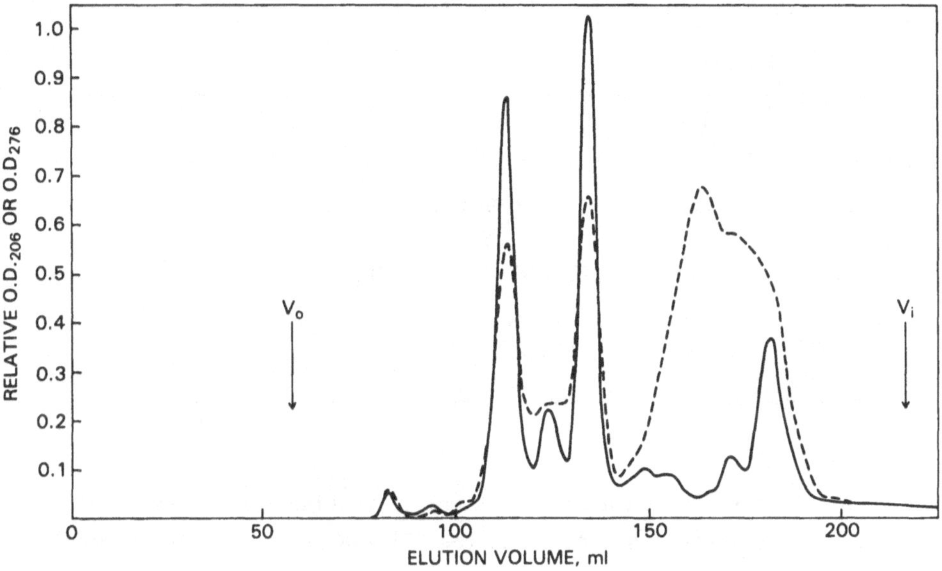

Fig. 3. Sephadex G-75 superfine chromatography of heat-stable fraction of fibrinogen digested by trypsin to the extent of 59.7 bonds cleaved per molecule. Solid line, effluent monitored at 276 nm, broken line at 206 nm. Ordinates of the latter are scaled down 8-fold.

without aromatic residues were liberated during the digestion, beside the two main peaks. With plasmin digestion a complication arises. The low molecular weight peptides are poorly soluble in the solvent used and turbidity develops in their region of elution, causing monotonous rise of the absorption that obscures the individual peaks. Nevertheless, these runs can be used to monitor the elution of the main peak and thus accomplish its isolation. The elution volumes of the well defined peaks were used to calculate the Stokes radii (11). The data are assembled in Table I. This table also contains the molecular weights derived from SDS-PAGE runs and amino acid analyses, and specific optical densities for the various fragments. The gel-electrophoresis patterns used for this purpose are shown in Figs. 4 and 5. As expected these patterns are

Table I: Some Physico Chemical Properties of the Fragments

Fragment	Molecular Weight SDS-PAGE	Molecular Weight Amino Acid Analysis	Stokes Radius Å	Specific $O.D._{280}$
Plasmic P1	24,500	25,594	26.5	1.46
Tryptic P1a	9,800	9,014	24.0	2.57
Tryptic P1b	6,600	5,982	16.0	2.20

equivalent to the elution diagrams of gel filtration runs recorded at 276 nm. It should be mentioned that the SDS-PAGE runs were performed on preparations chosen from those depicted in Fig. 1 and duplicating closely those used for gel filtration studies as far as the degree of digestion goes.

Both the gel filtration and gel electrophoresis data show that initially with both trypsin and plasmin high molecular weight fragments are formed, approaching the size of the maximal length of cleavable chain segment, which extends from residue 199, the C-terminal end of the α-chain segment in fragment D (12), to residue 610, the end of the α-chain. These, however, are short lived and because of this never reach stiochiometric proportions. With plasmin, rapidly a relatively stable fragment is obtained by this process, that will be labeled plasmic fragment P1 in the following. With trypsin, two lower molecular weight components appear that are much less resistant than the plasmic one to further enzymatic action. These were denoted tryptic fragment P1a and P1b. By chosing the proper digestion time these fragments can be obtained with relatively minor high molecular weight contaminants. The plasmic digest shown on Fig. 5 and the tryptic one shown in Figs. 2 and 3 was chosen for further purification. The proper cuts were made to exclude the flanking contaminants and pooled samples were lyophilized and recycled on the Sephadex G-75 superfine column under identical

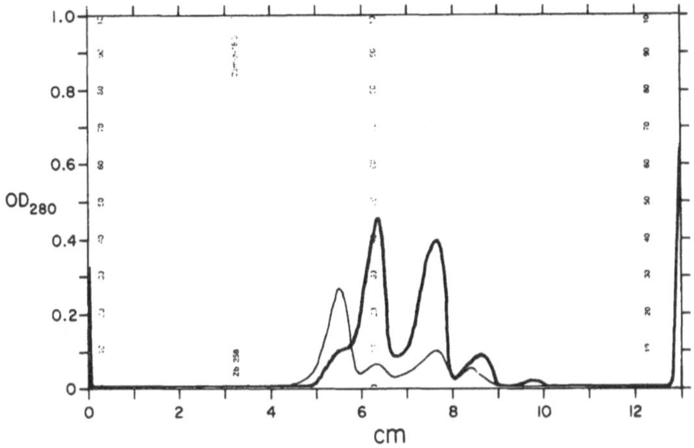

Fig. 4. SDS-polyacrylamide gel-electrophoresis patterns of heat-stable supernatants of fibrinogen digested by trypsin. Thin line, digestion to 14.0 moles, heavy line to 42.5 moles NaOH consumed per mole of fibrinogen in the pH stat. Direct scanning of gels at 280 nm.

conditions to the first fractionation. The recycled materials appeared homogeneous on gel filtration. These were hydrolyzed and their amino acid composition was determined. The data are assembled in Table II. From the spectra of the original solutions and the peptide concentrations determined from the amino acid analyses the specific optical densities of these fractions were determined and are listed in Table I. The spectral data served also to determine the tyrosine and tryptophan content of these fractions. The amino acid compositions compared with those of segments of the known sequence of the α-chain (13, 14) enabled the localization of these peptide fragments within the original α-chain.

Discussion

The C-terminal two-thirds of the α-chain of human fibrinogen has a peculiar sequence (13, 14). There are 7 tryptophans in the sequence of residues 276-391 and none further along the chain. Since these are the

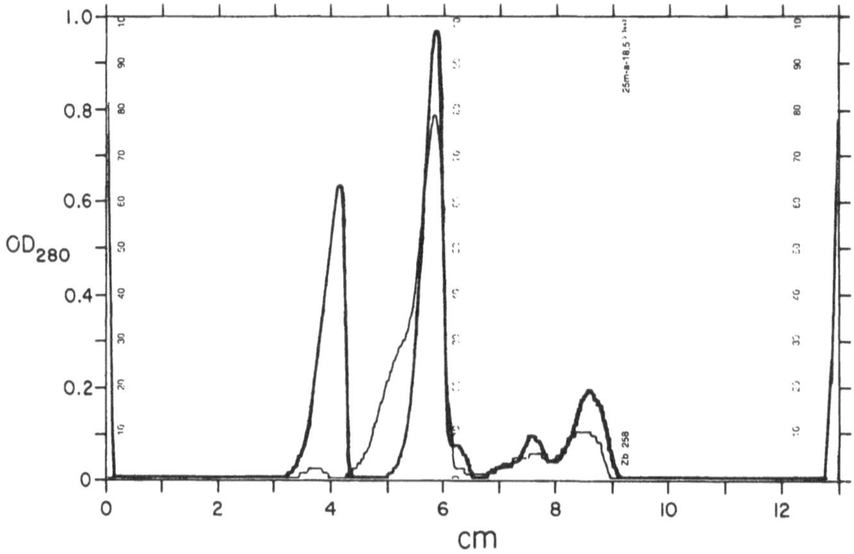

Fig. 5. SDS-polyacrglamide gel-electrophoresis of heat-stable supernatants of fibrinogen digested by plasmin. Thin line, digestion to 14.5 moles, heavy line to 29.3 moles NaOH per mole of fibrinogen consumed in the pH stat. Direct scanning at 280 nm. First peak on the left side is fragment E.

most hydrophobic of all amino acid residues, it is likely that this portion of the chain is in a folded conformation resistant to proteolytic attack. In addition, the primary structure also provides a similar effect. There are 6 potentially susceptible arginine and lysine residues in this sequence, but 4 are blocked by a following proline and with one the sequence is Pro-Arg-Asn-Pro. These facts suggest that the chain in this region may form a more compact structure, resistant to proteolytic enzymes, by a virture of both the primary sequence and the secondary structure. Indeed, as reported by others (4,5,8) and also evident in the work described here, plasmin cleaves off a fragment of

Table II: Amino Acid Composition of the Fragments

Residues in Chain Segments

Amino Acid	Plasmic P1 (240-491)		Tryptic P1a (253-348)		Tryptic P1b (349-406)	
	Found	Calulated	Found	Calculated	Found	Calculated
Asp	21.7	21	9.3	8	6.6	6
Thr	30.5	31	12.5	11	4.8	4
Ser	42.2	47	24.7	26	11.5	13
Glu	20.6	20	4.7	4	5.7	5
Gly	48.0	47	28.8	25	12.1	10
Ala	6.6	5	2.7	2	2.2	2
1/2 Cys	0.7	2	---	---	---	---
Val	9.5	10	---	---	2.9	3
Met	1.1	2	---	---	---	---
Ile	3.3	3	---	---	0.5	---
Leu	5.6	6	---	---	1.5	---
Tyr	2.0	2	0.4	1	0.2	---
Phe	4.7	3	---	---	1.9	2
His	4.7	4	0.2	---	1.4	2
Lys	9.4	10	1.2	1	0.9	---
Arg	14.9	14	4.5	4	3.5	3
Pro	19.4	18	11.5	10	4.8	5
Trp*	7.8	7	3.9	4	2.4	3

*From spectroscopy

approximately 26,000 MW from this chain that is resistant to further degradation by this enzyme. The fragment starts at Met 240 (8) and should extend to Arg 491 to reach the above molecular weight. The amino acid composition reported in Table II is in agreement with this assumption. There is an abundance of arginine and lysine residues in the second half of this sequence, nevertheless cleavage at these by plasmin is very slow. The Stokes radius of this fragment given in Table I also suggests a compact conformation, since it is close to that of chymotrypsinogen (22.5Å), a nearly spherical, tightly coiled protein. The fragment forms very rapidly, therefore, the Arg 239-Met 240 bond must be in an exposed situation. The whole segment of 38,500 MW, residues 240-610, liberated by cleavage at this point, is observed with very early digests, but it is degraded fast in several steps. Thus, the segment formed by residues 492-610 must have loose conformation, with bonds freely accessible to the enzyme.

Degradation by trypsin is slower but all the above mentioned fragments occur in the process. The first peak shown for example in the early trypsin digest in Figs. 2 and 4 has a MW of 26,000, identical to that of the plasmic fragment. However, tryptic digestion does not stop at this point, but continues until the two resistant fragments made of residues 253-348 and 349-406 are reached. These have virtually all their susceptible bonds blocked by proline residues. There are only two Arg-Asn bonds that are not, however, with these a proline is the second neighbor. Thus, the sequence peculiarities are sufficient to explain the formation of these fragments, but this does not exclude the possibility that they are also in a conformation other than a random coil.

Although the data reported in this paper suggest a more compact conformation for these fragments than a random coil further studies are necessary to prove this point in a definitive way.

References

1. Doolittle, R.F.,: Adv. Protein Chem. 27, 1-109 (1973).
2. Mihalyi, E.: Application of Proteolytic Enzymes to Protein Structure Studies, 2nd Edition, Vol. I-II, CRC Press, Boca Raton, Florida, (1978).
3. Mills, D., Karpatkin, S.: Biochem. Biophys. Acta: 271, 163-173 (1972).

4. Mosesson, M.W., Galanakis, D.K., Finlayson, J.S.: J. Biol. Chem. 249, 4656-4664 (1974).
5. Ferguson, E.W., Fretto, L.J., McKee, P.A., J. Biol. Chem. 250, 7210-7218 (1975).
6. Nussenzweig, V., Seligmann, M., Pelmont, J., Grabar, P.: Ann. Inst. Pasteur: 100, 377-389 (1961).
7. Takagi, T.,Doolittle, R.F.: Biochemistry: 14, 5149-5156 (1975).
8. Harfenist, E.J., Canfield, R.E.: Biochemistry 14, 4110-4117 (1975).
9. Sgouris, J.T., Inman, J.K., McCall, K.B., Hyndman, L.A., Anderson, H.D.: Vox Sang. 5, 357-376 (1960).
10. Mihalyi, E., Weinberg, R.M., Towne, D.W., Friedman, M.E.: Biochemistry 15, 5372-5381 (1976).
11. Laurent, T.C., Killander, J.: J. Chromatog. 14, 317-330 (1964).
12. Doolittle, R.F., Cassman, K.G., Cottrell, B.A., Friezner, S.J., Takagi, T.: Biochemistry 16, 1710-1715 (1977).
13. Watt, K.W.K., Cottrell, B.A., Strong, D.D., Doolittle, R.F.: Biochemistry 24, 5410-5416 (1979).
14. Henschen, A., Lottspeich, F., Hessel, B.: Hoppe-Seyler's Z. Physiol. Chem. 360, 1951-1956 (1979).

BIODISTRIBUTION OF HUMAN FIBRINOGEN-DERIVED PEPTIDES IN RABBITS

Job Harenberg, Gerd Stehle, Stephan Waibel

Medical University Clinic, Departments of Internal Medicine and Clinical Pharmacology, Bergheimerstr. 58, 6900 Heidelberg, FRG

Michael Eisenhut and Hans-Joachim Hermann

Department of Nuclear Medicine
University of Heidelberg, FRG

Introduction

Only limited data are available regarding the distribution and elimination of fibrinopeptides, which are cleaved from fibrinogen during the coagulation and fibrinolytic process. Recently, we have shown that human fibrinopeptide A is rapidly distributed in the vascular system, the liver and the kidneys after intravenous injection and then excreted almost undegraded via glomerular filtration into the urine in rabbits. A small part of the A-peptide, however, is degraded by carboxypeptidases in the proximal tubule cells and then excreted into the urine or reabsorbed into the plasma. Halflives of the radiolabeled A-peptide were determined as 0.5, 3.5, 15 and 90 min for four phases of elimination. The longer halflives are probably mediated by the redistribution of the peptide from the organ into the plasma (1).

We now report the results of the distribution and elimination of fibrinopeptides and in comparison of ACTH using radio-labeled substances and scintigraphic documentation.

Fibrinogen - Structural Variants and Interactions
© 1985 Walter de Gruyter & Co., Berlin · New York - Printed in Germany

Materials and Methods

Synthetic deaminotyrosine fibrinopeptide A (FPA) and peptide Bβ 15-42 were obtained from ImCo, Stockholm. ACTH was purified using the HPLC method (2). The peptides were labeled with 123-iodine by the chloramine T method as described by Hunter and Greenwood (3). Subsequently, the peptides are designated as 123-I-FPA, 123-I-Bβ 15-42 and 123-I-ACTH. 123-I was obtained from the Kernforschungszentrum Karlsruhe. The reaction mixture was separated on a 10 x 150 mm column with Sephadex G10. The iodinated peptides were eluted in the fraction 7-15. About 90 % of the iodine was incorporated into the peptides to be used for application. The specific activity was 0.1 mCi/μg peptide.

Healthy rabbits weighing 2.5 - 3.5 kg were used in the experiments. Each animal was investigated only once. The animals were anesthetized with ketamine hydrochloride 30 mg/kg body weight i.v. and 4 mg/kg body weight i.m. 20 min prior to the onset of the investigation. Anesthesia was maintained with half the initial anesthetic dose every 60 min. For those rabbits undergoing nephrectomy, a midline incision was made, followed by bilateral ligation of the kidney hilus.

Freshly labeled peptide (0.5 mCi) was injected i.v. into 15 normal and 15 nephrectomized rabbits. The rabbits were immobilized in a supine position during the study. The radioactive distribution of the labeled peptide was studied by scintigraphy using a gamma camera connected to a highly sensitive, low energy collimator. The scintigraphic data were stored on a magnetic disc with a Dycom 80 computer system (Elscint). The acquisition rate was 1 frame/10 s for the first 5 min and then 1 fram/5 min for the following 120 min. The field of the gamma camera included the region between the neck and the urinary bladder of the rabbit. The heart, liver, gallbladder, kidneys, and the urinary bladder were marked by ROI (region of interest) technique for kinetic evaluation over each organ. All data were corrected according to the halflife of 123-iodine (13.2h).

Results

The biodistribution of 123-I-FPA and 123-I-Bβ 15-42 shows that radioactivity was taken up primarily in the cardiovascular system, liver, gallbladder, kidneys, and urinary bladder. No radioactivity was detected in the lungs, spleen, gastrointestinal tract or thyroid gland. Within 1.5 min after injection, the heart, large arteries and liver became visible. Ten min after the injection, a considerable amount of radioactivity was detected in the kidneys and urinary bladder. After 30 min, the gallbladder showed radioactivity. No activity was found in the area of the thyroid gland. One hour after application of the radiolabeled peptide, almost all radioactivity was in the urinary bladder (Fig. 1 and 2).

The biodistribution of radiolabeled ACTH shows a rapid uptake of radioactivity into the thyroid gland within 10 min. Radioactivity as detected also after an initial distribution in the cardiovascular system, in the liver, the kidneys and the two adrenal glands. No activity was found in the area of the spleen, the gastrointestinal tract or other organs. Radioactivity was rapidly excreted into the urine. One hour after the application of the radiolabeled peptide, almost all radioactivity was in the urinary bladder (Fig. 3).

In nephrectomized animals, the radiolabeled peptides FPA, Bβ 15-42 and ACTH demonstrated a similar behavior. The radioactivity remained in the cardiovascular system, the liver and for ACTH in the thyroid gland. Some radioactivity was detected

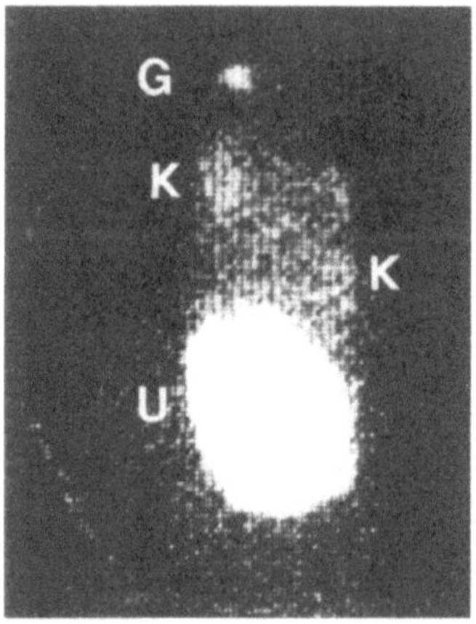

Fig. 1

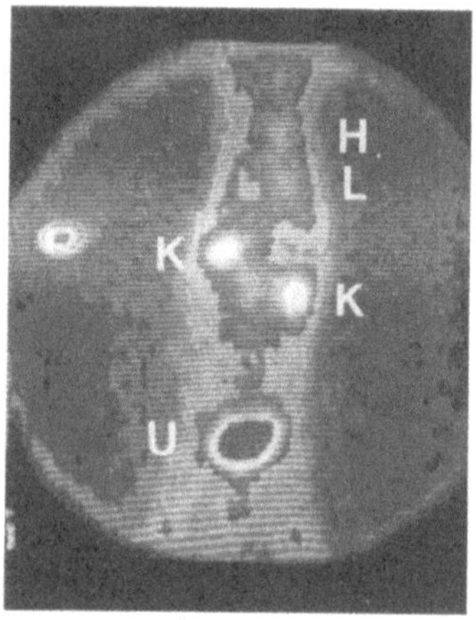

Fig. 2

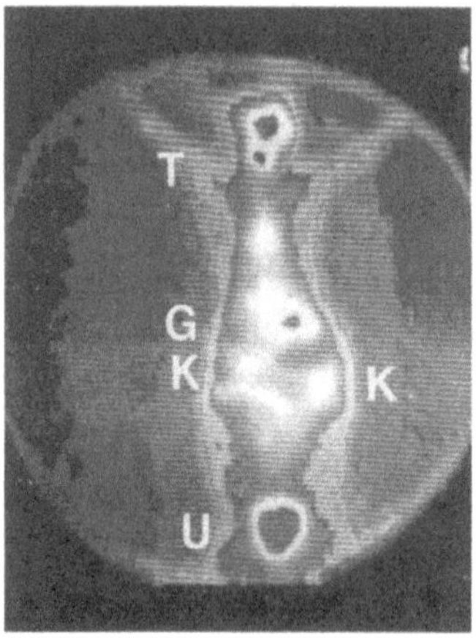

Fig. 3

Figures 1-6: The scintigraphic pictures 60 min after the administration of FPA (Fig.1,4), Bβ 15-42 (Fig. 2,5) and ACTH (Fig.3,6) in normal (Fig.1-3) and nephrectomized rabbits (Fig.4-6) are shown. Abbreviations: T = thyroid gland, H = heart, L = liver, G = gall bladder, K = kidney and U = urinary bladder.

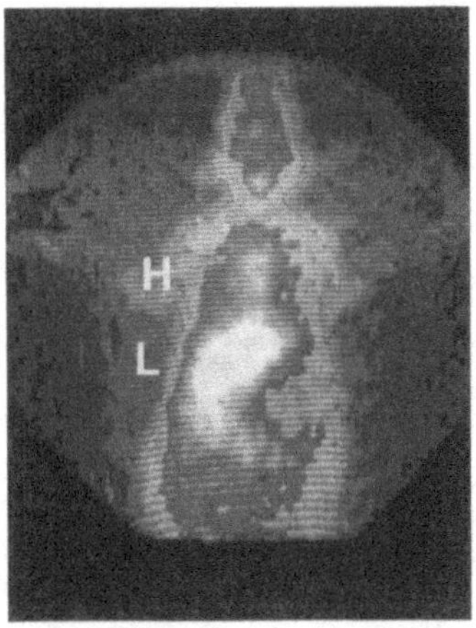

Fig. 4

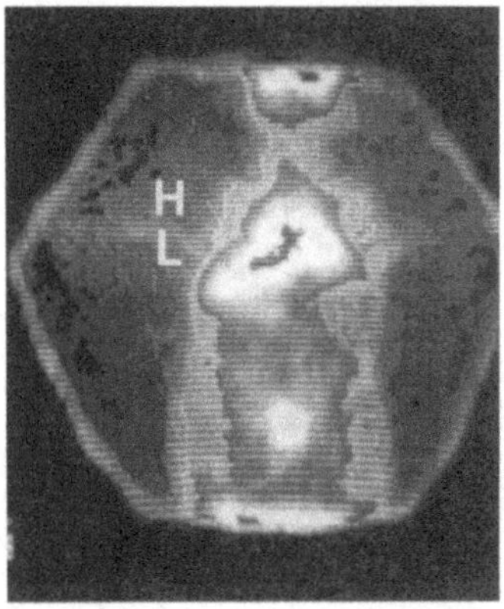

Fig. 5

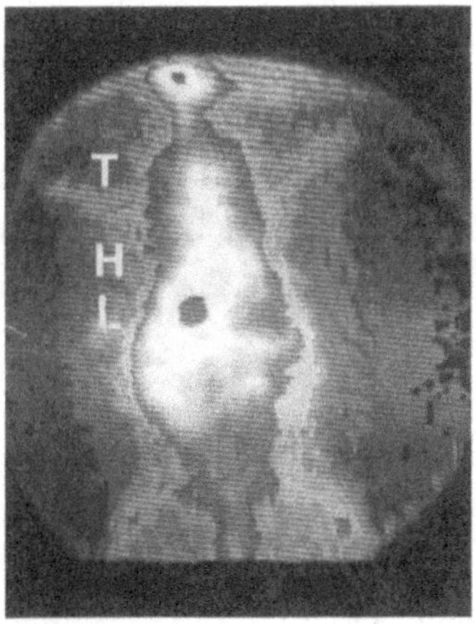

Fig. 6

Figures 1-6: See opposite page.

in the brain area. Over all other organs, no increase of radioactivity was measured throughout the observation period of 1 h. The scintigraphic pictures 1 h after the application of the peptides are shown in Fig. 4 - 6.

Discussion

Experiments by Nossel et al. on healthy volunteers indicated a plasma halflife for FPA of 3 - 5 min after i.v. bolus injection of homologous peptides (4). Recently, results of experiments on rats (5) and rabbits (1) have suggested that this rate probably reflects the distribution and equilibration of FPA in the cardiovascular system. The rapid components of distribution have been divided into two halflives of 30 s and 3.5 min in rabbits (1). A slow component could also be divided into two compartments with halflives of 15 and 90 min. These halflives were mainly influenced by distribution phenomena between liver, kidney and the blood compartment (1). The present results indicate that these observations of the distribution of FPA are not specific for this peptide, but that a similar behavior can be found for other small peptides such as Bβ 15-42 and ACTH.

After administration of labeled ACTH, iodine was taken up by the thyroid gland, indicating hydrolysis of the iodinated amino acid (6). In our study, radioactivity was found within 10 min over the area of the thyroid, indicating hydrolysis of the labeled iodine from ACTH. In contrast after application of the labeled FPA and Bβ 15-42, no radioactivity was detected over the thyroid gland using scintigraphic documentation. This finding supports the previously made assumption that FPA is not degraded at the amino terminal end, because the deaminotyrosine group has been synthesized at the N-terminal end of the peptide.

IN VIVO DISTRIBUTION OF HUMAN FIBRINOPEPTIDES

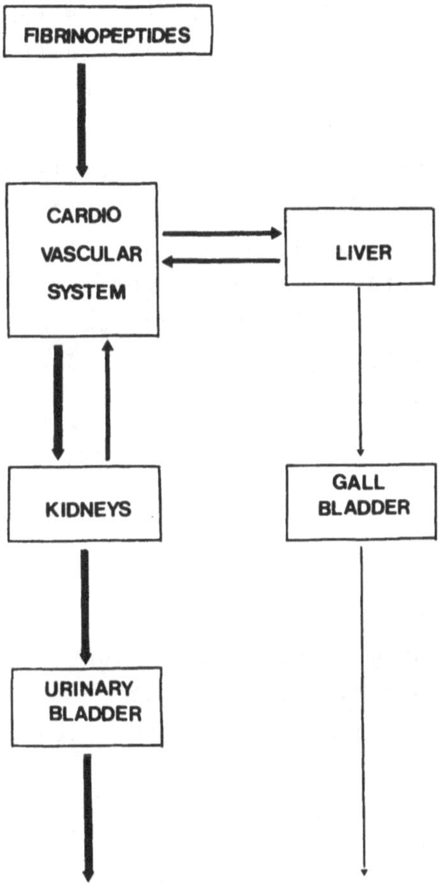

Figure 7: Distribution and elimination pattern of human radiolabeled Fibrinopeptide A and Bβ 15-42 in rabbits according to scintigraphy.

The experiments in nephrectomized rabbits were undertaken to determine the role of the kidney in the excretion and distribution of the FPA, Bβ 15-42 and ACTH. The results clearly demonstrate that no other compartments than the cardiovascular system, the liver and the kidney and for hormones the receptor organs exist for the peptides. Furthermore, it was shown that the peptides are excreted almost completely into the urine, since no other elimination processes could be detected in these experiments (Fig. 7). Small peptides and hormones are predominantly excreted as almost unchanged peptides(see 1). The lack of measurable FPA concentration in the urine (7) may be attributed to interference of other components of the urine with the assay system.

References

1. Harenberg, J., Stehle, G., Waibel, S., Hermann, H.J., Eisenhut, M., Zimmermann, R.: Thromb. Res. 32, 1 (1983).
2. Eisenhut, M.: unpublished observation.
3. Hunter, W.M., Greenwood, F.C.: Nature 194, 495 (1962).
4. Nossel, H.L., Younger, L.R., Wilner, G.D., Procupez, T., Canfield, R.E., Butler, V.P.: Proc.Nat.Acad.Sci.USA 68, 2350 (1971).
5. Lane, D.A., Siodlak, M., Thompson, E., Allen-Mersh, T.G.: Thromb. Res. 26, 73 (1982).
6. Franco-Saenez, R., van Gemert, M., Saffran, M.: Hormone Res. 15, 44 (1981).
7. Alkjaersig, N., Fletcher, A.P.: Blood 60, 148 (1982).

STRUCTURE-FUNCTION STUDIES ON A VASOACTIVE PENTAPEPTIDE DERIVED FROM PLASMIN DEGRADATION OF HUMAN FIBRIN(OGEN)

Rolf Wallin, Katarina Saldeen and Tom Saldeen

Institute of Forensic Medicine, University of Uppsala, Sweden

INTRODUCTION

A vasoactive pentapeptide derived from plasmin degradation of human fibrinogen or fibrin has been isolated (1). This peptide, which increases microvascular permeability, has been called 6A and has the sequence Ala-Arg-Pro-Ala-Lys and corresponds to residues 43-47 of the human fibrinogen Bβ chain.

The aim of the present study was to examine the structural requirements for biological activity of this peptide.

MATERIALS AND METHODS

Material. Pentapeptide 6A and 33 analogues of this pentapeptide were synthesized by the Merrifield solid phase technique, purified and analyzed as described in detail previously (2,3). A synthetic counterpart of 6A is as potent as the natural peptide in increasing microvascular permeability (2). Mepyramine maleate was a gift from Smith, Kline and French Laboratories Ltd, UK. ^{125}I-labelled human serum albumin was bought from Institutt for Energiteknikk, Kjeller, Norway. Indomethacin was a gift from Dumex, Helsingborg, Sweden. The carboxypeptidase N inhibitor 2-mercaptomethyl-3-guanidino-ethylthio-propionic acid was a gift from Dr. J.W. Ryan, Miami, Florida. All chemicals were of analytical or reagent grade.

Microvascular permeability in rat skin was studied in male Sprague-Dawley rats weighing 210-235 g. After the dorsal fur had been shaved, ^{125}I-labelled human serum albumin was injected intravenously. Each test substance, dissolved in 0.1 ml of buffer saline, was then injected intracutaneously in at least four different places in the dorsal skin of at least four rats. Four injections of buffer in each animal served as controls. Skin circles of uniform size surrounding each injection site were cut out after 30 min. Each piece of skin was weighed and the radioactivity was counted in a

Fibrinogen - Structural Variants and Interactions
© 1985 Walter de Gruyter & Co., Berlin · New York - Printed in Germany

gamma spectrometer. The influence of the test samples on the microvascular permeability was expressed as the "permeability quotient" as follows:

$$\text{permeability quotient} = \frac{\text{average content of } ^{125}\text{I-albumin in test spot}}{\text{content of } ^{125}\text{I-albumin in control spots from the same rat}}$$

<u>Histamine release from rat mast cells</u>. Adult male Sprague-Dawley rats (body weight 250-350 g) were killed by decapitation under ether anesthesia. Mixed cells of which about 10 % were mast cells were obtained by infusion of 8 ml of Sørensen´s phosphate buffer intraperitoneally and 4 ml intrathoracically, followed by gentle massage and removal of the intraabdominal and intrathoracic mast-cell-rich fluid with a pipette. The fluid was centrifuged at 350 g for 5 min at room temperature. The cells were washed once in Sørensen´s buffer, centrifuged, and suspended in an appropriate volume of the buffer. The cell suspensions were preincubated for 5-10 min at 37°C before the experiment was begun. 0.2 ml of the test substance was transferred to a glass test tube. 1.8 ml of the cell suspension was added to each tube and the mixture was incubated for exactly 10 min at 37°C. The reaction was interrupted by rapid cooling in an ice-bath. The tubes were then centrifuged at 350 g for 5 min at 0°C and the supernatants were removed for fluorometric assay of histamine. To the cell sediment of each tube was added 0.2 ml of 0.1 M HCl and the mixture was boiled for 5 min. A 1.8 ml aliquote of Sørensen´s buffer was then added to each mixture. Fluorometric assay of these samples gave the remaining amounts of histamine in the cells. Histamine was determined fluorometrically and the amount of histamine released was expressed in per cent of the total amount of histamine present in the cells at the start of the experiment. All determinations were made in duplicate.

<u>The effect of the peptides on angiotensin-converting enzyme</u>. The peptides were tested for their abilities to inhibit the activity of partially purified human urinary angiotensin-converting enzyme (4). The enzyme activity was measured in terms of the rate of hydrolysis of ^{3}H-hippuryl-His-Leu-(0.1 μCi; 25 Ci/nmol). The substrate and a given peptide, all in 100 μl of 0.05 M Hepes buffer, pH 8.0, plus 0.1 M NaCl and 0.75 M Na_2SO_4, were mixed

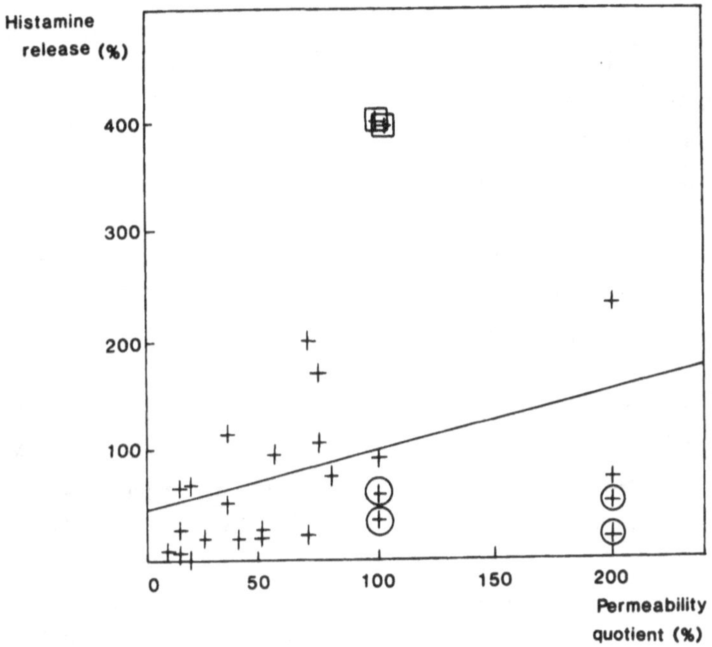

FIGURE 1: Correlation between histamine release and permeability quotients for the various analogues (r = 0.4) □ = peptides 13 and 24 ○ = peptides 19,28,30 and 32.

with an equal volume of enzyme in the same buffer and the reaction was allowed to proceed for 15 min at $37^{o}C$ (5). The quantity of enzyme used sufficed, in the absence of an inhibitor, to catalyze the hydrolysis of 12 % of substrate under the specified conditions. The concentration of peptide 6A or a related peptide required to reduce the enzyme activity by half was termed I_{50}.

RESULTS

The results are presented in Tables I-III. C-terminal amides are indicated by (NH_2), L-amino acid residues are capitalized and D-amino acid residues are denoted by lower-case letters. The permeability quotients and the I_{50} values for inhibition of angiotensin-converting enzyme have been reported previously (3,6) and are included for comparison with the effect of the

TABLE I

Peptide no.	Sequence	Permeability quotient(%)	Histamine release(%)	I_{50} ACE [M]
6A	Ala-Arg-Pro-Ala-Lys	100	100	1.9×10^{-5}
1	Gly-Ala-Arg-Pro-Ala-Lys	50	25	2.9×10^{-6}
2	ala-Arg-Pro-Ala-Lys	15	25	8.3×10^{-6}
3	C_2H_5CO-Arg-Pro-Ala-Lys	15	5	2.3×10^{-6}
4	Phe-Arg-Pro-Ala-Lys	40	20	4.5×10^{-7}
5	Arg-Pro-Ala-Lys	100	90	1.8×10^{-5}
6	Arg-Ala-Pro-Ala-Lys	15	65	2.3×10^{-6}
7	arg-Pro-Ala-Lys	10	5	4.6×10^{-5}
8	Arg-pro-Ala-Lys	10	nd	2.9×10^{-4}
9	Arg-Pro-Lys	70	nd	1.3×10^{-4}
10	Arg-Pro-Gly-Lys	30	nd	2.5×10^{-5}
11	Arg-Pro-pro-Lys	35	50	7.0×10^{-4}
12	Arg-Pro-Pro-Lys	75	105	1.6×10^{-4}
13	Arg-Pro-Lys-Lys	100	400	1.4×10^{-4}
14	Arg-Ala-Pro-Lys	20	70	2.6×10^{-5}
15	Arg-pro-Lys	15	0	$> 10^{-3}$
16	Arg-Pro-Ala-Ala-Lys	80	75	2.0×10^{-5}
17	Ala-Lys-Pro-Ala-Lys	70	20	8.0×10^{-6}
18	Ala-Nle-Pro-Ala-Lys	20	0	2.5×10^{-6}
19	Ala-Lys-Pro-Ala-Arg	200	15	5.4×10^{-6}
20	Ala-Arg-Gly-Ala-Lys	10	25	1.8×10^{-4}
21	Ala-Arg-ΔPro-Ala-Lys	50	nd	5.9×10^{-6}
22	Ala-Arg-Ala-Pro-Lys	35	115	2.3×10^{-4}
23	Ala-Arg-Pro-ala-Lys	70	200	1.6×10^{-4}
24	Ala-Arg-Pro-Lys-Lys	100	400	2.6×10^{-5}
25	Ala-Arg-Pro-Gly-Lys	55	95	3.4×10^{-6}
26	Ala-Arg-Pro-Pro-Lys	75	170	1.4×10^{-4}
27	Ala-Arg-Pro-Ala-Nle	25	20	2.5×10^{-6}
28	Ala-Arg-Pro-Ala-Arg	100	60	7.8×10^{-6}
29	Ala-Arg-Pro-Ala-Lys (NH_2)	200	230	9.5×10^{-5}
30	Ala-Arg-Pro-Ala-lys	100	30	nd
31	Ala-Arg-Pro-Ala-Nle (NH_2)	200	75	1.6×10^{-5}
32	Ala-Arg-Pro-Ala-lys (NH_2)	200	50	$>10^{-3}$
33	Ala-Arg-Pro-Ala-Lys-Gly	50	20	2.8×10^{-5}

Peptide	Sequence	Permeability quotient (%) Peptide alone	Peptide + mepyramine	Peptide + carboxypeptidase inhibitor
6A	Ala-Arg-Pro-Ala-Lys	100	<10^{X}	165^{X}
13	Arg-Pro-Lys-Lys	100	65^{X}	180^{X}
19	Ala-Lys-Pro-Ala-Arg	200	10^{X}	245
24	Ala-Arg-Pro-Lys-Lys	100	55^{X}	130^{X}
28	Ala-Arg-Pro-Ala-Arg	100	15^{X}	140^{X}
30	Ala-Arg-Pro-Ala-lys	100	<10^{X}	125
32	Ala-Arg-Pro-Ala-lys (NH_2)	200	30^{X}	250

Table II. Effect of mepyramine maleate (mepyramine) 1 µg/spot and a carboxypeptidase inhibitor, 2-mercaptomethyl-3-guanidino-ethylthio-propanoic acid, 500 nmol/spot, on microvascular permeability in rat skin after intracutaneous injection of peptide 6A, four analogues with a low capacity to release histamine from mast cells and two peptides with a high capacity. Permeability quotient due to 40 nmol 6A = 100 %. X = significantly different from peptide alone.

analogues on release of histamine from mast cells.

Peptide 6A in a concentration of 4×10^{-4} M released about 30 % of the histamine from mast cells. In Table I the effect of the analogues at the same concentration on histamine release is given in per cent of the effect of 6A itself. The analogues induced histamine release to varying degrees, some being more and some less potent than the native peptide.

In Fig. 1 the correlation between histamine release and the permeability quotients for the various analogues are given; the correlation coefficient being 0.4. Peptides 13 (Arg-Pro-Lys-Lys) and 24 (Ala-Arg-Pro-Lys-Lys) had a very strong histamine-releasing capacity, and were four times more potent than peptide 6A. Peptides 19 (Ala-Lys-Pro-Ala-Arg), 28(Ala-Arg-Pro-Ala-Arg),

	Permeability quotient
6A	4.1 ± 0.4
Indo + 6A	2.9 ± 0.3 X
6 A + mepyramine	1.5 ± 0.4 X
Indo + 6A + mepyramine	1.4 ± 0.3 X

Table III. Effect of intravenously injected indomethacin (Indo) (4 mg/kg) on the permeability increasing effect of 6A in rats treated or non-treated with mepyramine maleate (mepyramine) (1 μg/spot), (n = 4). X = significantly different from 6A alone.

30 (Ala-Arg-Pro-Ala-lys) and 32 (Ala-Arg-Pro-Ala-lys (NH_2) all had a very low capacity to release histamine in comparison with their ability to induce increased microvascular permeability in rat skin.

The influence of the H_1-blocker mepyramine maleate and of a carboxypeptidase inhibitor on the permeability-increasing effects of peptide 6A and these analogues is shown in Table II. Mepyramine maleate in the dosage used, 1 μg/spot, strongly inhibited this effect of 6A and of the four analogues with a low capacity to induce histamine release from mast cells, but seemed to be less effective in inhibiting the effect of the two peptides with a high capacity to induce release of histamine. The carboxypeptidase inhibitor increased the effect of these latter two peptides and of peptide 6A but only of one of the other four analogues. When these six peptides were excluded, the correlation coefficient between permeability and histamine release rose to 0.7 and the probability of correlation was 1.

The effect of indomethacin on the permeability-increasing action of peptide 6A in rats treated and not treated with mepyramine maleate is seen in Table III. Pretreatment with indomethacin significantly decreased the permeability increase due to 6A alone, but had no effect in rats given mepyramine maleate together with 6A.

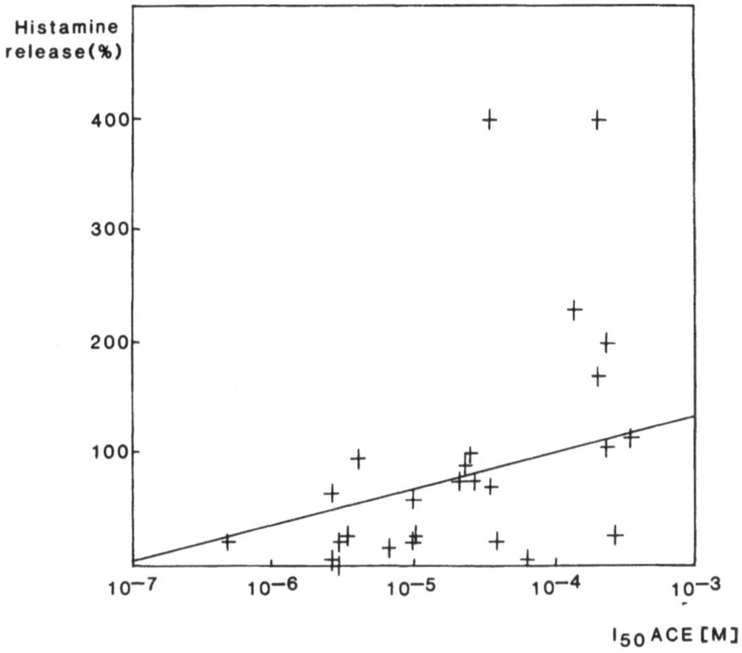

FIGURE 2: Correlation between histamine release and inhibitory effect on angiotensin converting enzyme for the various analogues (r = 0.29)

Indomethacin had no influence on the release of histamine from mast cells induced by peptide 6A. Thus, in one experiment the average release of histamine due to 5 x 10^{-4}M of 6A was 16 % of the total amount of histamine in the cells. Addition of 200 μM indomethacin to the cells 15 min before the addition of 6A resulted in a release of 17 % of this total amount.

Peptide 6A is a moderate inhibitor of angiotensin-converting enzyme, I_{50} being 1.9 x 10^{-5} M. The I_{50} for the various analogues varied between 4.5 x 10^{-7} and >10^{-3}. There was no correlation between the inhibitory effect of angiotensin-converting enzyme and histamine release (r = 0.29) (Fig. 2) or effect on vascular permeability (r = 0.13) (Fig. 3).

DISCUSSION

For the activity of the vasoactive pentapeptide derived from human fibrin-

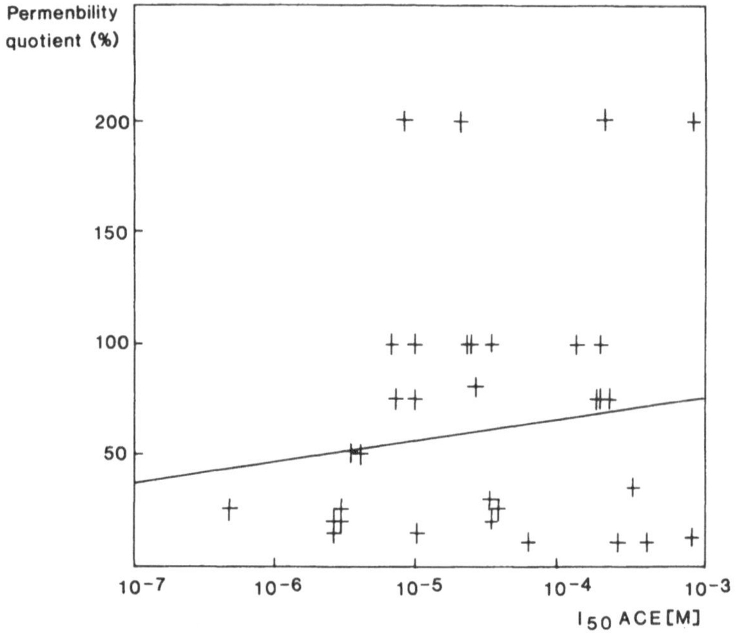

FIGURE 3: Correlation between permeability quotients and inhibitory effect on angiotensin converting enzyme for the various analogues (r = 0.13).

(ogen) studied in the present investigation there are certain structural requirements. A tetrapeptide with basic amino acids at both ends and a proline residue adjacent to the N-terminal amino acid is essential for high activity on permeability and the restrictions seem to reside within the N-terminal part of the molecule, whereas the C-terminal region can be extensively modified without loss of activity (2,3).

As to the inhibitory effect on angiotensin-converting enzyme Ala' can be omitted without loss of activity and replacement of Ala' by Phe yields an inhibitor over 40 times more potent than peptide 6A (6). The basic charge of Arg and the positive charge of Lys are not essential and D-form residues in the C-terminal positions greatly reduce the potency. In fact peptide 32, Ala-Arg-Pro-lys (NH_2), which has a very strong effect on permeability, is the weakest of the inhibitors of angiotensin-converting enzyme. This indicates that the inhibition of this enzyme is not a major mechanism underlying the effect of the pentapeptide on microvascular permeability.

Release of histamine, on the other hand, seems to play an important role in this latter effect. A correlation was found between the effect of the various analogues on the release of histamine from mast cells and their influence on microvascular permeability, although two of the analogues were unproportionately more potent as releasers of histamine and four of them less potent, than was expected from their effect on microvascular permeability. Both potent analogues had the C-terminal sequence Lys-Lys, and it is well known that peptides with a basic charge are potent releasers of histamine (7). The reason for the lower potency of these analogues in their effect on microvascular permeability may lie in their enzymatic degradation in vivo, as the carboxypeptidase inhibitor strongly increased their effect on permeability. A lower capacity of carboxypeptidase for in vivo degradation might be one explanation for the finding that the four peptides with the C-terminal sequences -Arg,-Arg,-D-Lys and D-Lys (NH_2), respectively rather than L-lys, had a more potent effect on microvascular permeability than could be expected from their effect on release of histamine, as the carboxypeptidase inhibitor seemed to be less effective on these peptides. The fact that almost all of their activity on microvascular permeability could be inhibited by the H_1-blocker mepyramine maleate indicates that histamine is a major mediator of the permeability increase by these four analogues also.

Inhibition of angiotensin-converting enzyme might well be a contributory cause of the effect of the peptide on microvascular permeability. Thus, of nine analogues which were more potent as angiotensin-converting enzyme inhibitors than peptide 6A itself, seven induced a stronger increase in vascular permeability than was expected from their influence on release of histamine. On the other hand, of 12 analogues which were less potent as inhibitors of angiotensin-converting enzyme than peptide 6A, only one had a more potent effect on microvascular permeability than on histamine release and all the other were less potent. Peptides with stronger effects as inhibitors of angiotensin-converting enzyme may be less rapidly degraded in vivo and induce release of histamine over a longer period of time.

That pretreatment with indomethacin decreased the permeability suggests that in addition to histamine an arachidonic acid product might also be involved in the permeability increase. The effect of indomethacin should not be due to less release of histamine, as indomethacin did not reduce the 6A induced release of histamine from mast cells. We found in previous studies that peptide 6A induced release of prostacyclin from bovine mesenteric arteries and dilation of these vessels, effects which could be inhibited by indomethacin and were not due to histamine (8). Prostacyclin induces an increase in blood flow which may potentiate the permeability enhancement due to histamine. The fact that indomethacin did not further decrease the permeability in rats treated with mepyramine maleate indicates that prostacyclin alone does not increase permeability, but that it has to be combined with another mediator such as histamine.

ACKNOWLEDGEMENTS

This research was supported by the Swedish Medical Research Council and the Torsten and Ragnar Söderberg Foundation.

REFERENCES

1. Belew, M., Gerdin, B., Porath, J., Saldeen, T.: Thrombosis Research 13, 983-994 (1978).
2. Belew, M., Gerdin, B., Larsson, L.E., Lindeberg, G., Ragnarsson, U. Saldeen, T., Wallin, R.: Biochim. Biophys. Acta 632, 87-94 (1980).
3. Gerdin, B., Lindeberg, G., Ragnarsson, U., Saldeen, T., Wallin, R.: Biochim. Biophys. Acta 757, 366-370 (1983).
4. Ryan, J.W., Chung, A., Martin, L.C., Ryan, U.S.: Tissue & Cell 10, 555-563 (1978)
5. Ryan, J.W., Chung, A., Ryan, U.S.: Environmental Health Perspectives 35, 165-170 (1980).
6. Saldeen, T., Ryan, J.W., Ragnarsson, U.: Kinins - III (Ed. Fritz, H. et al.) Plenum Publishing Corporation, 829 -834 (1983).
7. Johnson, A.R., Erdös, R.R., Proc. Soc. Exp. Biol. Med. 142, 1252-1256, (1973).
8. Andersson, R.G.G., Saldeen, K., Saldeen, T.: Thrombosis Research 30: 213-218 (1983).

PURIFICATION AND CHARACTERISTICS OF A VASOACTIVE PEPTIDE DERIVED FROM ELASTASE DEGRADATION OF HUMAN FIBRIN(OGEN)

Rolf Wallin, Katarina Saldeen, Makonnen Belew, Kjell Ohlsson and Tom Saldeen

Institute of Forensic Medicine and Department of Biochemistry, University of Uppsala, Department of Pharmacology, University of Linköping and Department of Experimental Medicine, Malmö General Hospital, University of Lund, Sweden.

INTRODUCTION

The presence of leucocytes around extravascular fibrin deposits suggests that the leucocyte elastases might be partly responsible for the extravascular degradation of fibrin. Previous studies have shown that degradation of fibrin and fibrinogen by plasmin leads to the release of several peptides which increase vascular permeability (1) and induce oedema, for example in the lungs (2).

The result of the present investigation show that small peptides released from fibrinogen after degradation by leucocyte elastases increase vascular permeability markedly.

MATERIALS

Human fibrinogen, grade L, was kindly supplied by AB Kabi, Stockholm, Sweden. Human leucocyte elastases were isolated from extracts of lysosome-like granules of human leukaemic myeloid cells. Bio-Gel® P-6 (200-400 mesh) was bought from Bio Rad Laboratories, Richmond, California, USA. Pyridine-washed cellulose (Munktell´s No. 410) for column zone electrophoresis and Munktell´s chromatography paper for electrophoresis are products of Grycksbo Pappersbruk AB, Sweden. Sephadex® and Sepharose® gels and chromatographic columns were obtained from Pharmacia Fine Chemicals AB, Uppsala, Sweden. DIAFLO® ultrafiltration membranes (type PM 10) were obtained from Amicon

---

fdp = fibrinogen degradation products
lmw-fdp = low molecular weight (less than 10 000 daltons) fibrinogen degradation products

Fibrinogen - Structural Variants and Interactions
© 1985 Walter de Gruyter & Co., Berlin · New York - Printed in Germany

Corp., Lexington, Mass., USA and millipore filters (type PG) from the Millipore Filter Corp., Bedford, Mass., USA. High-performance liquid chromatography was carried out on an Altex Model 312 Low Cost Gradient Research Chromatograph from Altex, Berkeley, California, USA with a Spherisorb® ODS 5μ column from Jones Chromatography Ltd., Llanbradock, Mid Glamorgan, England. Bradykinin, the peptide Arg-Pro-Pro-Gly-Phe-Ser-Pro-Phe-Arg was a gift from Sandoz AG, Basel, Switzerland. Substance P, Arg-Pro-Lys-Pro-Gln-Gln-Phe-Phe-Gly-Leu-Met-NH2 was purchased from Bachem Feinchemikalien AG, Bubendorf, Switzerland. Peptide 6A (Ala-Arg-Pro-Ala-Lys) was synthesized as described in detail previously (3). It is a vasoactive peptide derived from plasmin degradation of human fribrin(ogen) and corresponds to residues 43-47 of the Bβ-chain.

^{125}I-labelled human serum albumin was bought from Institutt for Energiteknikk, Kjeller, Norway. The carboxypeptidase N inhibitor 2-mercaptomethyl-3-guanidino-ethylpropionic acid was a gift from Dr. J.W. Ryan, Miami, Florida, USA. Mepyramine maleate was a gift from Smith, Kline & French Laboratories Ltd., Herts, UK. Indometahcin was a gift from Dumex, Helsingborg, Sweden. All buffers used were Millipore-filtered and the chemicals used were of analytical reagent grade.

METHODS

Human fibrinogen was made plasminogen-free by affinity chromatography on Lysine-Sepharose prior to use (13). It was then exhaustively dialysed against 0.1M ammonium acetate, ph 8.5 prior to enzymatic degradation. The human leucocyte elastases were isolated by a combination of gel filtration affinity chromatography and preparative agarose gel electrophoresis (4). The purified isoenzymes were lyophilized and kept frozen at -70^0 C until used.

Fibrinogen degradation products (fdp) were prepared by incubating the dialysed solution of fibrinogen (see above) and leucocyte elastases (mixed in a molar ratio of 100:1) at 37^0 C for 48 h. The lysis was stopped by rapid cooling of the reaction mixture in an ice bath. The lysate was then ultrafiltered on a DIAFLO PM10 membrane. The retenate & diffusate obtained was lypphilized. The diffusate (containing peptides of molecular weight less than 10,000 daltons, i.e. lmw-fdp fraction, accounted for about 20 % of the starting material as estimated from absorbance measurements at 280 nm and contained all of the measurable biological activity.

Microvascular permeability was studied in a rat skin model essentially as described by Udaka et al (5) which was slightly modified by us previously (6). The leakage of ^{125}I-human serum albumin into a standardized area of skin surrounding intracutaneously injected test samples was determined. The effect on microvascular permeability was expressed as the "permeability quotient" (Q) to the following formula:

$$Q = \frac{\text{Average content of } ^{125}\text{I-albumin in test spot}}{\text{Content of } ^{125}\text{I-albumin in control spots from the same rat}}$$

The final Q was calculated as the mean of three or, normally, four spots. When estimating subtle changes in Q values, significance was tested with Student's t-test.

Histamin release from rat mast cells was assayed by a procedure we described recently (7,8).

Detection of peptides. The distribution of peptides (lmw-fdp) in column effluents was established by absorbance measurements at 230 and 280 nm and by the ninhydrin method after alkaline hydrolysis (9). The latter method was also used for quantitating peptides in pooled fractions.

Amino acid analysis was performed on a Durrum D-500 amino acid analyser as described previously (1).

Molecular-sieve chromatography, paper electrophoresis and column zone electrophoresis were performed as described in detail previously (1).

High performance liquid chromatography. The lyophilized peptide fractions were dissolved in a minimum amount of distilled water & 200 µl of the solution was applied to a Spherisorb® ODS. 5µm column (Ø 4.5 mm, length 25 cm) equilibrated with 0.1M AmOAc, pH 6.8. A linear gradient was immediately started, ranging from 0-75 % ethanol in the same buffer. The gradient rate was 1 % ethanol per minute, The flow rate was 1 ml/min and fractions of 1 ml were collected. The distribution of material in the effluent was monitored by continuous absorbance measurements at 230 nm and the ninhydrin method on the individual tubes, The fractions obtained were pooled, diluted 10-fold with distilled water, lyophilized at least twice and their vasoactivity determined. Suitable aliquots of the active fractions were submitted to amino acid analysis.

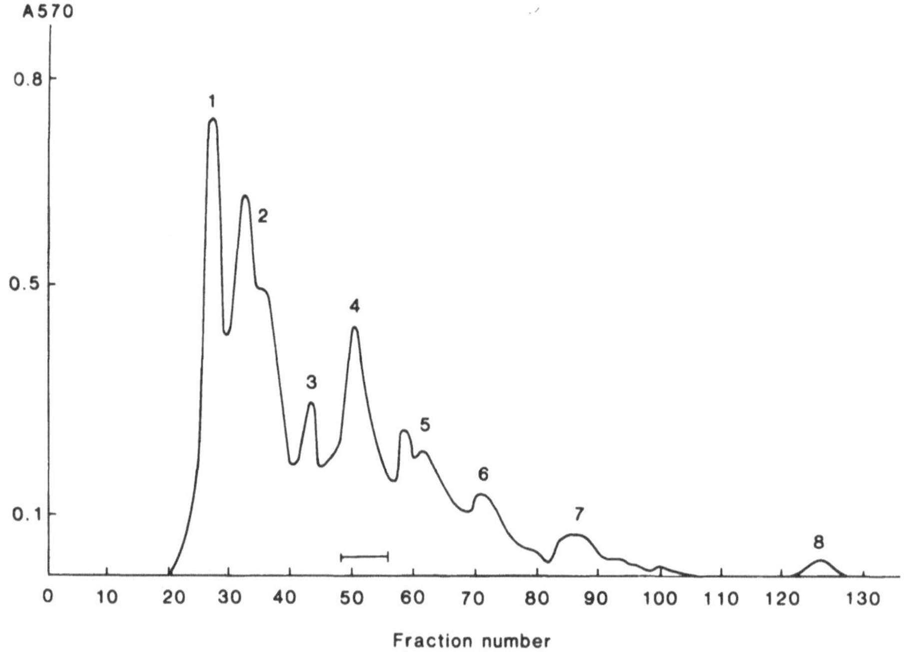

FIG. 1. Chromatography of the lmw-fdp on Bio-Gel P-6. The lyophilized peptides were dissolved in the minimum amount of distilled water prior to application, Column dimensions; 5 x 80 cm. Buffer 0.1M ammonium acetate. pH 6.8. Flow rate: 90 ml/h. Fraction volume = 17 ml; By far the highest specific activity was found in the pooled fraction 4.

RESULTS

Purification

As indicated above, all the measurable biological activity was found in the lmw-fdp fraction. The most potent peptide in this mixture was purified according to the following procedure. Fig. 1. shows the elution pattern obtained upon chromatography of the lmw-fdp fraction on a column of Bio-Gel P-6. Although most of the fractions showed a detectable amount of vasoactivity, fraction 4 gave by far the highest specificity activity. This fraction was further purified by column zone electrophoresis at pH 5.0 and 1.9. Fig. 2. shows the separation obtained after electrophoresis at pH 5.0. Most of the activity was localized in sub-fraction 4 A. Upon further electrophoresis of pool 4 A at pH 1.9. (see Fig. 3.) a further separation into two distinct peaks (4A:I & 4A:II) was obtained. The larger of the two peaks (peak 4A:I) was totally inactive while peak 4A:II contained all of the applied activity. Homogenity test of sub-fraction 4A:II by HPLC indicated the presence of only 1 component.

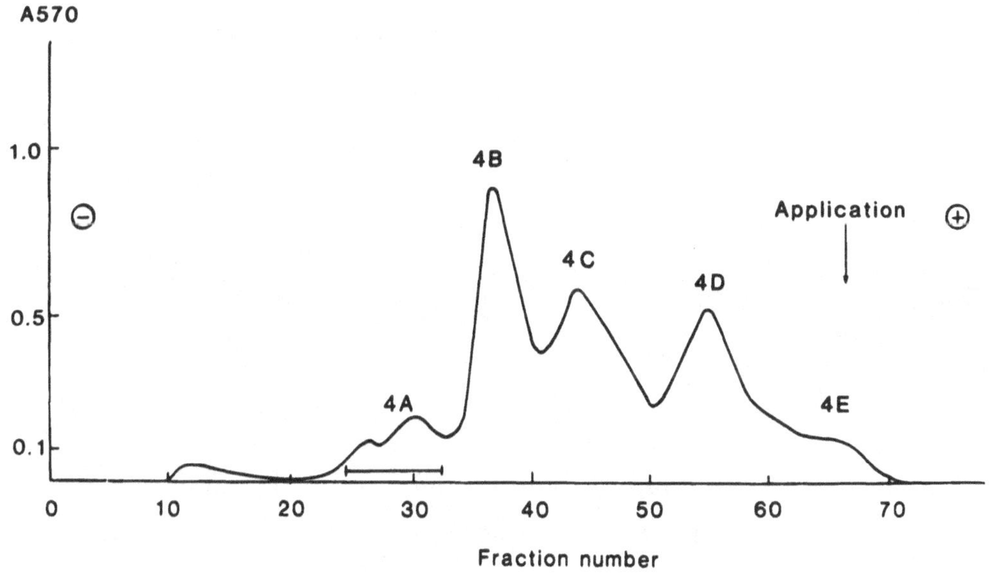

FIG. 2. Column zone electrophoresis of fraction 4 (from the run shown in Fig. 1). Column dimension ÷ 1 x 90 cm (Vo = 61 ml). Electrode buffer: 0.05 M pyridine/0.05 M acetic acid, pH 5.0. Duration of run: 20 h at 1000 V and 6 mA. The run was made at + 8^0 C.

Amino acid analysis showed that 4A:II is composed of 14 amino acid residues. On the basis of the known sequence of the fibrinogen molecule, and the amino acid composition of sub-fraction 4A:II, it was deduced that this peptide corresponds to residues 30-43 of the Bβ -chain with the sequence Arg-Pro-Ala-Pro-Pro-Pro-Ile-Ser-Gly-Gly-Gly-Tyr-Arg-Ala.

Physiological activity

The increase in vasular permeability caused by the purified Bβ 30-43 peptide and by some other known vasoactive peptides, which are structurally similar to this peptide, is shown in Fig. 4. The Bβ-chain 30-43 peptide was at least 10-15 times more potent than peptide 6A (see ref. 1) and about half as active as bradykinin. With respect to its effect on inducing the release of histamine from mast cells the Bβ 30-43 peptide was about 50 % and 15 % as effective respectively as bradykinin and substance P (Fig. 5).

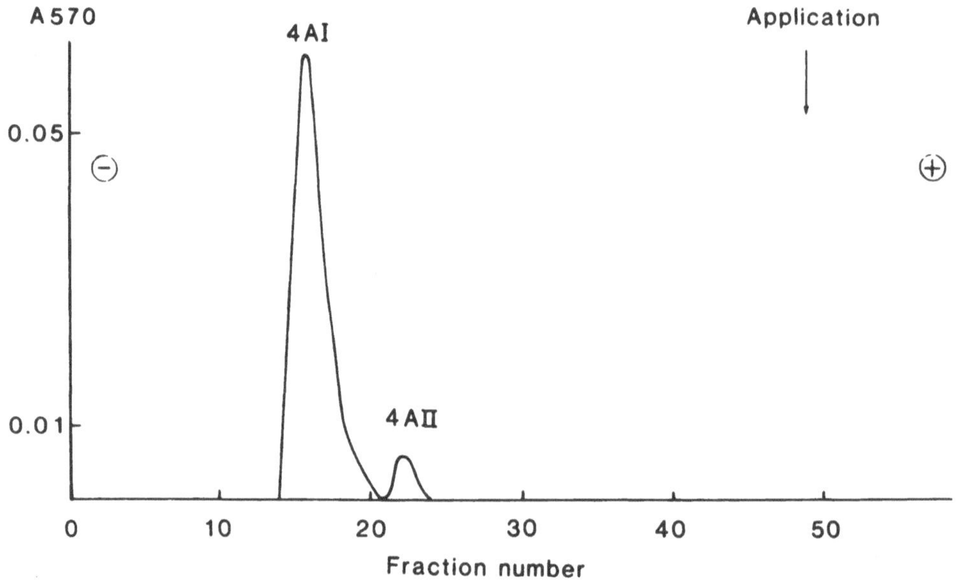

FIG. 3. Further purification of subfraction 4A (from the run shown in FIG. 2.) by column zone electrophoresis at pH 1.9. Column dimensions: 1 x 95 cm (Vo = 69 ml). Electrode buffer: 0.54 M formic acid/1.30 M acetic acid. pH 1.9. Duration of run: 20 h at 1000 V and 12 mA. The run was made at + 8^{0} C.

The effect of an antihistamine drug, the H_1-blocker mepyramine maleate, on the vasoactivity of the above 3 peptides was also investigated. The result (see Fig. 6 and Table 1) indicate that injection of mepyramine maleate, at a concentration of 1μg/spot, and each of the peptides simultaneously into the skin significantly inhibited the permeability increase caused by the Bβ 30-43 peptide. bradykinin and substance P. The results in Table I further indicate that a carboxypeptidase inhibitor potentiated the microvascular permeability of the Bβ 30-43 peptide while indomethacin reduced its vasoactivity. We have also observed that addition of indomethacin (final concentration 200 μM) to the mast cells prior to incubation with the Bβ 30-43 did not influence the release of histamine. However, under the same conditions, the activity of substance P was significantly reduced.

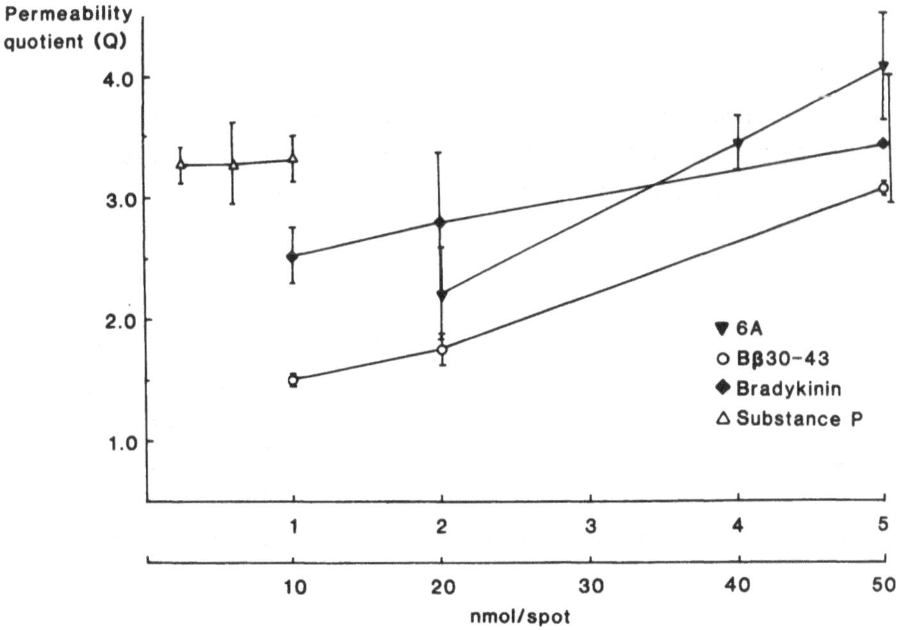

FIG. 4. Dose-response curves for the effect of four vasoactive peptides on microvascular permeability in rat skin after intracutaneous injection, (The expanded scale on the abscissa, 1-50 nmol/spot, is applicable only for peptide 6A.

DISCUSSION

Accumulation of fibrin in various organs, e.g. the lungs, is sometimes associated with increased vascular permeability and the development of oedema. It has been suggested that local release of fibrin degradation products might contribute to the increase in vascular permeability (2). We have previously isolated two peptides (viz. 6A and 6D, see ref. 2) released from fibrin(ogen) upon degradation with plasmin. Both peptides significantly increase vascular permeability.

The existence of leucocytes around fibrin deposits and the significantly high amounts of leucocyte elastases in the lungs of patients with pulmonary oedema suggested that the degradation of fibrin(ogen) by such elastases might lead to the production of lmw-fdp with vasoactive properties. This investigation showed that this was indeed the case och led to the isolation of one peptide with especially high permeability-increasing effect. It was at least 10-15 times more potent than the previously isolated peptides 6A and 6D (1) derived from plasmin degradation of fibrin(ogen).

As concerns the mechanisms underlying such permeability-increasing effects, the ability to induce the release of histamine from mast cells appears to be a major causative factor. It also appears that basic amino acids (10) proline (8) and aromatic amino acids situated in the vicinity of the C-terminal end (11) of such peptides seem to have a functional role for inducing histamine release from mast cells. It is relevant to point out that the Bβ 30-43 peptide, substance P and bradykinin have the sequence Arg-Pro- at their amino terminal end. According to Fewtrell et al. (11) Arg-Pro at the N-terminal seems to play an important role for inducing the release of histamine. In view of identical sequence at the N-terminal of these three peptides, it is reasonable to conclude that Arg at the N-terminus of the Bβ 30-43 peptide also plays an important role for its function in this respect.

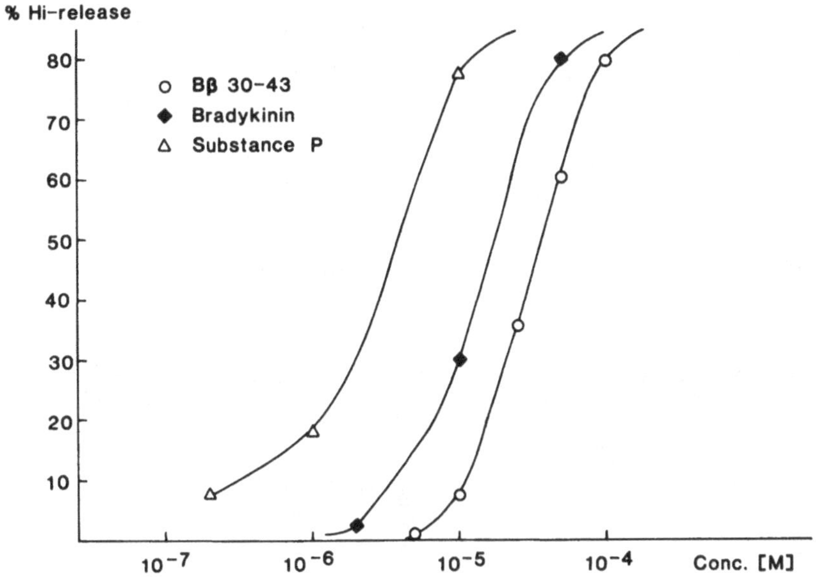

FIG. 5. Release of histamine (% Hi-release) from mast cells induced by three vasoactive peptides.

Indomethacin significantly decreased the permeability increasing effect of the Bβ 30-43 peptide but not its ability to induce histamine release from mast cells. This might suggest that in addition a product of arachidonic acid is probably involved in the vasoactivity of this peptide. However, the histamine release due to substance P was decreased by indomethacin which is in agreement with earlier observations of Mazurek et al. (12). The finding that the carboxypeptidase inhibitor potentiated the effect of Bβ 30-43 on micro-vascular permeability suggests that this peptide is to a certain degree inactivated by carboxypeptidase in vivo.

It is well conceivable that high local concentrations of this peptide occur during extravascular degradation of fibrin or fibrinogen in various organs, e.g. in the lungs, and might contribute to the enhanced vascular permeability.

Peptide	Injected amount (nmoles)	Permeability quotient (Q)			
		Peptide alone	Peptide + mepyramine	Peptide + carboxypeptidase inhibitor	Peptide + indo
Bβ 30-43	5	3.2 ± 0.2	1.6 ± 0.5^{x}	3.7 ± 0.4^{x1}	2.3 ± 0.4^{x}
6A	40	4.1 ± 0.4	1.2 ± 0.1^{x}	3.9 ± 0.4^{x2}	2.9 ± 0.3^{x}
Bradykinin	1	2.6 ± 0.2	1.7 ± 0.2^{x}	nd	2.0 ± 0.2^{x}
Substance P	1	3.4 ± 0.3	2.4 ± 0.3^{x}	nd	nd

Table I. Effect of injection of mepyramine maleate (mepyramine) (1μg/spot), a carboxypeptidase inhibitor, 2-mercaptomethyl-3-guanidino-ethylthiopropanoic acid (500 nmol/spot), and indomethacin (indo) (4 mg/kg intravenously), on microvascular permeability in rat skin after intracutaneous injection of different peptides. X = significantly different from peptide alone. nd = not determined. 1) Injection of 2 nmol/spot. Peptide alone $Q = 2.7 \pm 0.3$ 2) Injection of 20 nmol/spot. Peptide alone $Q = 2.7 \pm 0.2$.

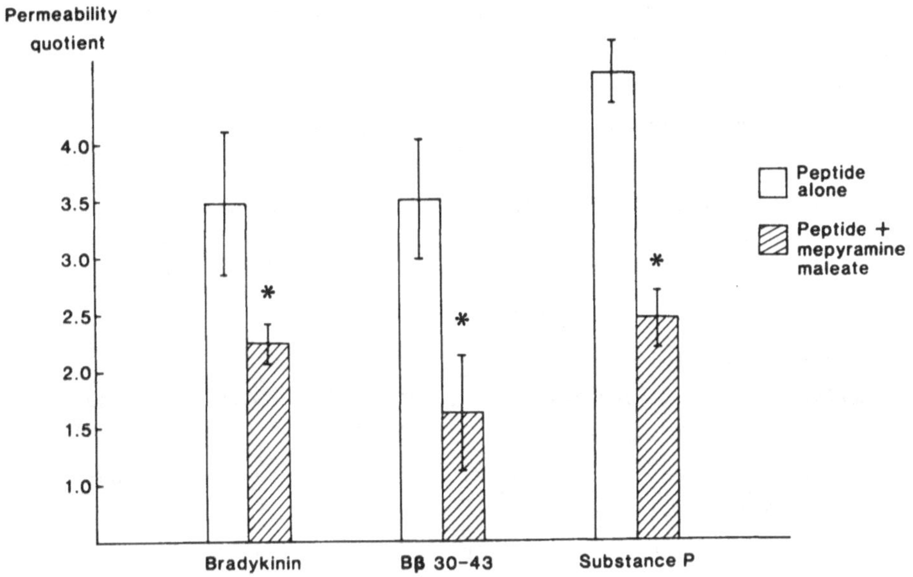

FIG. 6. The influence of 1 μg of mepyramine maleate per spot on the permeability-increasing effects of peptides in rat skin after intracutaneous injection, Injected amounts: Bradykinin 2 nmol/spot, Bβ 30-43 5 nmol/spot and substance P 1 nmol/spot.

ACKNOWLEDGEMENTS

This research was supported by the Swedish Medical Research Council and the Torsten and Ragnar Söderberg Foundation and the Fortia Fund, Uppsala University.

REFERENCES

1. Belew, M., Gerdin, B., Porath, J. and Saldeen, T.: Thrombos. Res. 13: 983-994 (1978).
2. Saldeen. T.: Microvasc. Res. 11, 227-259 (1976).
3. Belew, M., Gerdin, B., Lindeberg, G., Porath, J., Saldeen, T. and Wallin, R., Biochim. Biophys. Acta 621, 169-178 (1980).
4. Ohlsson, K. and Olsson, I.: Eur. J. Biochem. 42, 519-527 (1974).

5. Udaka, K., Takeuchi, U. and Movat, H.C.: Proc. Soc. Exp. Biol. Med. 133, 1384-1387 (1973).
6. Gerdin, B., and Saldeen, T.: Thrombos. Res. 13, 995-1006 (1978)
7. Eriksson, M., Saldeen, K., Saldeen, T., Strandberg, K. and Wallin, R.: Int. J. Microcirc: Clin. Exp. 2, 337-345 (1983).
8. Wallin, R., Saldeen, K., Saldeen, T.: In this volume.
9. Hirs. C.H.W., Moore, S. and Stein. W.H.: J. Biol. Chem. 219, 623-642, (1956).
10. Johnsson, A.L. and Erdös, E.G.: Proc. Soc. Exp. Biol. Med. 142, 1252-1256 (1973).
11. Fewtrell, C.M.S., Foreman, J.C., Jordan. C.C,. Oehme, P., Renner, H. and Stewart. J.M.: J. Physiol. 330, 393-411 (1982).
12. Mazurek, N., Pecht. I., Teichberg, V.I., and Blumberg, S.: Neuropharmacology, 20: 1025-1027 (1981).
13. Deutsch, D.G., Mertz, E.T.: Science 170, 1095-1096 (1970).

RELATION OF CROSSLINKED TO NON CROSSLINKED FIBRIN DERIVATIVES IN TUMOR ASCITES COMPARED TO CIRRHOSIS ASCITES

Reimar Hafter, Werner Klaubert+, Rotraud Gollwitzer++,
Rene von Hugo, Henner Graeff
Frauenklinik der Technischen Universität München, +Medizinische Klinik III der Ludwig-Maximilians-Universität München-Grosshadern, ++Max-Planck-Institut für Biochemie, Martinsried bei München, Federal Republic of Germany

Introduction

The differentiation of effusions, whether of malignant or non malignant genesis, is of great clinical interest. Despite the availability of several well established methods, like cytological evaluation, detection of carcinoembryonic and non-carcinoembryonic tumor markers, there is still a need for further differential information. Astedt and coworkers (1) have emphasized as early as 1972 the high specifity of fibrin degradation products (FDP) for ascitic fluid in malignancy. The FDP reflect the combined enzymatic action on the fibrinogen molecule. Tumor cells produce in vivo as well as in cell cultures procoagulant and fibrinolytic activity (2). Part of the FDP may arise also from crosslinked fibrin network deposited around the tumor surface. As we have shown earlier (3,4,5) a considerable portion of FDP is comprised of crosslinked high molecular weight fibrin derivatives with molecular weights up to 500 K and more.

This study reports the investigation of ascitic fluid from patients with ovarian cancer and with alcohol induced liver cirrhosis in regard to their levels of crosslinked high molecular weight fibrin derivatives, fibrinogen, fibrin(ogen) degradation products, fibrinopeptide A, F-CB3 related antigen (α chain fragment Aα 241-474), fibronectin and total protein.

Fibrinogen - Structural Variants and Interactions
© 1985 Walter de Gruyter & Co., Berlin · New York - Printed in Germany

Furthermore, fluids of benign ovarian cysts were compared. The aim of the study was to evaluate whether crosslinked fibrin derivatives or one of the other parameters might be suitable as tumor marker for effusions.

Materials and Methods

Twenty-one samples of ascitic fluid from patients with ovarian cancer, nine samples of ascitic fluid from patients with alcohol induced liver cirrhosis, and eight samples of fluids from benign ovarian cysts were investigated.

All specimens were aspirated by percutaneous puncture or during laparotomy. An anticoagulant solution containing 0.13 mol/l sodium citrate, 0.06 mol/l tris, 0.05 mol/l EDTA (disodium salt) and 2000 KIU/ml aprotinin (Trasylol, Bayer) adjusted to pH 7.4, was added (1:10). The samples were then immediately centrifuged.

Clottable fibrinogen was measured according to Hörmann and Gollwitzer (6). Fibrin related antigen (FRA) was measured by radial immunodiffusion on M-Partigen plates (Behringwerke) with fibrinogen standards. Since FRA is comprised of different fragments (fibrinogen, low and high molecular weight FDP) in varying amounts and diffusion behaviors, only a crude overall value can thus be obtained.

F-CB3 related antigen was measured radioimmunologically (7,8). Labelling was performed by the lactate peroxidase method (9). A modified binding assay was applied and calculation performed by use of a computerized spline function program. Normal range in plasma is up to 20 pmol/ml.

Fibrinopeptide A (FPA) was measured radioimmunologically using the bentonite method (Byk Mallinckrodt). In our laboratory procedure the normal range in plasma is less than 10 ng/ml (=0.01 µg/ml). Determination of the ratio of crosslinked to non crosslinked fibrin (and fibrinogen) was performed by heat precipitation with 0.5 ml samples to which was added 4.5 ml 0.05 M phosphate buffer, pH 5. The samples were heated for 10 min at 60° C and the precipitate harvested by centrifugation, washed twice with saline and incubated with 0.3 ml 8 M urea and 0.05 ml beta-mercaptoethanol at 60° C until complete dissolution occurred (approx. 2h). The samples were subjected to SDS-PAGE and the banding pattern densitometrically scanned. The areas under the peaks of the dimeric γ-chains were set in relation to the areas under the monomeric γ-chain (see Fig.1).

Fibronectin was measured by laser nephelometry (10). Normal range in plasma is 0.25 - 0.3 mg/ml. Total protein was measured using biuret reagent. Normal range in plasma is 60 - 80 mg/ml.

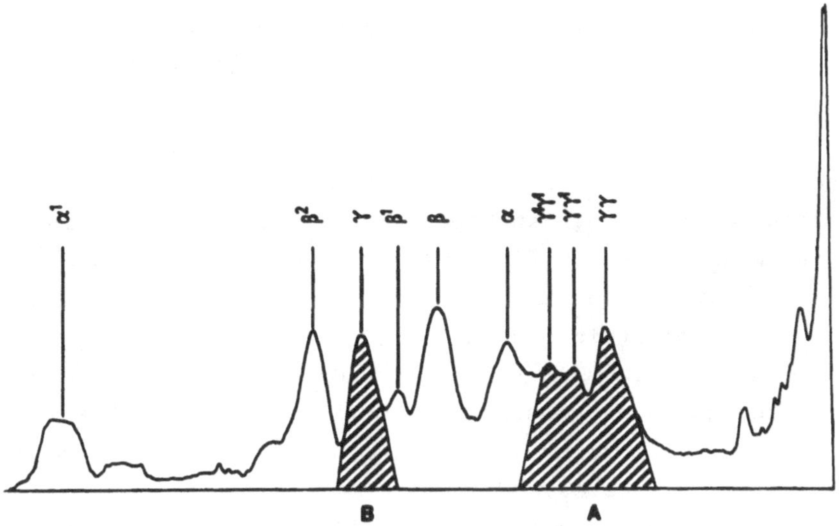

Fig. 1

Densitometric scan of subunit polypeptide chains from fibrin derivatives (and fibrinogen) of a tumor ascites sample following heat precipitation and SDS-PAGE (7.5 % gel). The ratio of dimeric γ-chains to monomeric γ-chains (A:B) as an expression of crosslinked to non crosslinked fibrin and fibrinogen was 2.4 in this sample

Statistical evaluation: The non-parametrical U-Test of Mann-Whitney for unpaired samples was applied.

Results

Table I presents the mean values and respective ranges in tumor ascites compared to cirrhosis ascites and to the fluid content of benign ovarian cysts. The levels of clottable fibrinogen, fibrinopeptide A and total protein were found to be approximately twice as high in tumor ascites as in cirrhosis ascites. The mean levels of FRA (fibrinogen and FDP) and F-CB3 related antigen are five times higher than in cirrhosis ascites.

The mean ratio of crosslinked fibrin to non crosslinked fibrin and fibrinogen in tumor ascites was 1.63 ± 0.97 compared to 0.64 ± 0.32 in cirrhosis ascites. From these data and the different levels of FRA in tumor and cirrhosis ascites it can be estimated that the content of crosslinked fibrin in tumor ascites is about ten times higher than in cirrhosis ascites (means 1.66 ± 0.41 mg/ml versus 0.16 ± 0.25 mg/ml). It must be emphasized that these values are only approximations due to uncertainties in the FRA determination (see Methods) and problems in the estimation of the ratios between crosslinked and non crosslinked fibrin derivatives. The same holds true for the value ranges, which extend from 1.1 to 2.7 mg/ml for tumor ascites and 0.04 to 0.9 mg/ml for cirrhosis ascites. In any case these approximations seem to show that there is no overlap in the values of crosslinked fibrin between tumor and cirrhosis ascites.

Highly significant were the differences in fibronectin content. The levels were six times higher in tumor ascites (0.20 ± 0.07 mg/ml) compared to cirrhosis ascites (0.03 ± 0.02 mg/ml). The single values of each group were in different ranges.

Table 1

Fibrinogen, Fibrin Derivatives and Fibronectin in Effusions (mean ± S.D.)

		Ascitic Fluid Malign. Tumor n = 21 (1)	Ascitic Fluid Liver Cirrhosis n = 9 (2)	Significances p 1 vs. 2	Fluid of Benign Ovarian Cysts n = 8
Fibrinogen (clotable)	mg/ml	0.22 ± 0.15	0.12 ± 0.09	< 0.05	Ø
	range	0.1 - 0.6	0.01 - 0.3		
	nmol/ml	0.65 ± 0.44			
FRA	mg/ml	2.68 ± 0.67	0.41 ± 0.65	< 0.01	0.38 ± 0.38
	range	1.82 - 4.40	0.1 - 2.26		0.15 - 1.10
F-CB3	µg/ml	60 ± 40	13 ± 7	< 0.01	0.83 ± 0.89
	nmol/ml	2.47 ± 1.77	0.57 ± 0.32		0.04 ± 0.04
	range	0.51 - 6.9	0.21 - 1.0		
FPA	µg/ml	0.39 ± 0.19	0.22 ± 0.16	n.s.	0.04 ± 0.01
	range	0.10 - 0.78	0.06 - 0.52		
	nmol/ml	0.26 ± 0.13	0.15 ± 0.1		0.03 ± 0.01
Ratio of Crosslinked to Non Crosslinked Fibrin(ogen)		1.63 ± 0.97	0.64 ± 0.32	< 0.05	Ø
Fibronectin	mg/ml	0.20 ± 0.07	0.03 ± 0.02	< 0.01	0.04 ± 0.01
	range	0.07 - 0.37	0.002 - 0.055		
	nmol/ml	0.45 ± 0.17	0.07 ± 0.06		0.1 ± 0.025
Total Protein	mg/ml	48 ± 9	28 ± 10	< 0.01	32 ± 23
	range	31 - 70	18 - 52		20 - 54

In fluids of benign ovarian cysts, no clottable fibrinogen was found, although in six of eight samples a certain amount of FRA, and in all samples up to 0.8 μg/ml F-CB3 related antigen and FPA up to 0.04 μg/ml could be detected. The total protein content ranged between 2.0 and 5.4 mg/ml, its mean value was as high as that for cirrhosis ascites. Crosslinked fibrin could not be demonstrated in any one of the cyst punctures.

Discussion

Ascites formation with ovarian cancer has generally been attributed to obstruction of the lymphatic drainage system by tumor cells. There is sufficient experimental evidence that fibrin and platelets interact with cancer cells _in vitro_ and that hemostatic changes occur during tumor cell growth and dissemination _in vivo_ (11).

In tumor ascites the high ratio of FRA versus clottable fibrinogen strongly indicates an enhanced turnover of fibrinogen in this fluid. This is in accordance with observations of Svanberg and Astedt (2) who demonstrated high amounts of FDP in tumor ascites compared to significantly smaller amounts in effusions of patients with liver cirrhosis.

The high fibrinolytic activity in tumor ascites compared to cirrhosis ascites is also well documented by the five times higher levels of F-CB3 related antigen. By this assay the very early split products of the α-chain of fibrin(ogen) are measured (7,8).

The FPA levels in both ascitic fluids are significantly higher than in normal plasma. In tumor ascites the level of FPA was found to be about twice as high as in cirrhosis ascites. The high molar relation of F-CB3 to FPA could suggest that some portion of the FDP might result from fibrinogenolysis rather than from fibrinolysis. However, the ratio of crosslinked fibrin to non crosslinked fibrin(ogen) in tumor ascites compared to cirrhosis ascites favours the assumption of fibrinolysis at least in tumor ascites. Tumor ascites contains approximately ten times higher levels of crosslinked fibrin derivatives than cirrhosis ascites.

Fragments DY and DD could be demonstrated (4,5) as prominent crosslinked derivatives, while the nature of others could be elicited as crosslinked X-Oligomers (3,4,5). Their structural compositions could be deduced from molecular weight determinations to be likely XY, DXD, DXY, YXY and DXX (4,12). Similar banding patterns could also be demonstrated from ascitic fluid of patients with liver cirrhosis (4,13). The crosslinked fibrin derivatives originate from the combined action of coagulative and fibrinolytic activity. This concept was sustained by in vitro experiments (4). Under simultaneous action of thrombin, factor XIIIa and plasmin on fibrinogen in presence of calcium electrophoretical patterns comparable to those from tumor ascites, but even richer in banding pattern, could be demonstrated. Experiments with plasma samples (13) resulted in an even stronger correlation in banding pattern of fibrin derivatives and hypothesize a similar genesis as in ascitic fluid.

In addition, it must be noted that enhanced crosslinking could also result from release of tissue transaminase from tumor cells (14). Hence the marker function of crosslinked fibrin derivatives in ascitic fluid could be associated with a tumor generated component. This might indicate a tumor specific pathophysiological mechanism in the development of tumor induced ascites, different from the mechanism of cirrhotic ascites development.

Furthermore, significantly higher levels of fibronectin were found in tumor ascites compared with cirrhosis ascites. These findings are in accordance with similar results reported recently (15). The fibronectin content observed in tumor ascites ranges fourty percent lower than in plasma of tumor patients and/or of healthy subjects (16). It seems reasonable that fibronectin content in ascites results mainly from plasma transudation. On the other hand, fibronectin has a high affinity to fibrin (17), hence there may be a correlation with the high levels of fibrin derivatives in tumor ascites. In addition, it is known that tumor cells have a high turnover of surface membrane proteins (18). One can speculate that an enhanced release of fibronectin from the surface of cancer cells might also account for the high levels of fibronectin in tumor ascites. In cirrhosis ascites the low level of fibronectin corresponds to reduced plasma levels in patients with severe liver cirrhosis with presence of ascites (19). Since no overlap could be observed between the individual values of tumor and cirrhosis ascites, fibronectin determination might also be of diagnostic help to differentiate between malignant and non malignant effusions.

References

1. Astedt, B., Svanberg, L., Nilsson, I.M.: Lancet II, 1312 (1972).
2. Svanberg, L., Astedt, B.: Cancer 35, 1382-1387 (1975).
3. Hafter, R., Graeff, H.: In: Fibrinogen - Recent Biochemical and Medical Aspects, A. Henschen, H. Graeff, F. Lottspeich (eds) Walter de Gruyter, Berlin-New York, 291-305 (1982).
4. Hafter, R., Klaubert, E., Graeff, H.: In: Fibrinogen - Structure, Functional Aspects, Metabolism. F. Haverkate, A. Henschen, W. Nieuwenhuizen, P.W. Straub (eds) Walter de Gruyter, Berlin-New York, 215-225 (1983).
5. Graeff, H., Hafter, R.: Seminars Thromb. Hemost. 8, 57-68 (1982).
6. Hörmann, H., Gollwitzer, R.: Hoppe Seyler's Z. Physiol. Chem. 346, 21-41 (1966).
7. Gollwitzer, R., Timpl, R.: FEBS Lett. 58, 269-272 (1975).
8. Gollwitzer, R., Hafter, R., Timpl, R., Graeff, H.: Thromb. Res. 11, 859.868 (1977).
9. David, G.S., Reisfeld, R.A.: Biochem. 13, 1014-1021 (1974)

10. Pott, G., Meyering, M., Voss, B., Karges, H.E., Sieber, A.: J. Clin. Chem. Clin. Biochem. 18, 893-895 (1980).

11. Donati, M.B., Mussoni, L., Poggi, A., de Gaetano, G., Garattini, S.: Europ. J. Cancer 14, 343-347 (1978).

12. Francis, C.W., Marder, V.J.: Seminars Thromb. Hemost. 8, 25-35 (1982).

13. Hafter, R., Klaubert, W., Gollwitzer, R., von Hugo, R., Graeff, H.: Thromb. Res. 35, 53-64 (1984).

14. Laki, K.: In: Onkohämostaseologie, H. Gastpar (ed), F.K. Schattauer, Stuttgart, 61-71 (1976).

15. Schölmerich, J., Volk, B.A., Köttgen, E., Ehlers, S., Gerok, W.: Z. Gastroent. 20, 517-518 (1982).

16. Stathakis, N.E., Foruntas, A., Trianos, E.: J. Clin. Pathol. 34, 504-508 (1981).

17. Ruoslathi, E., Vaheri, A.: J. Exp. Med. 141, 497-501 (1975).

18. Black, P.H.: Advances Cancer Res. 32, 75-199 (1980).

19. Matsuda, M., Yamanaka, T., Matsuda, A.: Clin. chim. Acta 118, 191-199 (1982).

VIII. INTERACTION WITH PLASMINOGEN AND ITS ACTIVATOR

STUDY OF THE INTERACTION BETWEEN PLASMINOGEN AND FIBRINOGEN DEGRADATION PRODUCTS USING IMMUNOENZYMOLOGICAL ASSAY

Jeannette SORIA[1-2], Claudine SORIA[3], Fred DUNN[3], Katsumi DEGUCHI[3], Massoud MIRSHAHI[1], Roger LIJNEN[4], Wilhem NIEUWENHUIZEN[5], Fritz HAVERKATE[5], Meyer SAMAMA[2], Jacques CAEN[3]

1. Service de Médecine Interne et Oncologie (Pr. Bernadou), Hôtel-Dieu, Paris, France.
2. Laboratoires Centraux de Biochimie (Pr. Fabiani) et Hématologie (Pr. Samama), Hôtel-Dieu, Paris, France.
3. Laboratoires Centraux d'Hématologie (Pr. Caen), et de Biochimie (Pr. Rousselet), Hôpital Lariboisière, Paris, France.
4. Department Medical Research, Campus Gasthisberg, Katholieke Universiteit Leuven, Leuven, Belgium.
5. Health Research Division, TNO, Gaubius Institute, Leiden, The Netherlands.

SUMMARY

The interaction of plasminogen with fibrin(ogen) plays an important role in the regulation of the fibrinolytic system. It is well known that the lysine binding sites in the plasminogen A chain are mainly responsible for the interaction with fibrinogen, whereas the complementary sites in fibrin(ogen) are not yet characterized. The aim of this study is to determine which parts of fibrinogen interact with the lysine binding sites of plasminogen.

This work carried out using two different immunoenzymological assays :

1. Binding of either Glu- Lys- or Mini- plasminogen to immobilized fibrinogen, fragment D Cate, fragment D EGTA and fragment E.

The interaction between plasminogen and fibrinogen degradation

Fibrinogen - Structural Variants and Interactions
© 1985 Walter de Gruyter & Co., Berlin · New York - Printed in Germany

products evidenced by this technique is related to the presence of the lysine binding sites on plasminogen because it is found that lys-plasminogen bound more to fragments D and E than Glu-plasminogen, whereas Mini-plasminogen did not bound at all to these fragments.
2. Competitive inhibition binding of plasminogen to immobilized fragments D or E by increasing amounts of fragments D and E in solution.
Our results suggest that the fibrinogen molecule contains several complementary sites to lysine binding sites of plasminogen, located both in D and E domains of fibrinogen. Due to the fact that D and E in solution inhibit plasminogen binding to either fragment D or fragment E coated-polystyrene, it is proposed that the complementary sites of plasminogen in D and E fragments are similar.

INTRODUCTION

In the last few years, the interaction between plasminogen and fibrin has been extensively studied and it has been clearly demonstrated that the interaction plays an important role in the regulation of the fibrinolytic system.
It has been shown that the lysine binding sites in plasminogen are involved in the interaction with fibrin, whereas the complementary sites in fibrin (ogen) have not yet been described (1).The aim of the present study was to investigate which part of fibrinogen interacts with the lysine binding sites of plasminogen. This work was performed using an immunoenzymological assay.
Two types of reaction were used :
1.Direct binding of Glu- Lys- or Mini- plasminogen to fragment D or fragment E-coated-polystyrene.
The amount of plasminogen which bound to the immobilized fragments was determined by immunoenzymological assay using antiplasminogen.
2.Competitive inhibition binding : serial dilutions of fragments D or E

were incubated with plasminogen, and then the mixture was assayed for residual binding on fragment D or fragment E-coated plates.

MATERIAL AND METHODS

96 well polystyrene microtiter plates for ELISA (Dynatech Laboratory, Paris).

Fibrinogen, human Glu- and Lys-plasminogen were from Kabi (Stockholm).

Mini-plasminogen was obtained by elastase digestion of plasminogen and separation of the fragments by gel filtration on Sephadex G 75 and affinity chromatography on lysine Sepharose as described by Sottrup-Jensen et al. (2).

Fragment E was isolated according to the technique of Nilhen (3).

Fragment D_1 and D_3 were prepared according to (4, 5, 6)

Buffer and reagent used for immunoenzymatic assay :

- coating-buffer : 0.1 M phosphate buffer, pH 7.4,
- PBS Tween-buffer, pH 7.6 was made up with 8.0 g. of NaCl, 0.2 g of KH_2PO_4, 2.9 g. of Na_2HPO_4, 12 H_2O, 0.5 ml tween 20 in one liter of distilled water, then 1 g. of bovine albumin was added to it.

Reagent for the determination of peroxidase activity : 40 mg of Orthophenylene diamine (OPD) were dissolved in 100 ml, pH 5 buffer consisting 0.05 M citric acid an 0.1 M Na_2HPO_4, then 40 µl of 30 % perhydrol were added. (7)

Antisera used were prepared in the rabbit (anti-D and anti-E, Stago Laboratories), antiplasminogen (Behring Laboratories).

Peroxidase-labeled sheep immunoglobulin anti-rabbit Ig was purchased from Cappel.

<u>Technique used to determine the interaction between various kinds of plasminogen and fragment D or fragment E.</u>

Two types of reaction were carried out :

1. <u>Direct-plasminogen binding on fragment D or fragment E coated-polystyrene</u> (Figure 1).

Experiments were performed as follows :

a) - Preparation of fibrinogen (or fragment D or fragment E) coated plates :

- Coating-procedure :

96 wells polystyrene microplates (Dynatech-Microelisa) were coated by adding to each well 100 µl of fibrinogen or fibrinogen fragments at 0.5 and 1 µg/ml in the coating-buffer, and incubated overnight at 37°C.
In order to determine the quantity of protein bound to polystyrene wells, after incubation the solution of each well was collected and kept for analysis.
Then the plates were washed four times with 0.15 M NaCl containing 0.05 % tween 20. This step was automated using a Dynatech automatic washer.

- Determination of the amount of fibrinogen (or fragment D or fragment E) bound to polystyrene :

The difference between the concentration of fibrinogen (or fragment D or fragment E) added and the concentration of these products remaining in the supernatant after coating, represents the quantity of products which is bound to polystyrene.
The quantity of non-absorbed fibrinogen (fragment D or fragment E) present in the supernatant of each well was determined by Elisa procedure (8)
For this assay immunoglobulin from specific antisera (antifibrinogen, anti-D and anti-E purchased from Stago) were isolated the technique of Fine et al. (9). Peroxidase-labelling of immunoglobulins was performed by the two step technique of Avrameas. (10)

b) - Plasminogen binding to coated-plates :

The plasminogens were disolved in PBS containing 0.05 % tween 20 in order to avoid nonspecific adsorption. 100 µl of plasminogen solution containing either 4, 2, 1 and 0.5 µg/ml of glu- Lys or Mini-plasminogen were added to each well.
After two hours incubation at 37 °C, the plates were washed four times with 0.15 M NaCl containing 0.05 % tween 20.

c) - Determination of the amount of plasminogen bound to either fibrinogen, fragment D or fragment E coated-wells :

It was carried out by a two step immunoenzymatic procedure :

- In a first step, antiplasminogen bound to immobilized plasminogen, and in a second step, the antiplasminogen bound was revealed using a peroxidase-labeled anti-rabbit Ig. This analysis was performed as follows :

- to each well were added 100 µl of antiplasminogen diluted 1/1,000 in PBS, 0.05 % tween 20. After two hours incubation at 37°C, the plates were washed four times with 0.15 M NaCl containing 0.05 % tween 20.

- then, 100 µl of peroxidase-labeled anti-rabbit IgG diluted 1/1,000 in the same buffer were added. After two hours incubation at 37°C, the plates were washed four times and peroxidase-labeled anti-IgG bound to the wells was determined by adding into each well 200 µl of orthophenylene diamine (OPD) solution : the colour development was stopped after exactly 90 or 180 seconds by adding 50 µl of 2.5 N sulfuric acid. The absorbance was measured at 492 nm using an auto-reader (Dynatech). In order to determine the amount of plasminogen bound to fibrinogen, fragment D or fragment E, a dose-response curve was established by plotting the bound immuno-reactive enzyme (OD at 492 nm) versus increasing amounts of plasminogen added to fibrinogen, fragment D or fragment E-coated wells.

2. <u>Competitive inhibition binding</u>

The inhibition of plasminogen binding to fragment D or fragment E-coated plates by increasing concentrations of D and E in the mixture was tested. The analysis was performed as shown above except that plasminogen was previously incubated for 18 hours at 4°C with increasing amounts of fragments D or E.

RESULTS

When the plates were coated with 0.5 and 1 µg/ml fragments D or E, the total amount was bound to polystyrene because after incubation no fragment D nor E was found by Elisa procedure in the supernatant of coating.

1. Interaction between Glu- Lys- or Mini-plasminogen and fibrinogen degradation products :

Using buffer containing tween 20, the ligand was not interacting directly with plastic, since in controls carried out with uncoated plates, there was no binding of plasminogen to polystyrene.
The dose-response curves of lys-plasminogen bound to fragment D Cate, D_{EGTA} or E-coated plates, versus increasing concentrations of lys-plasminogen added were shown in Fig. 2. It appeared that lys-plasminogen interacted with fragment E, and also with fragment D_{EGTA} and with fragment D Cate. As it can be seen in Fig. 3, mini-plasminogen did not bind either to fragment D or to fragment E, whereas lys-plasminogen interacts with D and E fragments more than glu-plasminogen does.

2. Competitive inhibition binding :

Both fragment D ans fragment E prevented the binding of lys-plasminogen to either fragment D or E-coated wells. The slope of inhibition obtained with increasing concentrations of fragment D or fragment E was very similar wheter using fragment D or fragment E coated plates (Fig. 4a and 4b).

DISCUSSION

In order to study the part of fibrinogen which interacts with plasminogen, we have used an immunoenzymological assay to analyse the interaction between plasminogen and fibrinogen degradation products (D Cate, D_{EGTA} or E). Because of the absence of binding of mini-plasminogen (which is devoid of lysine binding sites) to fibrinogen or fragment D or fragment E-coated plates, we can assume that only interaction between lysine binding sites in plasminogen and complementary sites in fibrinogen are evidenced by this assay. From the data presented in Fig. 2 and 3 (direct binding of plasminogen to fragment D and fragment E-coated plates) it appeared that both purified fragment D and fragment E interact with plasminogen.

Fibrinogen degradation products interact more with lys-plasminogen than with glu-plasminogen. These results are in accordance with the hypothesis that the NH_2 terminal part in glu-plasminogen masks partially a lysine binding site in glu-plasminogen as already mentioned by Thorsen (11) and by Wiman and Wallen (12).

It appeared that the interaction of plasminogen with fragment E is similar to that observed with fragment D when considering equimolecular amount of D and E-coated plates. Data with fragment E were analysed on the basis of two equivalent binding sites per molecule. In addition the plasminogen binding to fragment D Cate (D_1) and D_{EGTA} (D_3) was compared. D Cate was obtained by fibrinogen digestion in the presence of calcium and it contained an intact carboxy-terminal end of the γ chain. Fragment D_{EGTA} was prepared by degradation of fibrinogen in the presence of EGTA. In comparison with D Cate, D_{EGTA} lacks a 13,000 MW piece at the carboxy-terminal end of its γ chain. (4, 5, 6).

It appeared (Fig. 2 and 3) that the difference between plasminogen binding to D_{EGTA} (D_3) and that to D Cate (D_1) was not significant.

According to the fact that a conformational change may occur during immobilization of fibrinogen degradation products on polystyrene, inducing a modification of the binding site for plasminogen, a control was carried out using a competitive inhibition binding assay. In the test, the inhibitory effect of the binding of plasminogen to both D and E-coated plates was studied by adding to plasminogen increasing amounts of either fragment D or fragment E in solution : the results shown in Fig. 4 revealed that both D and E fragments in solution inhibited the interaction of plasminogen to either D or E-coated plates. It appeared that no significant difference existed between the inhibitory effect induced by D and E fragments when taking into account their molecular concentration.

In conclusion, our results suggest that the fibrinogen molecule contains several complementary sites to the lysine binding sites of plasminogen, located both in D and E fragments. These complementary sites in D and E domains seem to be very similar because both D and E fragments in solution inhibit plasminogen binding to either fragment E or fragment E-coated plates.

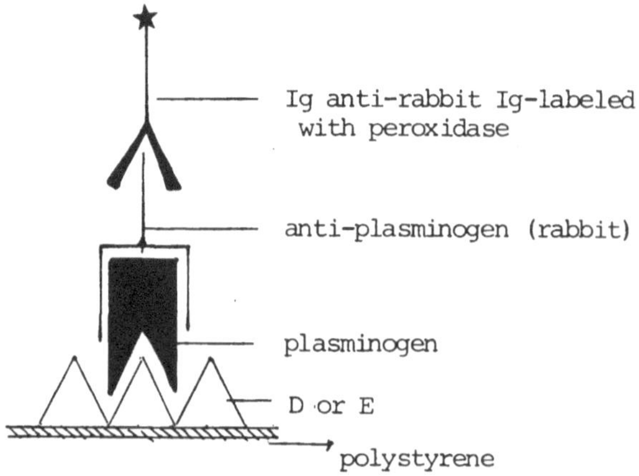

Figure 1 : Direct plasminogen binding assay to fragment D or fragment E.

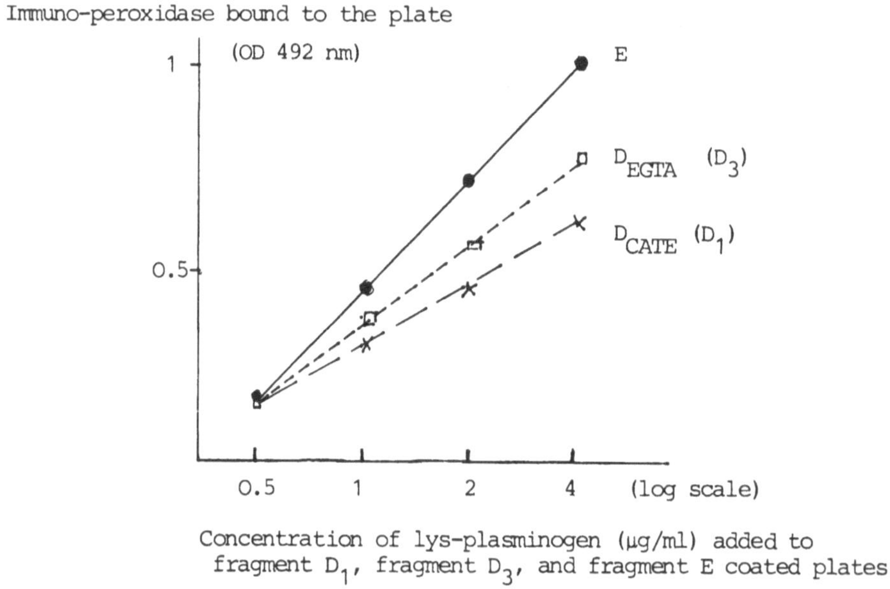

Figure 2 : Interaction of lys-plasminogen with fragment D_1 (1.07×10^{-8}M) or fragment D3 (1.25×10^{-8}M) or fragment E* coated wells (4×10^{-8}M).

* Molar concentration of fragment E was calculated on the basis of two equivalent sites per molecule.

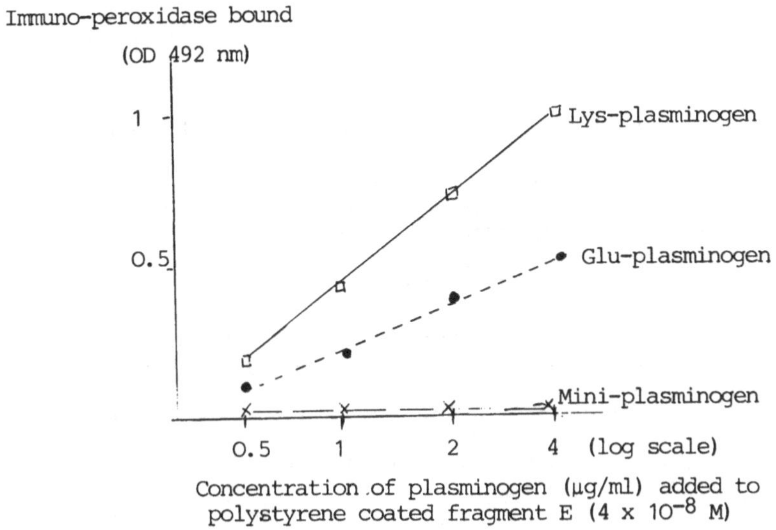

Figure 3 : Interaction of Lys- Glu- or Mini-plasminogen to immobilized fragment E.

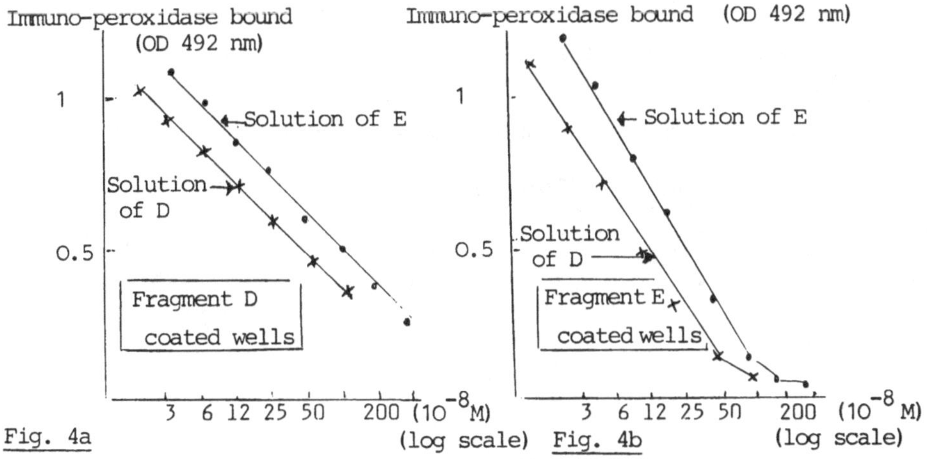

Concentration of D or E in solution added to plasminogen before incubation with fragment D or fragment E coated wells (fragment E on the basis of two equivalent binding sites per molecule).

Figure 4 : Inhibition binding of plasminogen to immobilized fragment D or fragment E induced by fragments D and E in solution.

REFERENCES

1. Rickli, E.E., Otavsky, N.L. : Eur. J. Biochem. 59, 441-447 (1975).

2. Sottrup-Jensen,L., Claeys, H., Zajdel, M., Petersen, T.E., Magnusson, S. : Progress in Chemical Fibrinolysis and Thrombolysis. Davidson JF, Rowan RM, Samama M, Desnoyers P (Eds), 3, 191-209 (1977).

3. Nihlen, J.E.:Thromb. Diathes. Haemorrh. 18, 89-100 (1967).

4. Van Ruyven-Vermeer, I.A.M., Nieuwenhuizen, W., Haverkate, F., Timan, F. : Hoppe-Seyler's Z. Physiol. Chem. 360, 633-637 (1979).

5. Haverkate, F., Timan, G. : Thromb. Res. 10, 803-812 (1977).

6. Nieuwenhuizen, W., Vermond, A., Haverkate, F. : Biochim. Biophys. Acta 667, 321-327 (1981).

7. Wolters, G., Kuijpers, L., Kacaki, J., Schuurs, A. : J. Clin. Path. 29, 873-879 (1976).

8. Engwall, E., Carlsson, H.E. : 1st International Symposium on Immuno-enzymatic techniques. INSERM, Symp. n°2. Feldman (Ed), North Holland publishers Company, Amsterdam, 135-147 (1976).

9. Fine, J.M., Steinbuch, M.A. : Rev. Eur. Et. Clin. Biol. 15, 1115-1122 (1970).

10. Avrameas, S. : Immunochemistry 6, 825-831 (1969).

11. Thorsen, S. : Biochim. Biophys. Acta 339, 55-65 (1975).

FIBRIN AND PLASMINOGEN STRUCTURES INVOLVED IN THE TISSUE-TYPE PLASMINOGEN ACTIVATOR CATALYZED ACTIVATION OF PLASMINOGEN

Jan Verheijen, Willem Nieuwenhuizen, Daan Traas, Glenn Chang, Bep Hoegee-de Nobel

Gaubius Institute, Health Research Division TNO,
Herenstraat 5d, 2313 AD Leiden, The Netherlands

Introduction

Fibrin degradation in vivo is mainly performed by plasmin. Plasmin is present in plasma in an inactive zymogen form, plasminogen. Plasminogen contains five so-called kringle structures which are believed to play a role in plasminogen to fibrin interaction. Four of these five kringles can be removed from plasminogen by elastase digestion. The resulting 442-val-plasminogen or mini-plasminogen, harbours an intact active site but lacks four of its five kringles.
The conversion of plasminogen to plasmin (activation) is catalyzed by highly specific proteolytic enzymes, plasminogen activators. These plasminogen activators hydrolyze only one (560-arg-561-val) peptide band in plasminogen, thus forming active plasmin. For fibrinolysis in vivo the tissue-type or extrinsic plasminogen activator (t-PA) seems to be most important (for review see ref 1).
It has been known for some time that fibrin has a remarkable role in fibrinolysis. It is the substrate for plasmin, but it also accelerates plasmin formation from plasminogen catalyzed by t-PA. Thus fibrin accelerates its own degradation (2). Recently we have shown that insoluble polymeric fibrin is not a prerequisite for this process. Also soluble monomeric fibrin and some fibrin(ogen) degradation products have stimulating

Fibrinogen - Structural Variants and Interactions
© 1985 Walter de Gruyter & Co., Berlin · New York - Printed in Germany

properties on plasminogen activation by t-PA. Especially the plasmic fragments Y, D EGTA and D-dimer and the CNBr-fragment FCB-2 (Hol-DSK) (3,4) have stimulating properties.
The mechanism of this stimulation is not yet resolved. In this paper the role of plasminogen-fibrin interactions in this process was studied by comparing the activation of native 1-glu-plasminogen and 442-val-plasminogen (mini-plasminogen) in the presence of fibrin(ogen) fragments. It is concluded, that two different types of site contribute to the stimulating properties of fibrin. Both, but at least one type, may be exposed in D-dimer and only one in CNBr-fragment FCB-2. The site present in FCB-2 probably interacts with plasminogen in a kringle 1-4 dependent way.

Materials and Methods

- A cyanogen bromide digest of fibrinogen was prepared as described (5). The dialyzed mixture was called CNBr-digest. From this digest fragment FCB-2 (Hol-DSK) was purified as described (3).
- Fibrin monomer was prepared either by dissolving a non cross-linked fibrin clot in 3 mol/l NaBr pH 5.5 (3) or by fibrin formation in the presence of antipolymerizing Dcate fragments (3).
- Fragments Y, Dcate, D EGTA, D-dimer and E were prepared by limited plasmic degradation of fibrin(ogen) and purified as published (6,7).
- Two-chain tissue-type plasminogen activator was purified from melanoma cell culture medium according to Rijken et al. (8).
- 1-Glu-plasminogen was isolated from fresh human blood plasma using lysine-sepharose (9). After removal of 6-aminohexanoic acid by ammoniumsulfate precipitation the plasminogen was dissolved in 0.05 mol/l Tris.HCl, 0.1 mol/l NaCl, pH 5.7 an equal volume of glycerol was added and the solution was stored at -20°C.

- 442-Val-plasminogen (mini-plasminogen) was prepared from 1-glu-plasminogen essentially as described by Sottrup-Jensen et al. (10). The purity of plasminogen preparations was checked with SDS-polyacrylamide gel electrophoresis and N-terminal amino acid determinations (11).
- The effect of fibrin(ogen) fragments on plasminogen activation was measured in an indirect spectrophotometric assay (3, 5). In a total volume of 250 µl was present: 0.05 IU/ml t-PA, 0.13 µmol/l plasminogen (either glu or mini), 0.30 mmol/l S-2251, 0.10 mol/l Tris.HCl pH 7.5, 0.1% (v/v) Tween 80 and a varying amount of the fragments to be tested. Assays were carried out in 96 well polystyrene microtitration plates at 25°C and absorbance measurements at 405 nm were done with an eight channel photometer (Titertek, Multiskan).
- Fibrinogen (grade L) and S-2251 (H-D-Val-Leu-Lys-pNA) were obtained from Kabi, Stockholm, Sweden.

Results and Discussion

In previous studies from this laboratory it has been shown that the accelerating effect of fibrin on t-PA catalyzed plasminogen activation was not simply caused by a concentrating effect of t-PA and plasminogen on the insoluble fibrin. Soluble fragments, especially D-dimer (3) and cyanogen bromide fragment FCB-2 (4) also have pronounced stimulating properties on the activation of plasminogen by t-PA.
In our earlier work mixtures of 1-glu- and 77-lys-plasminogen were used. The rates of activation and/or the interactions of glu- and lys-plasminogen may differ. For this reason these experiments were repeated with pure native 1-glu-plasminogen. The results shown in figure 1 were essentially the same as with the less defined plasminogen preparations used before, indicating no important difference between glu- and lys-plasminogen. The accelerated increase in para-nitroanilin formation is not caused by an acceleration of S-2251 hydrolysis by plasmin as

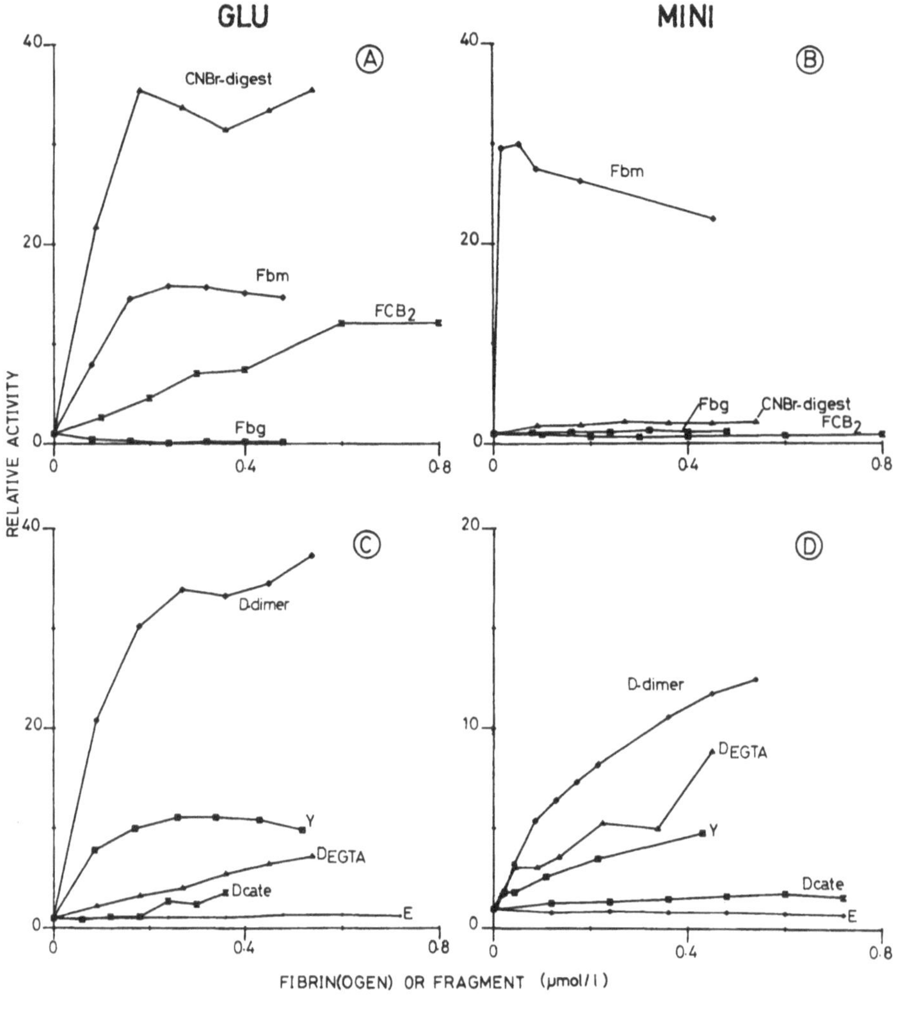

Figure 1. Effect of fibrin, fibrinogen and fragments on t-PA catalyzed activation of 1-glu-plasminogen and 442-val-plasminogen.

t-PA (0.012 IU) was incubated in a system described in Materials and Methods. Varying concentrations of fibrin monomer, fibrinogen or fragments were added as indicated. After different durations of incubation the absorbance change at 405 nm was measured and plotted against the square of the incubation duration. The slopes of these ΔA versus t^2 plots were used as a measure for the rate of plasminogen activation. A and C 0.13 µmol/l 1-glu-plasminogen and B and D 0.13 µmol/l 442-val-plasminogen (mini-plasminogen).

was shown before (3).
Consequently, plasminogen activation is accelerated. For this acceleration different molecular mechanisms can be proposed:
1. Only t-PA interacts with fibrin or some fibrin(ogen) fragment and t-PA becomes more effective in activation.
2. Only plasminogen interacts with fibrin or some fibrin(ogen) fragment and can more easily be activated by t-PA.
3. Both t-PA and plasminogen have an interaction with fibrin and this leads to effects 1 and 2 at the same time. In this model also the concentration of both molecules on the fibrin may play an additional role. In this paper the role of plasminogen-fibrin interactions in the stimulation of plasminogen activation is studied. The role of kringles 1-4 of plasminogen was studied from the differences in activation behaviour between native glu-plasminogen and mini-plasminogen. The results of the experiments with both kinds of plasminogen are shown in figure 1 and summarized in table I. Fibrinogen, frag-

Table I
Stimulation of 1-glu-plasminogen and 442-val-plasminogen (mini--plasminogen) activation by fibrin, fibrinogen and some of their fragments

	glu-plasminogen	mini-plasminogen
Fibrinogen	-	-
D-cate	-	-
E	-	-
Fibrin monomer	+	+
Y	+	+
D EGTA	+	+
D-dimer	+	+
FCB-2	+	-
CNBr-digest	+	-

This table is a summary of the results mentioned in figure 1.
a (+) indicates stimulation by the substance;
a (-) indicates no stimulation.

ment Dcate and fragment D do not affect glu-plasminogen or mini-plasminogen activation by t-PA. Fibrin-monomer, fragment Y, D EGTA and D-dimer all stimulate both glu-plasminogen and mini-plasminogen activation. However, the CNBr-fragment FCB-2 (and the whole CNBr-digest) stimulates glu-plasminogen activation but does not affect mini-plasminogen activation.

These results are not compatible with mechanism 1, i.e. a more effective t-PA by interaction with fibrin or a fragment. This would have given rise to accelerated mini-plasminogen activation. This mechanism is therefore excluded. Both, mechanisms 2 or 3 take into consideration an interaction of plasminogen with a fibrin domain and remain feasible mechanisms. An interaction, at least between plasminogen and FCB-2 is also suggested by another contribution in this volume (Nieuwenhuizen et al.).

Based on the present data two possible tentative models for stimulation of plasminogen activation by t-PA can be proposed, illustrated in a simplified cartoon form in figure 2. In model A fibrin is supposed to have at least two independent different binding sites for plasminogen. For binding to one of these sites (the one in the FCB-2 domain not stimulating mini-plasminogen) the kringle 1-4 region of plasminogen is required. The

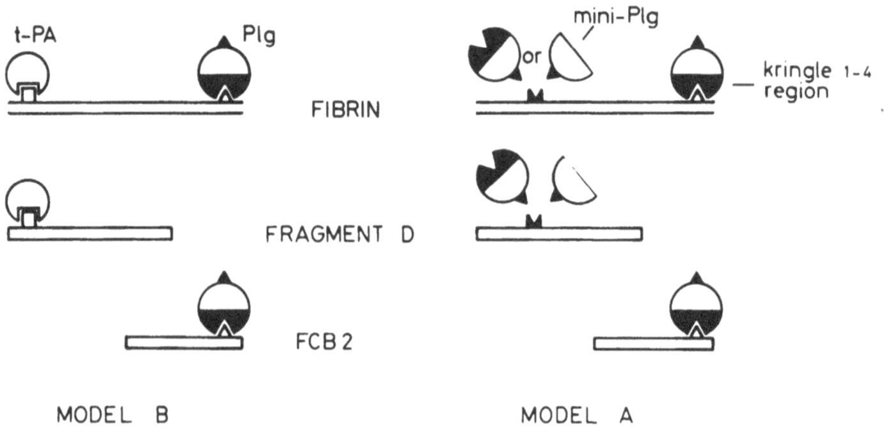

Figure 2. Schematic representation of possible interactions involved in stimulation of plasminogen activation.

other site (exposed in D-dimer and stimulating both mini-plasminogen and glu-plasminogen) binds plasminogen in a kringle 1-4 independent way. Both interactions lead to stimulation of plasminogen activation.

In model B fibrin is thought to have a plasminogen binding site in the FCB-2 domain which interacts with the kringle 1-4 region of plasminogen. Furthermore, a t-PA binding site is present in the D-domain (D EGTA, D-dimer).

Model B is the most attractive, since plasminogen binding to fibrin (12) and some fragments (Soria et al., this volume) as well as t-PA binding to fibrin (13) have been observed.

Recently, it has also been shown that t-PA bound to fibrin behaves differently towards synthetic peptide substrates as compared with free t-PA (14). In both models A and B, additional interacting, but non-stimulating sites, may be present.

The role of fibrin in its own breakdown is very complicated and more experiments, for instance direct binding studies with mini-plasminogen and kinetic studies, will be necessary to fully unravel this.

References

1. Collen, D.: Thrombos. Haemostas. 43, 77-89 (1980).
2. Wallén, P.: Thrombosis and Urokinase (Paoletti, R., Sherry, S. eds.) pp. 91-102, Academic Press, New York (1977).
3. Verheijen, J.H., Nieuwenhuizen, W., Wijngaards, G.: Thromb. Res. 27, 377-385 (1982).
4. Nieuwenhuizen, W., Verheijen, J.H., Vermond, A., Chang, G. T.G.: Biochim. Biophys. Acta 755, 531-533 (1983).
5. Verheijen, J.H., Mullaart, E., Chang, G.T.G., Kluft, C., Wijngaards, G.: Thrombos. Haemostas. 48, 266-269 (1982).
6. Nieuwenhuizen, W., Gravesen, M.: Biochim. Biophys. Acta 668, 81-88 (1981).
7. Van Ruyven-Vermeer, I.A.M., Nieuwenhuizen, W., Haverkate, F., Timan, T.: Hoppe-Seyler's Z. physiol. Chem. 360, 633-637 (1979).

8. Rijken, D.C., Collen, D.: J. Biol. Chem. 256, 7035-7041 (1981).

9. Robbins, K.C., Summaria, L.: Methods in Enzymology (Lorand, L. ed.) Vol. XLV, part B, chapter 22, Academic Press, New York (1976).

10. Sottrup-Jensen, L., Claeys, H., Zajdel, M., Petersen, T.E., Magnusson, S.: Progress in Chemical Fibrinolysis and Thrombolysis (Davidson, J.F., Rowan, R.M., Samama, M.M., Desnoyers, P.C. eds.) pp. 191-209, Raven Press, New York (1978).

11. Gray, W.R.: Methods in Enzymology (Hirsch, C.H.W. ed.) Vol. XI, pp. 139-151, Academic Press, New York (1967).

12. Wallén, P.: Fibrinolysis (Kline, D.L., Reddy, K.N.N. eds.) pp. 1-25, CRC Press Inc., Boca Raton Fla. (1980).

13. Thorsen, S., Glas-Greenwalt, P., Astrup, T.: Thromb. Diath. Haemorrh. 28, 65-74 (1972).

14. Rånby, M., Bergsdorf, N., Svenson, E., Wallén, P.: Haemostasis 11/S1, 74 abstr. (1982).

KINETICS OF THE TISSUE-TYPE PLASMINOGEN ACTIVATOR-MEDIATED ACTIVATION OF PLASMINOGEN. INFLUENCE OF CNBr FIBRIN(OGEN) FRAGMENT FCB-2 AND DIFFERENT FORMS OF PLASMINOGEN

Willem Nieuwenhuizen, Marijke Voskuilen, Daan W. Traas,
Bep Hoegee-de Nobel and Jan H. Verheijen

Gaubius Institute, Health Research Division TNO,
Herenstraat 5d, 2313 AD Leiden, The Netherlands

Introduction

Plasmin (E.C. 3.4.21.7) is considered as the main enzyme for the degradation of the fibrin matrix of a blood clot in vivo. It is generated by activation of its inactive precursor, plasminogen, which occurs in blood in a concentration of 1.5-2 μM (1). Native plasminogen (glu-plasminogen) has glutamic acid as the amino-terminal amino acid. It is easily converted to other forms with amino-terminals lysine, valine or methionine ("lys-plasminogen") by limited plasmin digestion (2,3). Upon limited elastase digestion several bonds, amongst which the Val 441-Val 442 bond are cleaved and this yields amongst other fragments Val_{442}-plasminogen (mini-plasminogen), lacking kringle structures 1-4 (4). The plasminogen activation can be exerted by tissue-type plasminogen activator (t-PA) (1,5,6), also designated as extrinsic plasminogen activator. Two forms of t-PA have been isolated i.e., a single-chain and a two-chain form, depending on whether or not the plasmin inhibitor aprotinin was added during the purification (6). Tissue-type plasminogen activator has a high affinity for fibrin (1) and also plasminogen interacts specifically with fibrin. The well-known accelerating effects of fibrin on the t-PA mediated plasminogen activation (7) may be based on one or both of these interactions.
Several investigators have attempted to analyse kinetically the

Fibrinogen - Structural Variants and Interactions
© 1985 Walter de Gruyter & Co., Berlin · New York - Printed in Germany

effects of fibrin of the t-PA mediated plasminogen activation. Some of them used radiolabelled fibrin, adsorbed to plastic walls. In such complicated systems fibrin acts both as a plasmin substrate and as an accelerator (8). As a consequence this complicates the interpretation of the kinetic data. Others (9) have used fibrin only as a stimulator and recorded the increase of plasmin activity with a chromogenic plasmin substrate. The use of insoluble fibrin, however, is a complicating factor, since incubation mixtures must be centrifuged before optical densities can be read. Recently, we have shown (10) that plasmin fibrin(ogen) degradation products, fragments Y, D EGTA and D-dimer, and also CNBr fragment FCB-2 (11) accelerate the plasminogen activation by t-PA. These fragments are soluble, in contrast with fibrin. This enables us to record the plasmin generation continuously in a homogeneous system and to determine the influence of these fibrin(ogen) fragments on the kinetic parameters of the activation reaction.
This report is an account of our studies on the influence of fragment FCB-2 on the kinetics of the two-chain t-PA-mediated activation of glu-, "lys"- and mini-plasminogen.

Materials and Methods

Two-chain t-PA was isolated from melanoma cell culture medium as described (6). Activities were expressed in urokinase equivalent units by comparison of fibrinolytic activity on plasminogen-enriched fibrin plates (12). Activities were converted to molar quantities using a specific activity of 90,000 IU/mg and a molecular weight of 70,000 (6,8).
Glu-plasminogen was prepared from fresh human plasma by affinity chromatography (13) on lysine-Sepharose, precipitated with ammoniumsulphate and dialyzed extensively against 0.1 M Tris. HCl, 0.01% (v/v) Tween 80, pH 7.5.
"Lys-plasminogen" was prepared from Cohn fraction III by a combination of affinity chromatography on lysine-Sepharose (13)

and ion-exchange chromatography on DEAE-Sephadex (3). SDS-polyacrylamide gel electrophoresis revealed a molecular weight of 85,000 which corresponds to "lys-plasminogen". End-group analysis by the dansyl method (14) revealed predominantly methionine as the amino-terminal amino acid.

Val_{442}-plasminogen (mini-plasminogen) was prepared and purified as described (4).

The concentration of all types of plasminogen was based on an $A^{1\%}_{1\ cm} = 16.1$ at 280 nm (2).

FCB-2 was prepared and purified as described before (11,15). Its concentration was based on Lowry determinations (16) using bovine serum albumin as a reference protein.

S-2251 (H-D-valyl-L-leucyl-L-lysine p-nitroanilide) was bought from KABI AB, Stockholm, Sweden and

Tween 80 from Baker Chemicals.

Assay system, principle

The principle of the assay system was that described by Drapier et al. (17) based on the work of Kosow (18) and adapted in this laboratory for t-PA assays (19). Briefly, t-PA, plasminogen and S-2251, the plasmin substrate are mixed and two reactions will occur i.e., t-PA activation of plasminogen to plasmin (kinetic parameters k_{cat} and K_m) and the hydrolysis of S-2251 by plasmin (kinetic parameters k_2 and K_{m2}), liberating p-nitroanilide (pNA). At S-2251 concentrations $\gg K_{m2}$, the reaction rate of the S-2251 hydrolysis equals k_2 [plasmin].

Drapier et al. derived the following equation:

$$[pNA]_t = [pNA]_o + k_2[plasmin]_o t + 0.5\ \gamma t^2$$

$$\text{where } \gamma = \frac{k_2 k_{cat}[t\text{-}PA][plasminogen]}{K_m + [plasminogen]}$$

$[pNA]_o$ and $[plasmin]_o$ are the pNA and plasmin concentration at zero time, respectively. These are usually very low, and can be corrected for by running blanks without t-PA. [pNA] is derived from the $A_{405\ nm}$ using $\varepsilon = 10{,}500\ M^{-1}cm^{-1}$.

Plotting of $[pNA]_t$ against t^2 yields a straight line with slope 0.5 γ. Plotting of γ^{-1} against $[plasminogen]^{-1}$ will yield a straight line with intercepts $\frac{1}{k_2 k_{cat} [t\text{-}PA]}$ and $-\frac{1}{K_m}$ on ordinate and abscissa, respectively.
Since k_2 is known (1326 min^{-1}) (personal observations in agreement with ref 20) and [t-PA] is chosen, this continuous assay system can be used to determine k_{cat} and K_m. As k_{cat} and/or K_m apparently change in the presence of FCB-2, it is also possible to determine the influence of FCB-2 on k_{cat} and K_m by making similar plots (γ^{-1} against $[plasminogen]^{-1}$) in the presence of FCB-2.

Experimental conditions

In total volumes of 0.250 ml 0.1 M Tris.HCl, 0.01% (v/v) Tween 80, pH 7.5, were present (final concentration): plasminogen (0.01-0.2 µM) and FCB-2 (0-700 nM).
In cases when FCB-2 was present, the samples were incubated for 2 hours before t-PA was added (see under Results and Discussion). The reactions were started by the addition of t-PA (0.31 nM), S-2251 (0.9 mM) and run at 37°C.
The increasing absorbance with time at 405 nm was continuously recorded with a Unicam Spectrophotometer SP 1700, equipped with a thermostated cuvet holder and a recorder AR 25. The tracings were replotted and treated as described above.

Results and Discussion

Plasminogen activation in the absence of FCB-2

Figure 1 shows a Lineweaver-Burk plot of the activation of glu-plasminogen with t-PA in the absence of FCB-2. A clear downward curvature is seen at plasminogen concentrations higher than 1 µM. At lower concentrations (below 1 µM) the plot is apparently linear and we decided to work in the plasminogen concentration range 0.01-0.2 µM. In that range K_m values are 0.053 µM and 0.030 µM for glu- and "lys-plasminogen", respect-

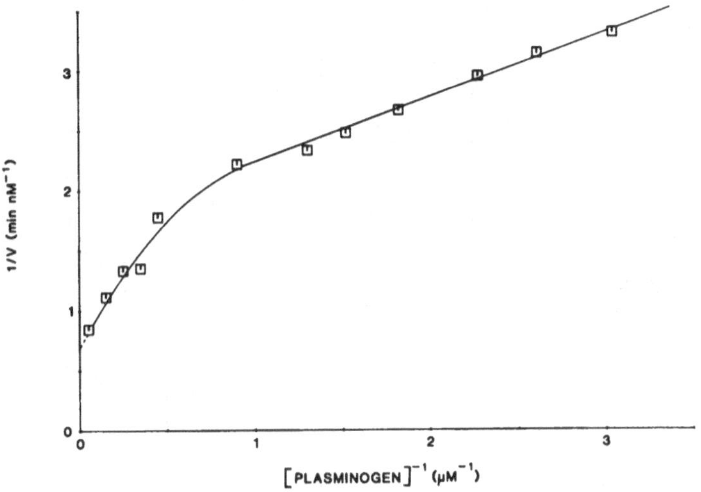

Figure 1. Lineweaver-Burk plot of the glu-plasminogen activation by tissue-type plasminogen activator in the absence of FCB-2 (details in text).

ively; the k_{cat} values are 0.12 and 0.37 min^{-1}, respectively. The deviation from Michaelis-Menten kinetics observed here for glu-plasminogen (figure 1), was also observed for lys-plasminogen by Rånby (9) at lys-plasminogen concentrations higher than 1 μM. As Rånby (9) pointed out, and as also illustrated by figure 1, erroneous results for K_m and k_{cat} will be obtained when plasminogen concentrations higher than 1 μM are used. Judging from the shape of the curve in figure 1, too high values for K_m and k_{cat} may be anticipated. This is in fact supported by results obtained by Wallén and Rånby in their earlier work (21) and more recently, by Hoylaerts et al. (8). The latter authors worked in the range 3.3-100 μM and 3.3-50 μM for glu- and lys-plasminogen, respectively. They found for glu-plasminogen K_m = 65 μM and k_{cat} = 3.6 min^{-1}; for lys-plasminogen K_m = 19 μM and k_{cat} = 12 min^{-1}. Working in the 5-50 μM concentration range Rånby (9) found K_m = 7.6 μM and k_{cat} = 0.47 min^{-1} for glu-plasminogen (table I).

Table 1

Kinetic parameters of plasminogen activation by two-chain t-PA

	glu-plasminogen		"lys-plasminogen"		reference
	without fibrin	with fibrin or FCB-2	without fibrin	with fibrin or FCB-2	
range (µM)	3.3-100	0.04-1	3.3-50	0.005-0.1	Hoylaerts
K_m (µM)	65	0.16	19	0.02	et al.
k_{cat} (min^{-1})	3.6	6	12	12	(8)
range (µM)			>> 1	>> 1	Wallén &
K_m (µM)			20	0.1	Rånby
k_{cat} (min^{-1})			6	24	(21)
range (µM)	5-50	0.02-5	< 1	0.002-0.16	Rånby
K_m (µM)	7.6	0.18	0.30	0.035	(9)
k_{cat} (min^{-1})	0.47	7.2	1.5	13.2	
range (µM)	0.01-0.2	0.01-0.2	0.01-0.2	0.01-0.2	this
K_m (µM)	0.053	0.003*	0.030*	0.010*	report
k_{cat} (min^{-1})	0.12	3.5**	0.37	13**	

*At 10 nM.

**At 350 and 700 nM FCB-2 for glu- and "lys-plasminogen", respectively.

Working in the plasminogen concentration range 0.06-0.9 µM, we found for glu-plasminogen 0.22 µM and 0.24 min^{-1} for K_m and k_{cat}, respectively. For "lys-plasminogen" these values were 0.38 µM and 1 min^{-1}. Our findings show that already in the plasminogen concentration range near 1 µM too high K_m and k_{cat} values are obtained.

Effect of preincubation on the plasminogen activation in the presence of FCB-2

In our early experiments we noted that the plots of [pNA] against time squared (see Materials and Methods) were not linear under the conditions used (figure 2). The curves exhibited an upward curvature and became apparently linear after 20 minutes reaction time. Upon preincubation of FCB-2 with plasminogen at room temperature, before addition of t-PA and S-2251 two effects were observed: the curves became more and more linear and their average slope increased, until, after 2-3 hours preincubation time, they were linear and had a maximum

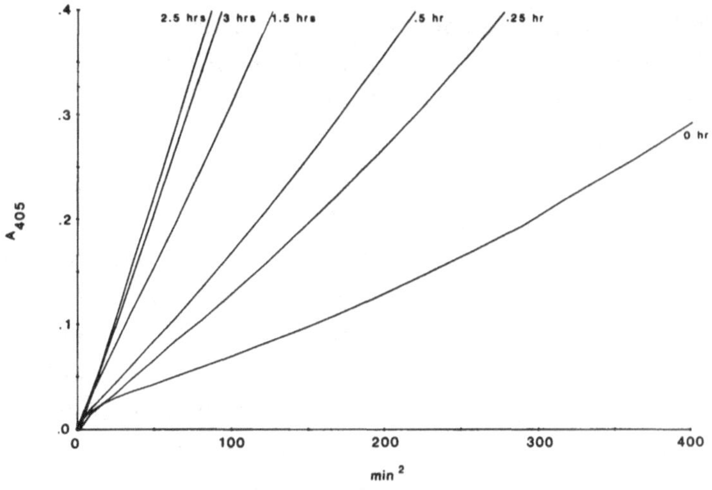

Figure 2. Kinetics of activation of plasminogen by tissue-type plasminogen activator in the presence of FCB-2; its dependence on preincubation time of plasminogen with FCB-2.

slope from the moment of addition of t-PA on ($t^2 = 0$ in figure 2). This phenomenon was observed both for glu- and lys-plasminogen.

As preincubation of FCB-2 with t-PA did not show this effect, our results suggest that, for a maximum rate of plasminogen activation, a complex between FCB-2 and plasminogen must be present and that the formation of this complex is relatively slow. The amount of complex seems to be rate-limiting under the conditions of the experiment. Hoylaerts et al. (8) concluded from their kinetic data that first t-PA and fibrin form a binary complex and that this binary complex forms a cyclic ternary complex with plasminogen, whereas our results indicate that first a complex between FCB-2 and plasminogen is formed. It must be noted, that the interaction rates of FCB-2 may not be representative for those of intact fibrin.

Kinetic parameters of mini-, glu- and lys-plasminogen; influence of FCB-2

Mini-plasminogen activation was not significantly affected by FCB-2. Both in the presence and absence of FCB-2 the same K_m and k_{cat} values were found i.e., 61 nM and 0.21 min^{-1}, respectively. These values do not differ much from those for glu-plasminogen without fibrin.

Figure 3 (left hand panel) shows that an abrupt drop in K_m occurs for glu-plasminogen from 53 nM to 3 nM and for lys-plasminogen (figure 3, right hand panel) from 30 nM to 10 nM at FCB-2 concentrations below 10 nM. Beyond this concentration range the K_m values do not change to a great extent. However, at concentrations above 200 nM (not shown) a gradual increase in K_m was seen. This is probably caused by competitive inhibition of FCB-2 on the S-2251 hydrolysis by plasmin. In the FCB-2 concentration range where the drop in K_m occurs, the k_{cat} values do not change significantly (figure 4). However, at higher FCB-2 concentrations, the k_{cat} values begin to increase and reach a maximum of 3.5 min^{-1} at [FCB-2] = 350 nM for glu-

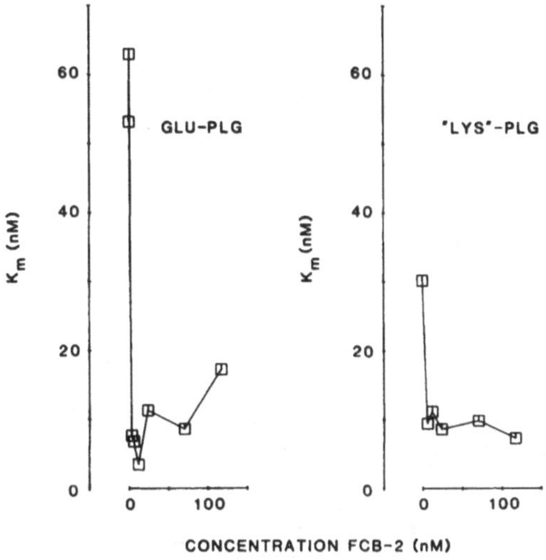

Figure 3. FCB-2 concentration. Dependence of K_m of the activation of glu-plasminogen (left hand panel) and lys-plasminogen (right hand panel) by tissue-type plasminogen activator.

plasminogen (figure 4, top) and 13 min^{-1} for lys-plasminogen at [FCB-2] = 700 nM for lys-plasminogen (figure 4, lower panel).

The mechanism for these phenomena is not clear at present. However, it is conceivable that the K_m drop is due to a high affinity complex between FCB-2 and plasminogen, whereas the increase in k_{cat} may be due to lower affinity interactions. Direct binding studies will be required to answer this question.

It must be noted that our K_m and k_{cat} values for lys-plasminogen activation in the presence of FCB-2 (in contrast with those in its absence) are similar to those obtained by Hoylaerts et al. (8) and by Rånby (9) (table I). This better agreement is probably due to the fact that all these values were determined in the same plasminogen concentration range well below 1 µM, where the activation follows Michaelis-Menten kinetics.

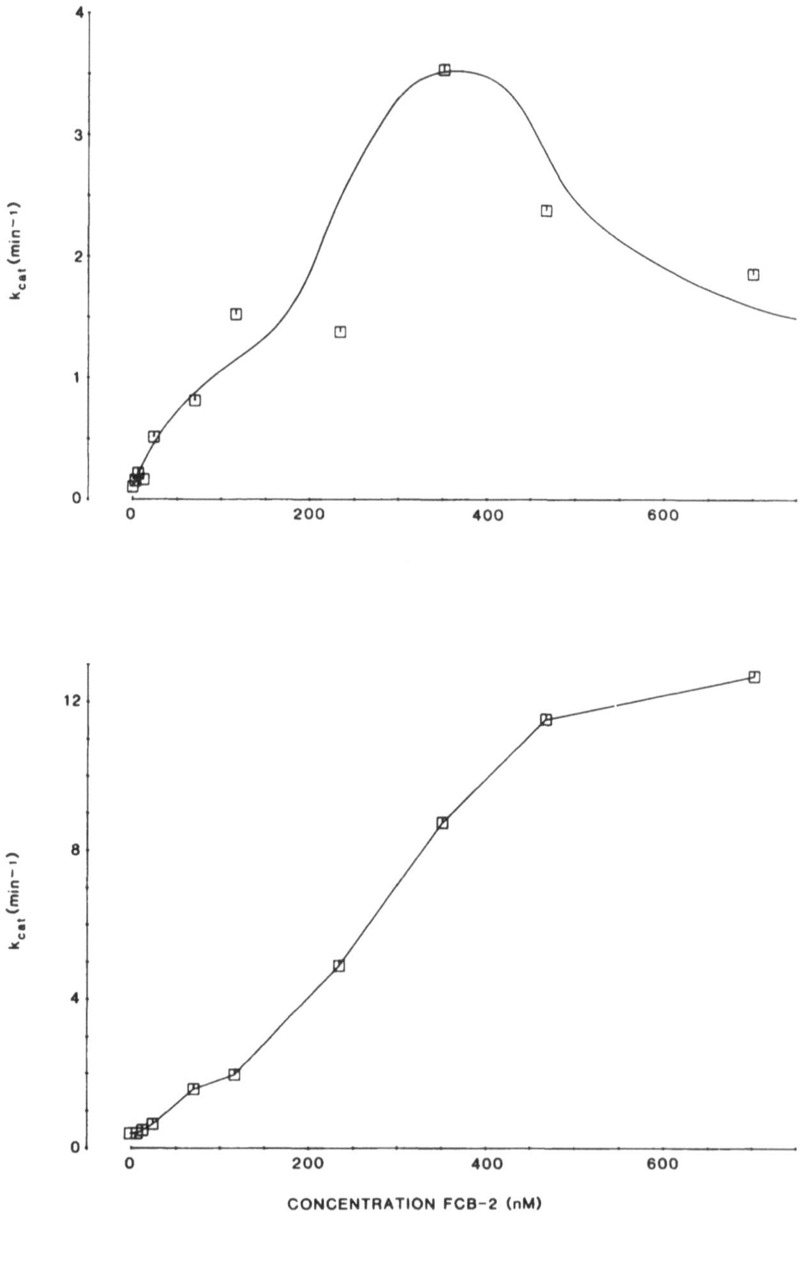

Figure 4. FCB-2 concentration dependence of k_{cat} of the activation of glu-plasminogen (top panel) and lys-plasminogen (lower panel) by tissue-type plasminogen activator.

Also the values for glu-plasminogen activation in the presence of FCB-2 agree better with the literature data (8,9) obtained in the presence than in the absence of fibrin (table I). Although there does exist a considerable drop in K_m's in the presence of fibrin, this drop has been overestimated (8). The main contribution to this drop is only apparent and is caused by the fact that the values (with and without fibrin) were determined in plasminogen ranges that differ two orders of magnitude (see above). In contrast, all values in this report have been determined in the same plasminogen concentration range. Furthermore, Hoylaerts et al. (8) have suggested that the K_m value for t-PA mediated plasminogen activation in the absence of fibrin is much higher than the plasminogen concentration in plasma. This would make the activation inefficient and fibrin would be required to trigger the fibrinolytic process. Regulation would occur at the fibrin level and regulation at the t-PA level e.g., by t-PA inhibitors (22), would not play an important role.

Our data confirm that there is a regulatory mechanism at the fibrin level but show also, that the K_m value for the activation is at least twenty times lower than the plasma plasminogen level (1.5-2 μM), even in the absence of fibrin. This indicates that plasmin formation can occur under these conditions if active t-PA is released into the blood and that regulation of the fibrinolytic process also at the t-PA level, may be essential for a good hemostatic balance.

References

1. Lijnen, H.R., Collen, D.: Seminars in Thrombosis and Haemostasis 8, 2-10 (1982).
2. Wallén, P., Wiman, B.: Biochim. Biophys. Acta 221, 20-30 (1970).
3. Wallén, P., Wiman, B.: Biochim. Biophys. Acta 257, 122-134 (1972).

4. Sottrup-Jensen, L., Claeys, H., Zajdel, M., Petersen, T.E., Magnusson, S.: Progress in Chemical Fibrinolysis and Thrombolysis, Vol. 3 (Davidson, J.F., Rowan, R.M., Samama, M.M., Desnoyers, P.C. eds.) pp. 191-209, Raven Press, New York (1978).

5. Rijken, D.C., Wijngaards, G., Zaal-de Jong, M, Welbergen, J.: Biochim. Biophys. Acta 580, 140-153 (1979).

6. Rijken, D.C., Collen, D.: J. Biol. Chem. 256, 7035-7041 (1981).

7. Wallén, P.: Thrombosis and Urokinase (Paoletti, R., Sherry, S. eds.) pp. 91-102, Academic Press, New York (1977).

8. Hoylaerts, M., Rijken, D.C., Lijnen, H.R., Collen, D.: J. Biol. Chem. 257, 2912-2919 (1982).

9. Rånby, M.: Biochim. Biophys. Acta 704, 461-469 (1982).

10. Verheijen, J.H., Nieuwenhuizen, W., Wijngaards, G.: Thromb. Res. 27, 377-385 (1982).

11. Nieuwenhuizen, W., Verheijen, J.H., Vermond, A., Chang, G. T.G.: Biochim. Biophys. Acta 755, 531-533 (1983).

12. Haverkate, F., Traas, D.W.: Thromb. Diath. Haemorrh. 32, 356-365 (1974).

13. Deutsch, D.G., Mertz, E.T.: Science 170, 1095-1096 (1970).

14. Gray, W.R.: Methods Enzymol. 25B, 121-143 (1972).

15. Blombäck, B., Blombäck, M., Henschen, A., Hessel, B., Iwanaga, S., Woods, K.R.: Nature 218, 130-134 (1968).

16. Lowry, O.H., Rosebrough, N.J., Farr, A.L., Randall, R.: J. Biol. Chem. 193, 265-275 (1951).

17. Drapier, J.C., Tenu, J.P., Lemaire, G., Petit, J.F.: Biochimie 61, 463-471 (1979).

18. Kosow, D.P.: Biochemistry 14, 4459-4465 (1975).

19. Verheijen, J.H., Mullaart, E., Chang, G.T.G., Kluft, C., Wijngaards, G.: Thrombos. Haemostas. 48, 266-269 (1982).

20. Claeson, G., Aurell, L., Karlsson, G., Friberger, P.: Progress in Chemical Fibrinolysis and Thrombolysis, Vol. 3 (Davidson, J.F., Rowan, R.M., Samama, M.M., Desnoyers, P.C. eds.) pp. 299-304, Raven Press, New York (1978).

21. Wallén, P., Rånby, M.: Prot. Biol. Fluids 28, abstr. 67 (1980).

22. Verheijen, J.H., Chang, G.T.G., Mullaart, E.: Thrombos. Haemostas. 50, abstr. 930 (1983).

IX. INTERACTION WITH CELLS

STRUCTURAL CHARACTERIZATION OF THE RECOGNITION SITE FOR PLATELET RECEPTORS ON HUMAN FIBRINOGEN

Jacek Hawiger, Marek Kloczewiak, Sheila Timmons, and Thomas J. Lukas

Division of Experimental Medicine, Department of Medicine, New England Deaconess Hospital and Harvard Medical School, Boston, Massachusetts 02215 and Department of Pharmacology, Vanderbilt University and Howard Hughes Medical Institute, Nashville, Tennessee 37232

Introduction

Fibrinogen is a protein with two distinct functions. On one hand it constitutes a substrate for enzymatic attack by thrombin, which results in polymerization of fibrin monomer and formation of a plasma clot (1, 2). On the other hand, fibrinogen is endowed with a remarkable propensity toward cellular receptors. It interacts not only with receptors on prokaryotic cells, such as staphylococci and streptococci (3, 4), but also with sites on eukaryotic cell fragments, i.e., platelets. As an adhesive macromolecule, fibrinogen is a key player in aggregation of platelets by linking them into aggregates. Very early in the aftermath of development of the *in vitro* method of platelet aggregation (5, 6), it was appreciated that this reaction requires fibrinogen (7, 8, 9). These reports were followed by a number of similar observations, culminating in the detailed analysis of the genetic disorder afibrinogenemia, with prolonged bleeding time and abnormal aggregation of platelets (10, 11). This "experiment of nature" provided key evidence for the role of fibrinogen in platelet function *in vivo* and *in vitro*. In fact, not only the fibrinogen which is present in plasma, but also that which is stored in platelet α granules, is essential for aggregation of platelets "activated" with

Fibrinogen - Structural Variants and Interactions
© 1985 Walter de Gruyter & Co., Berlin · New York - Printed in Germany

agents such as thrombin (12, 13). With these two functions - clotting and agglutinating - inprinted on its surface, fibrinogen is intimately involved in the formation of primary and secondary hemostatic plugs which seal off leaks in the blood circulatory system.

We will present information gathered during the last few years in our laboratory on the mechanism of the agglutinating function of fibrinogen in regard to human platelets. Starting with the induction of receptors which are not readily available on "non-activated" platelets, through analysis of the involvement of structural subunits and domains, and ending with pinpointing the main site on fibrinogen which recognizes platelet receptors, we will attempt to establish sufficient evidence for the molecular mechanism of platelet aggregation by fibrinogen. This evidence was greatly aided by synthetic peptides which we prepared and employed as inhibitors and promoters of platelet aggregation. To this end, a semi-synthetic molecule which can substitute for fibrinogen as a platelet aggregant will be presented to validate the function of the recognition site(s) identified by us on "natural" fibrinogen.

When Does Fibrinogen Interact With Platelets?

The mechanism of fibrinogen involvement in platelet aggregation remained unexplained, until Mustard and colleagues demonstrated that aggregation of platelets induced by ADP is accompanied by their binding of fibrinogen (14). Subsequent experiments of Marguerie and colleagues (15), Bennett and Vilaire (16), and ours (17) indicated that the main role of platelet agonists such as ADP, epinephrine, or thrombin is to induce exposure of binding sites for fibrinogen on the platelet membrane.

In our studies, intact resting platelets bound very little ^{125}I-fibrinogen (Fig. 1). After stimulation with 4.5 μM ADP, there was a four- to five-fold increase in the amount of ^{125}I-fibrinogen bound. This binding approached saturation at 300 μg of ^{125}I-fibrinogen, and equilibrium was achieved within 5 min. Approximately, 70% of the labeled fibrinogen could be displaced by a 20-fold excess of unlabeled fibrinogen, suggesting that binding was specific. Moreover, adenosine, AMP, and ATP in equimolar concentrations did not

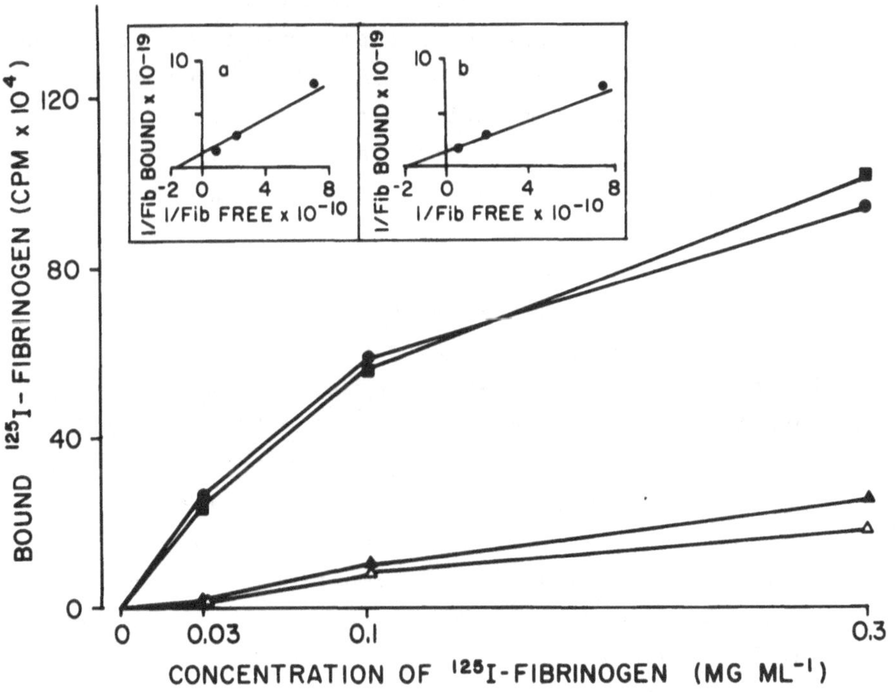

Fig. 1. Induction of binding of ^{125}I-labeled human fibrinogen to human platelets by ADP or thrombin. Platelets were treated with 4.5 μM ADP (■) or with thrombin (0.05 U/ml) followed by hirudin (0.1 U/ml) (●) prior to addition of ^{125}I-labeled fibrinogen. Non-specific binding was determined in both systems after adding a 20-fold excess of unlabeled fibrinogen (Δ). It was similar to the values obtained with untreated platelets suspended in buffer (▲). Inset "a" represents double reciprocal plot of "specific" binding (total observed binding less "non-specific" binding) calculated for ADP-treated platelets and inset "b" represents a double reciprocal plot of "specific" binding calculated for thrombin-treated platelets (from ref. 17). Reprinted by permission from Nature, 283, 195-197, Copyright (c) 1980 Macmillan Journals Limited.

increase the binding of ^{125}I-fibrinogen to platelets. The amount of ^{125}I-fibrinogen bound was calculated from a double reciprocal plot (17) of the binding data obtained after subtracting values for nonspecific binding (Fig. 1). Assuming uniform labeling of fibrinogen and a molecular weight of 340,000 (2), we estimated that 50,000 molecules of ^{125}I-fibrinogen per platelet were bound to ADP-stimulated platelets with the apparent dissociation constant of 1.5 X 10^{-7} M.

Because thrombin is almost instantly detectable when blood vessels are wounded (18, 19), we compared the effect of trace amounts of thrombin with that of ADP in inducing binding of ^{125}I-fibrinogen. After stimulation with thrombin (0.05 U/ml), and then inactivation of free thrombin with hirudin, we observed specific binding of ^{125}I-fibrinogen with remarkably similar characteristics to those after stimulation with ADP (Fig. 1). We estimated that 44,000 molecules of fibrinogen bound with an apparent dissociation constant of 1.8 X 10^{-7} M (Fig. 1). Thus, traces of thrombin can also induce mobilization of platelet-binding sites for plasma fibrinogen. The interaction of fibrinogen with "activated" platelets fulfills the criteria for specific, saturable, and reversible binding mediated by receptors of relatively high affinity (K_d = 2 X 10^{-7} M). Since the plasma concentration of fibrinogen is approximately 50 times higher, it is apparent that even when its plasma level decreases to 10% of the normal value, there is still a sufficient amount to saturate the available binding sites.

The binding of fibrinogen to platelets requires Ca^{++} because it is prevented or reversed by EDTA (14, 20), and involves platelet membrane glycoproteins IIb and IIIa (16, 21-23). Using monoclonal antibody against one of these glycoproteins, McEver and Majerus showed that about 40,000 molecules of this glycoprotein are available on one platelet (24). As it turns out, the concordance of these values - 40,000 molecules of platelet fibrinogen expressed on the platelet surface (13), 40,000 molecules of fibrinogen bound at equilibrium to ADP- or thrombin-treated platelets (17), and 40,000 molecules of glycoprotein IIb available to monovalent antibody (24) - is remarkable and suggests a 1:1 stoichiometry between fibrinogen and the glycoprotein IIb/IIIa complex on the membrane of "activated" platelets.

A genetic deficiency of glycoproteins IIb and IIIa, characteristic of Glanzmann's thrombasthenia, or treatment of platelets with monoclonal antibodies against the glycoprotein IIb/IIIa complex results in impaired binding of fibrinogen to platelets and their inability to aggregate (20-24). Isolated glycoproteins IIb/IIIa in the form of a Ca^{++}-held heterodimer bind fibrinogen (25). Thus, calcium-dependent stereochemical rearrangement of glycoproteins IIb and IIIa in the platelet membrane constitutes the formation of high affinity binding sites for fibrinogen. We have shown that this process is either prevented or reversed by cyclic AMP-mediated reactions, since an increase in cAMP levels induced by prostacyclin (PGI_2) or a non-prostanoid inhibitor, forskolin, correlates with inhibition of binding of fibrinogen to "activated" platelets, which is paralleled by inhibition of platelet aggregation (26).

Which Fibrinogen Chain Interacts With Platelet Receptors?

We became interested in determining the structural basis for fibrinogen's ability to interact with specific platelet receptors. Human fibrinogen is composed of three pairs of non-identical chains (α_2, β_2, γ_2) linked by a series of disulfide bonds and arranged in three main domains: one central E and two distal D (2). In general, three different strategies could be employed in determining the localization of a region on fibrinogen involved in recognition of platelet receptors.

First, gradual removal of portions of the molecule with plasmin and determination of the ability of the obtained fibrinogen derivatives to promote platelet aggregation can be followed. This approach was initially used by Niewiarowski and colleagues (27) who showed that plasmin-treated fibrinogen rapidly loses its ability to support platelet aggregation which was attributed to the loss of the carboxy-terminal segment of the α chain and the amino-terminal portions of the β chain. However, the critical role of the carboxy-terminal portions of the α chain was subsequently questioned by Holt, et al. (28).

Secondly, isolated fragments of fibrinogen (e.g., D, E, etc.) could be used to inhibit binding of ^{125}I-fibrinogen to platelets. This approach was used initially by Tomikawa, et al. (29) and then by Marguerie, et al. (30). However, it poses difficulties due to discreet differences in the preparations of fragments as will be discussed below.

We elected a different approach starting with the reduction of fibrinogen to dissociate individual subunits (α , β ,γ chains), then carboxy-methylation of the chains and their separation, and finally, chemical and enzymatic fragmentation of individual chains to shorter peptides. This approach was successfully employed by us earlier in a related project on determination of the fibrinogen recognition site for staphylococcal clumping receptor in collaboration with Russell Doolittle in La Jolla (3, 31).

Since fibrinogen is required for the aggregation of platelets by ADP (9-11), the basic test system was the aggregation of human platelets separated from plasma proteins and incubated with ADP and fibrinogen, or its subunits. Under such conditions, addition of fibrinogen in a concentration of 1.4 mg/ml (3.7 μM) induced platelet aggregation (Fig. 2). When isolated subunits of fibrinogen dispersed in ammonium bicarbonate, pH

7.8, were tested, γ chains gave a very strong reaction, fully comparable to the fibrinogen effect. Alpha chains in an equimolar concentration produced an intermediate effect, and β chains were completely inactive (32). The effect of γ chains was concentration-dependent in terms of aggregation parameters (Tmax) and was blocked by γ-chain specific antibody fragments which also inhibited fibrinogen-mediated platelet aggregation induced with ADP (32).

A Site Interacting with Platelet Receptors Exists on the Carboxy-terminal Segment of the γ Chain.

The procedure used previously (31) for isolation of a peptide inhibiting the staphylococcal clumping reaction with human

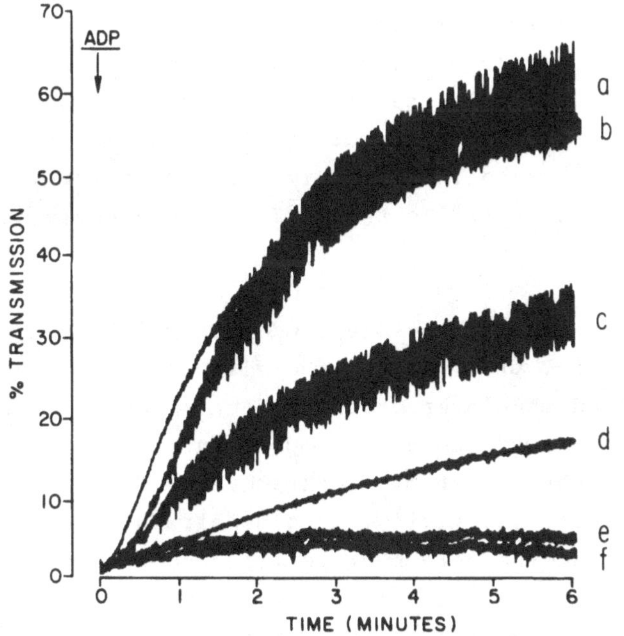

Fig. 2. The effect of fibrinogen and its isolated chains on aggregation of ADP-treated platelets. Human platelets (2×10^8/ml) separated from plasma proteins were incubated with human fibrinogen (3.7 μM) - curve b; reduced and carboxy--methylated fibrinogen (3.7 μM) - curve c; isolated α chain multimers (7.5 μM) - curve d; isolated β chain multimers (7.5 μM) - curve e; isolated γ chain multimers (7.5 μM) - curve a; and ammonium bicarbonate, pH 7.8, as diluent - curve f. Aggregation curves a and b representing the γ chain and fibrinogen are superimposed indicating their equipotent aggregating effect (from ref. 32).

fibrinogen was modified by us using high pressure liquid chromatography (HPLC) to achieve a higher yield of isolated peptide derived from the γ chain of human fibrinogen (33).

Amino acid composition of the purified peptide showed very good agreement with that of the 27-residue fragment located on the COOH-terminal of γ chain (33). The peptide obtained from the analytical column was sequenced in an automated sequencer and showed that the first 10 amino acids were Lys-Ile-Ile-Pro-Phe-Asn-Arg-Leu-Thr-Ile, again in agreement with the sequence reported by Henschen et al. (34) and Strong et al. (31).

The inhibitory effect of the 27-residue peptide on the binding of ^{125}I-fibrinogen to platelets was concentration-dependent. The inhibition curve reached saturation at 86 μM and the 50% inhibition of binding occurred at approximately 20 μM peptide concentration (33). The peptide completely inhibited the aggregation of platelets induced by ADP and fibrinogen. Trypsin digestion abolished the inhibitory activity of the isolated peptide tested in the ^{125}I-fibrinogen-platelet binding assay.

The 27-residue peptide has one bond susceptible to proteolysis by staphylococcal protease and the mixture of resulting peptides inhibited the interaction of fibrinogen with platelets (33) and with staphylococcal clumping factor (31). After separation of the digest by preparative HPLC, only the first peak had inhibitory activity, thus confirming our previous observation that the more hydrophilic carboxy-terminal portion of the parent 27-residue peptide is inhibitory (35). Amino terminal analysis of the pooled material containing the first active peak gave the sequence Gly-Glu-Glu corresponding to the residues 397-398-399 of the γ chain of human fibrinogen (34).

The isolated pentadecapeptide inhibited binding of ^{125}I-fibrinogen to human platelets treated with ADP. Inhibition of binding was concentration-dependent and was almost complete at 140 μM of the pentadecapeptide, but only 38 μM was required for 50% inhibitory concentration (IC_{50}) (Fig. 3).

For comparison, a synthetic pentadecapeptide of identical sequence (35) was examined and produced a similar concentration curve with IC_{50} of 28 μM. Both the isolated native and synthetic pentadecapeptides completely blocked platelet aggregation at a concentration of 200 μM, and 190μM respectively, when added prior to fibrinogen and ADP (Fig. 4).

Synthesis and Structure-function Relationship of Peptides Corresponding to the Carboxy-terminal Segment of the γ Chain.

It is apparent from the foregoing, that localization of the platelet receptor recognition site on human fibrinogen has been attributed to the γ and α chains of human fibrinogen (32)

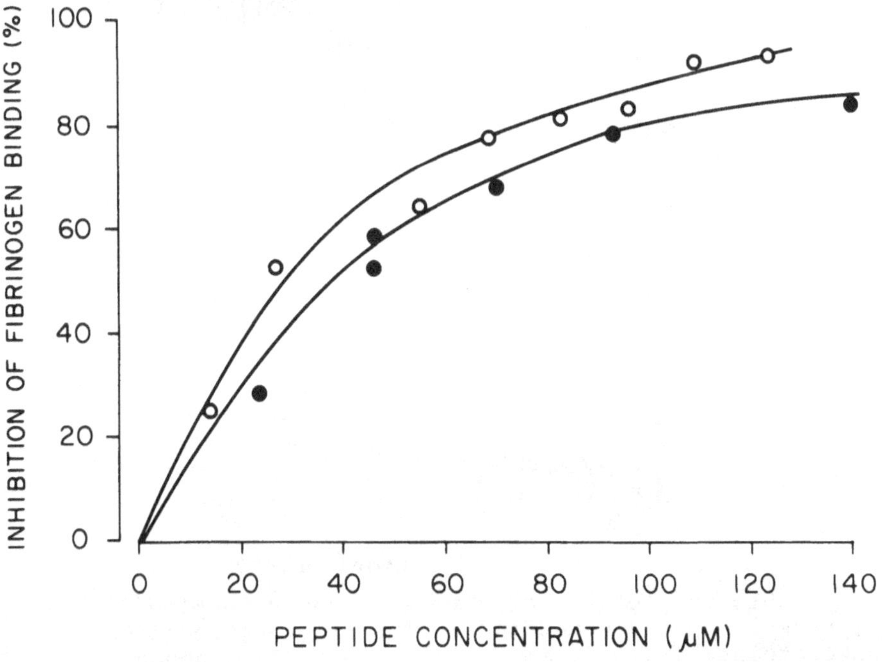

Fig. 3. Inhibition of binding of ^{125}I-fibrinogen (0.16 μM) to ADP-treated platelets by native pentadecapeptide (γ397-411) isolated from the γ chain (●) as compared to its synthetic analog (○) (from ref. 35).

and pinpointed by us to the last 15 amino acid residues on the carboxy-terminal segment of the γ chain (35). In the next study, the structure-function relationship of this segment of the γ chain was investigated in regard to its interaction with the platelet receptor. By synthesizing peptides of an increasing number of amino acid residues, patterned after the carboxy-terminal segment of the γ chain, the role of these residues in the interaction with the platelet receptor was compared. Thus, we sought to establish the minimal sequence of the carboxy-terminal segment of the γ chain required for its interaction with the platelet receptor and to determine which amino acid residues play a key role in its recognition.

The amino acid compositions of synthetic peptides containing 5-, 7-, 10-, 12-, and 15-amino acid residues of the sequence starting with the carboxy-terminal valine (γ411) are given in Table I. They were in excellent agreement with data predicted

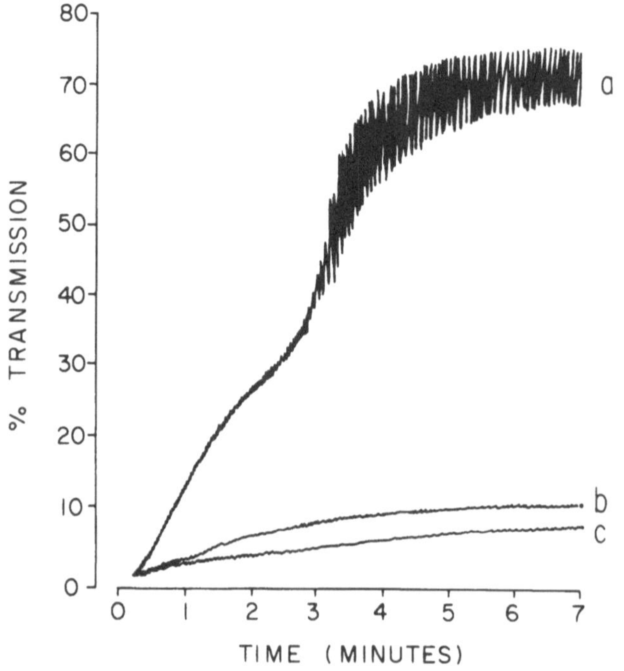

Fig. 4. Inhibitory effect of native pentadecapeptide (curve b) and synthetic pentadecapeptide (curve c) on aggregation of human platelets induced with ADP(5 μM) and fibrinogen (0.16 μM). Aggregation tracing "a" represents a system without tested peptides (from ref. 35).

Table I. The amino acid sequence of the carboxy-terminal segment of the human fibrinogen γ chain and designed sequences of synthetic peptides[+]

		IC_{50} (μM)[++]
γ384-411	M↓-K-I-I-P-F-N-R-L-T-I-G-E↓-G-Q̂-Q-H-H-L-G-G-A-K̂-Q-A-G-D-V-COOH (CNBr at M; SP at E; FXIII at Q and K)	20
γ397-411	G-Q-Q-H-H-L-G-G-A-K-Q-A-G-D-V-COOH	28
Y-γ397-411	Y-G-Q-Q-H-H-L-G-G-A-K-Q-A-G-D-V-COOH	32
C-Y-γ397-411	C-Y-G-Q-Q-H-H-L-G-G-A-K-Q-A-G-D-V-COOH	22
γ400-411	H-H-L-G-G-A-K↓-Q-A-G-D-V-COOH (Tr at K)	28
γ400-406 (fragment of γ400-411)	H-H-L-G-G-A-K	1000
γ402-411	L-G-G-A-K-Q-A-G-D-V-COOH	106
γ405-411	A-K-Q-A-G-D-V-COOH	102
γ407-411	Q-A-G-D-V-COOH	190

[+]The sequence of γ384-411 is based on data reported by Henschen et al. (1980); arrows show the positions of residues involved in CNBr cleavage and enzymatic reaction with Factor XIIIa (FXIII), staphylococcal protease (SP), and trypsin (Tr). [++]Concentration of peptide causing 50% inhibition of ^{125}I-fibrinogen binding. Reprinted with permission from Biochemistry, 23, 1767-1774, 1984. Copyright 1984 American Chemical Society.

from the known sequence of the γ chain (34). In addition to these peptides, two synthetic peptides containing tyrosine or cysteinyl-tyrosine at the amino terminal were prepared (36). The inhibitory potency of all seven synthetic peptides toward binding of ^{125}I-fibrinogen to the platelet receptor was determined. Pentadecapeptide (γ397-411) was inhibitory with IC_{50} = 28 μM, a value similar to a related peptide possessing an additional tyrosine residue (tyrosyl-γ397-411). Likewise, addition of a cysteine residue to this peptide to form a cystine-linked dimeric peptide cystinyl-(tyrosyl-γ397-411)$_2$ did not change significantly the inhibitory potency of such a peptide.

Dodecapeptide (γ400-411) had IC_{50} = 28 μM, in reference to ^{125}I-fibrinogen which is identical to that of the pentadecapeptide. The inhibitory potency of other peptides decreased with shortening of the peptide length. Decapeptide (γ402-411) was almost four times less potent (IC_{50} = 106 μM) than dodecapeptide (γ400-411). The heptapeptide (γ405-411) had IC_{50} = 102 μM similar to the decapeptide, but the pentapeptide (γ407-411) was less reactive (IC_{50} = 190 μM). Whereas, the dodecapeptide inhibited completely ^{125}I-fibrinogen binding at 80-100 μM, the pentapeptide required a much higher concentration (500 μM) for complete inhibition. Acetylation of dodecapeptide (γ 400-411) abolished its inhibitory activity. At a concentration of 80 μM, no inhibition of fibrinogen binding could be observed.

Since the dodecapeptide (γ 400-411) has one bond susceptible to trypsin, enzymatic degradation was done to see whether the resulting smaller peptides retained reactivity toward the platelet receptor. The inhibitory activity of these peptides toward binding of ^{125}I-fibrinogen to the platelet receptor was measured. None exhibited activity comparable to the starting dodecapeptide (γ400-411). In particular, the 7 residue amino-terminal peptide which corresponds, in terms of its composition, to γ400-406 has low inhibitory activity at a concentration of 1 mM.

In addition to the inhibition studies, we have demonstrated direct binding of ^{125}I-labeled tyrosyl-pentadecapeptide to platelet receptor (36). This binding was concentration-dependent and when plotted as bound ligand against the logarithm of free ligand according to Klotz (37) it produced a sigmoid curve with an inflection point corresponding to the dissociation constant (K_d) of 3×10^{-4} M (Fig. 5).

Observed binding was specific as determined by inhibition with unlabeled dodecapeptide (γ400-411) using ADP- or thrombin-stimulated platelets (36). The pentadecapeptide (γ397-411) and dodecapeptide (γ400-411) at equimolar concentrations (60 μM) fully inhibited ADP-induced γ chain dependent aggregation of platelets as judged by the slope and Tmax value of aggregation tracings (Fig. 6). The aggregation system contained 60 μM albumin. The inhibitory effect of the deca-

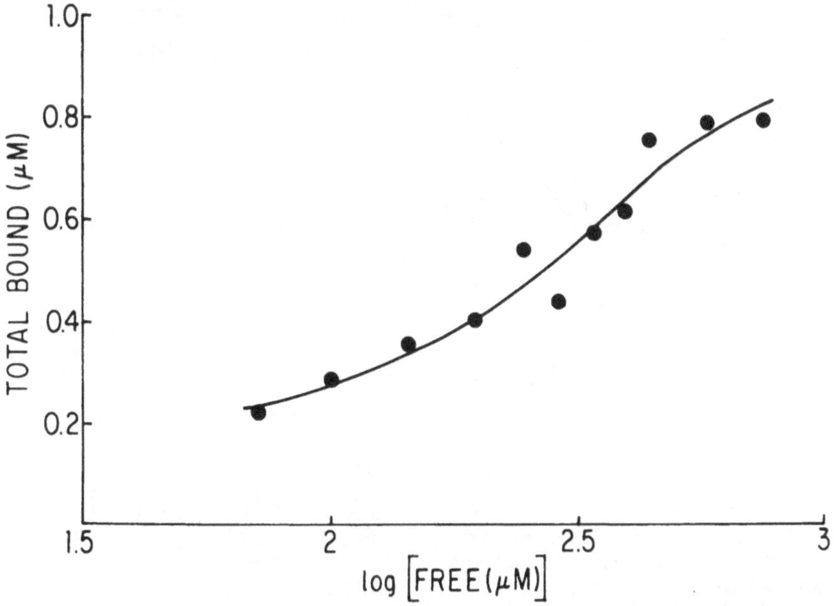

Fig. 5. Binding of ^{125}I-labeled tyrosyl-pentadecapeptide to platelets (1×10^8/0.5 ml) treated with ADP (10 μM). The binding data are presented as Klotz plot with inflection point corresponding to apparent K_d of 3×10^{-4}M (from ref. 36). Reprinted with permission from Biochemistry, 23, 1767-1774, 1984. Copyright 1984 American Chemical Society.

peptide (γ402-411, 60 μM) was apparently diminished when compared to the effect of the dodecapeptide. Heptapeptide (γ405-411) and pentapeptide (γ407-411) both used at concentrations of 60 μM inhibited platelet aggregation to a much lesser degree (Fig. 6).

Design of a Synthetic Pentadecapeptide-albumin Molecule Which Substitutes for Fibrinogen.

Proof that the pentadecapeptide represents the effector arm of human fibrinogen, essential for platelet aggregation, came from the experiment in which we conjugated a synthetic peptide cysteinyl-tyrosyl- γ397-411 to human serum albumin to prepare a polyvalent molecule mimicking fibrinogen (36). The peptide-albumin conjugate added to platelets separated from plasma

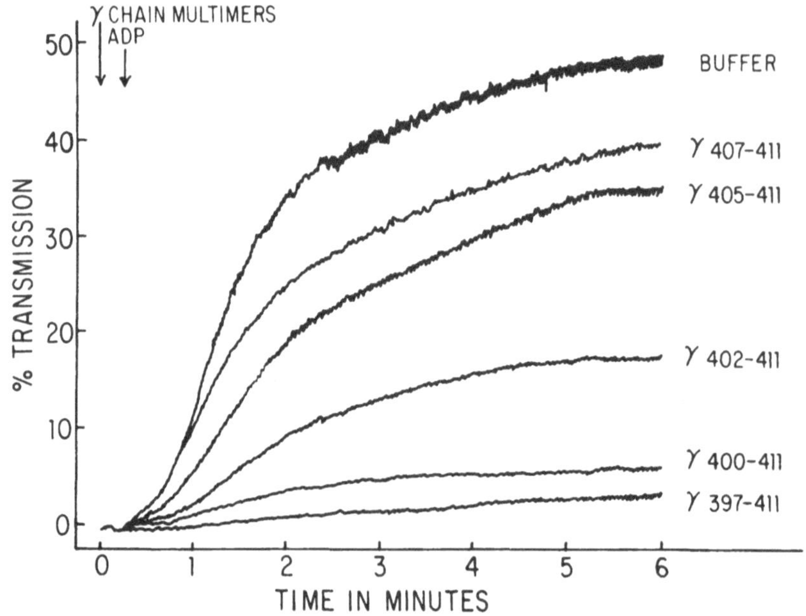

Fig. 6. Inhibitory effect of a series of synthetic peptide analogs of the carboxy-terminal segment of the γ chain on aggregation of human platelets induced with ADP (5 μM) and γ chain multimers (7 μM). Pentadecapeptide (γ397-411) and dodecapeptide (γ400-411) in concentration of 60 μM abolished platelet aggregation and pentapeptide (γ407-411) was least inhibitory at the same concentration (from ref. 36). Reprinted with permission from Biochemistry, 23, 1767-1774, 1984. Copyright 1984 American Chemical Society.

proteins caused platelet aggregation in the presence of 10 μM ADP (Fig. 7). In the absence of ADP the peptide-albumin conjugate did not aggregate platelets. Aggregation of ADP-treated platelets by a polyvalent synthetic peptide-albumin conjugate in concentration of 2.5 μM was prompt and biphasic. It was abolished by prior addition of unconjugated monovalent pentadecapeptide (Fig. 7) or dodecapeptide (results not shown in Fig. 7). A parallel control of cross-linked albumin, prepared in the same way, did not aggregate ADP-treated platelets.

The specificity of the interaction of the polyvalent synthetic pentadecapeptide-albumin conjugate with platelets was verified with monovalent Fab antibody fragments directed against the carboxy-terminal segment of the γ chain (γ385-411) isolated

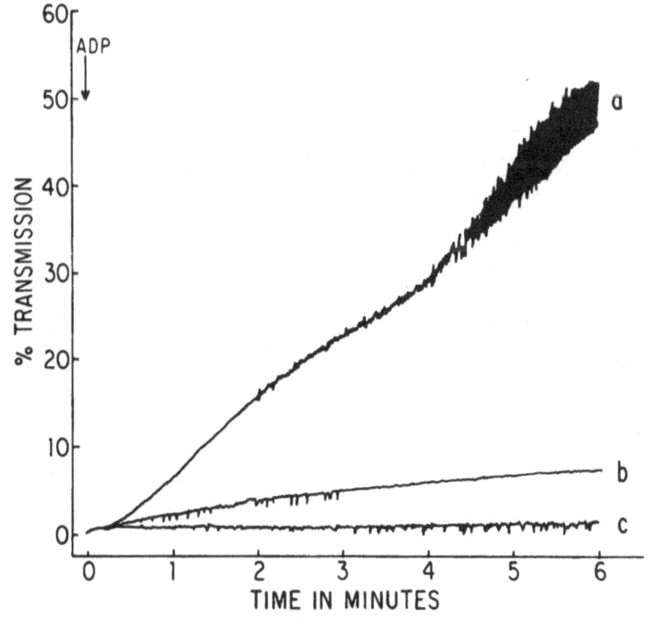

Fig. 7. Platelet-aggregation effect of a synthetic peptide-albumin conjugate (2.5 μM). Cysteinyl-tyrosyl-pentadecapeptide was coupled to human serum albumin in a molar ratio of 2:1 and added to human platelets treated with ADP (curve a). The aggregation was inhibited by addition of pentadecapeptide (60 μM) or of monospecific antibody Fab fragments against γ 385-411 isolated from the γ chain of human fibrinogen (from ref. 36). Reprinted with permission from Biochemistry, 23, 1767-1774, 1984. Copyright 1984 American Chemical Society.

from human fibrinogen. These antibody Fab fragments at a concentration of 1.8 μM inhibited binding of ^{125}I-fibrinogen (0.16 μM) to ADP-stimulated platelets (results not shown). They also inhibited substantially aggregation of ADP-treated platelets induced by a polyvalent conjugate of synthetic pentadecapeptide with albumin (Fig. 7).

The Platelet Receptor Recognition Site on γ Chain of Human Fibrinogen and its Polymerization Sites Are Not Related.

The elegant studies of Laudano and Doolittle (38), who designed a series of peptide analogues of the amino terminus of the α chain and of the β chain of human fibrin, indicate that some of these peptides are potent inhibitors of fibrin monomer polymerization resulting from thrombin's action on fibrinogen (38). With this information at hand, a tetrapeptide, Gly-Pro-Arg-Pro was employed by Harfenist and colleagues to prepare thrombin-degranulated platelets by preventing polymerization of fibrin monomer derived from platelet fibrinogen upon the action of thrombin (39). These authors did not observe an inhibitory effect of the "anti-polymerizing" tetrapeptide on binding of ^{125}I-fibrinogen to ADP-treated rabbit platelets. In contrast, Plow and Marguerie reported that Gly-Pro-Arg-Pro inhibited the interaction of fibrinogen with its platelet receptor (40). However, this inhibition did not exceed 60%, even at 1000-fold molar excess of tetrapeptide over fibrinogen. In our hands (Table II),the concentration of tetrapeptide causing 50% inhibition of binding of ^{125}I-fibrinogen (0.16 μM) was 2000 μM, thus making it unlikely that this tetrapeptide is a specific inhibitor of the fibrinogen interaction with platelets responsible for their aggregation as postulated by Plow and Marguerie (40). Since Olexa and Budzynski indicated that one of the polymerization sites of human fibrinogen encompasses the γ chain residues 374 to 411 (41), the possibility arose that the γ chain carboxy-terminal peptides, isolated and purified in our laboratory,

Table II. Inhibition of ^{125}I-fibrinogen binding to platelets and of fibrin monomer polymerization by fragments of fibrinogen and synthetic peptides

Fragment or Peptide	^{125}I-fibrinogen binding 50% Inhibitory Concentration µM	Fibrin Monomer Polymerization Inhibitory Activity§	
γ 385-411	20	+	(44.0)π
γ 397-411	28	O	(3.4)π
		O	(59.0)
γ 400-411	28	O	(5.0)
		O	(50.0)
Fibrinogen	0.7	++++	(4.8)
Fragment D_1	13*	+++	(3.4)
Fragment D_3	none	O	(3.4)
D_1-D_1	52τ	++++	(3.4)
Gly-Pro-Arg-Pro	2230÷	++++	(30.0)

*The highest concentration tested; τthis concentration gave 22% inhibition; ÷ 3.5 mM gave only 60% inhibition; §polymerisation inhibitory activity is expressed by range from O to ++++ where O indicates no inhibition and ++++ indicates complete inhibition; πnumbers in parentheses indicate concentration (µM) of fragments and peptides tested in polymerization system.

could influence the polymerization of fibrin monomer. Therefore, we examined whether fragments of fibrinogen which inhibit its binding to platelet receptors also interfere with polymerization of fibrin monomer. As shown in Table II, the pentadecapeptide isolated from the γ chain and encompassing residues γ397-411 and the 27-residue peptide containing residues γ385-411 did not have a demonstrable inhibitory effect on fibrin monomer polymerization (35). For comparison, we tested the inhibitory effect of larger fragments of fibrinogen and we observed complete inhibition of fibrin monomer polymerization at 3.4 μM of fragment D_1 or the dimer of fragment D_1 (D_1-D_1). Fragment D_3 was not inhibitory at the same concentration (Table II).

Fragment D_1 in its monomeric and dimeric form (D_1-D_1) inhibited fibrinogen binding at the highest concentrations studied, i.e., 13 μM and 52 μM producing 50% and 22% inhibition, respectively. This difference indicates that Factor XIII-catalyzed crosslinking of the carboxy-terminal segment of the γ chain diminishes its reactivity toward the platelet receptors. In contrast, fragment D_1-D_1 possesses the same inhibitory potency toward polymerization of fibrin monomers as fragment D_1. Among all of the tested materials only unlabeled fibrinogen, the pentadecapeptide, and the 27-residue peptide caused 90% or more inhibition of binding. The IC_{50} value for fragment D_1 (13 μM) obtained in our laboratory (35) differed from the IC_{50} value of 170 μM reported by Marguerie et al. (30). Since the IC_{50} value for unlabeled fibrinogen was identical in both studies, indicating similar conditions for platelet binding of ^{125}I-fibrinogen, the much weaker inhibition by fragment D_{100} reported by Marguerie et al. (30) suggests that these authors worked with a preparation of fragment D_{100} which was partially degraded at the carboxy-terminal segment of the γ chain (residues γ406-411) identified by us as the essential part of the platelet receptor recognition site (33). The lack of these 5 residues may not affect the apparent Mr of fragment D_{100} reported by Marguerie et al. (30).

A Model of Fibrinogen with Recognition Sites for Cellular Receptors.

Summarizing to this point, we were able to trim down the main recognition site for platelet receptors to the last twelve amino acids on the carboxy-terminal segment of the γ chain of human fibrinogen (Fig. 8). This site represents 3% of the γ chain and is overlapped with the slightly longer recognition site for the staphylococcal clumping receptor, pinpointed by Strong and colleagues to the last 15 residues (31). Although

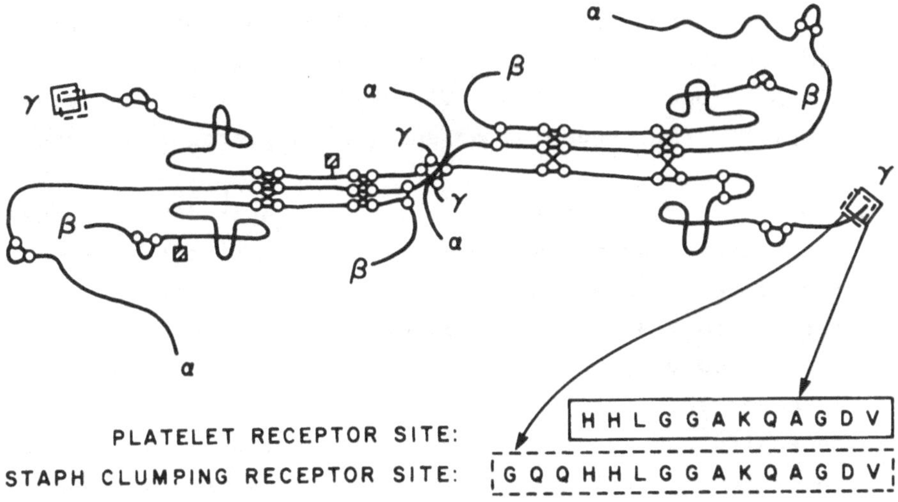

Fig. 8. Localization of platelet recognition sites (36) and staphylococcal clumping receptor sites (31) on the carboxy-terminal segment of the γ chain. The dimeric fibrinogen molecule is drawn according to the model proposed by Hoeprich and Doolittle (45) in which γ chains are arranged in antiparallel alignment. Two distally located recognition sites for cell receptors fulfill divalent requirements for cell agglutinating function of human fibrinogen.

the platelet receptor recognition site (γ400-411) and the staphylococcal clumping receptor site (γ397-411) on human fibrinogen share a very significant similarity, a glance at both interactions of fibrinogen with these receptors reveals significant differences. The staphylococcal clumping receptor binds fibrinogen instantly without any prior rearrangement and without the requirement for Ca^{++}. The receptor on platelets binds fibrinogen only when they are "activated" with ADP or other agonists (14-17). It appears that platelet receptors have a built-in "clutch" mechanism, preventing them from immediate and direct engagement in linkage with fibrinogen. This makes evolutionary sense from the standpoint of protecting the human circulation from continuous formation of platelet aggregates, which would happen if fibrinogen could bind at anytime to its receptor on platelets, as it can with staphylococci. To protect the mammalian circulation from continuous showers of platelet aggregates, nature endowed blood platelets with a "receptor clutch", which does not allow engagement unless platelets become "activated". The activating switch requires ADP supplied from exogenous or endogenous platelet sources.

The reactivity of the carboxy-terminal segment of the γ chain toward platelet receptors depends to a significant extent on two histidine residues (His-400 and His-401). Removal of these residues leads to a pronounced decrease in inhibitory activity. However, histidine residues are not absolutely required for the inhibitory activity. Other functional residues situated along the length of the dodecapeptide such as lysine-406 and valine-411 are also important. In contrast, no inhibitory effect of 5- and 7-residue peptides in the interaction with the staphylococcal clumping receptor was observed, while the longer peptides were inhibitory according to findings reported by Strong et al. (31).

It is important to note that simple occupancy of platelet

receptors by small peptides is not sufficient for aggregation unless an appropriately constructed divalent molecule is bound. Polyvalency was achieved by coupling the synthetic peptide, cysteinyl-tyrosyl-γ397-411, to human serum albumin. This endowed the conjugate with platelet aggregating function resembling that of the native human fibrinogen molecule. Because platelet aggregation induced by the conjugate was blocked by monovalent pentadecapeptide or dodecapeptide as well as by monospecific antibody Fab fragments directed against the carboxy-terminal segment of human fibrinogen γ chain (γ385-411), we conclude that the minimal structure of the γ chain encompassing residues 400-411 is involved in the aggregating function of human fibrinogen.

The platelet receptor reactive residues encompass a highly hydrophilic part of the γ chain, which was postulated by Chen and Doolittle (42) to be exposed to the surrounding solvent, thus being easily accessible to enzymes (Factor XIII, plasmin, and staphylococcal protease) and to staphylococcal clumping factor receptor (3, 31). At present, we do not know how much the functional reactivity of the platelet recognition site at the carboxy-terminal end of the γ chain is influenced by the additional sequence present in the γ' variant of human fibrinogen (43, 44).

In conclusion, our structure-function studies, utilizing a knowledge of the covalent structure of fibrinogen, led us to pinpoint the main site recognizing the platelet receptor to the twelve carboxy-terminal residues of the γ chain of human fibrinogen. The structural continuity of this segment is essential for optimal interaction of human fibrinogen with the platelet receptor. This site, present at the two distal ends of the fibrinogen molecule, fulfills the criteria for a divalent interaction which is essential for the platelet aggregating function of fibrinogen. Construction of an aggregating molecule by crafting two synthetic peptides to an "inert" albumin molecule endowed the conjugate with platelet aggregating properties mimicking that of native fibrinogen. The role of other sites on fibrinogen, located on the α chain, in enhancing and consolidating fibrinogen-platelet linkages, awaits further experimentation.

Acknowledgements

The authors' research was supported by U.S. Public Health Service research grants, HL30648, HL30649 and GM30861. We are indebted to Russell F. Doolittle and D. Martin Watterson for their generous advice and support, and to Kristina Lyscars for her excellent assistance in preparation of the manuscript.

References

1. Blomback, B., Hogg, D.H., Gardlund, B., Hessel, B., Kudryk, B.: Thromb. Res. (Suppl. II) 8, 329-346 (1976).
2. Doolittle, R.F.: Hemostasis and Thrombosis, editors A.L. Bloom and D.P. Thomas, Churchill Livingston, London 1981.
3. Hawiger, J., Timmons, S., Strong, D.D., Cottrell, B.A., Riley, M., Doolittle, R.F.: Biochemistry 21, 1407-1419 (1982).
4. Timmons, S., Kloczewiak, M., Hawiger, J.: Clin. Res. 31, 498A (1983).
5. Born, G.V.R.: Nature 194, 927-929 (1962).
6. O'Brien, J.R.: J. Clin. Path. 15, 452-455 (1962).
7. Caen, J., Inceman, S.: Nouv. Rev. Franc. Hemat. 3, 614-615 (1963).
8. Gugler, E., Luscher, E.: Ann. Paediat. 200, 125-127 (1963).
9. Brinkhous, K.M., Read, M.S., Mason, R.G.: Lab. Invest. 44, 335-342 (1965).
10. Inceman, S., Caen, J., Bernard, J.: J. Lab. Clin. Med. 68, 21-32 (1966).

11. Weiss, H.J., Rogers, M.: N. Engl. J. Med. 285, 369-374 (1971).

12. Morse, E.E., Jackson, D.P., Conley, C.L.: J. Clin. Invest. 44, 809-816 (1965).

13. Tollefsen, D.M., Majerus, P.W.: J. Clin. Invest. 55, 1259-1268 (1975).

14. Mustard, J.F., Packham, M.A., Kinlough-Rathbone, R.L., Perry, D.W., Regoeczi, E.: Blood 52, 453-461 (1978).

15. Marguerie, G.A., Plow, E.F., Edgington, T.S.: J. Biol. Chem. 254, 5357-5363 (1979).

16. Bennett, J.S., Vilaire, G.: J. Clin. Invest. 64, 1393-1401 (1979).

17. Hawiger, J., Parkinson, S., Timmons, S.: Nature 283, 195-197 (1980).

18. Teitel, J.M., Bauer, K.A., Lau, H.J.,Rosenberg, R.D.: Blood 59, 1086-1097 (1982).

19. Shuman, M.A., Majerus, P.W.: J. Clin. Invest. 58, 1249-1258 (1976).

20. Peerschke, E.I., Zucker, M.B., Grant, R.A., Egan, J.J., Johnson, M.M.: Blood 55, 841-847 (1980).

21. Kornecki, E., Niewiarowski, S., Morinelli, T.A., Kloczewiak, M.: J. Biol. Chem. 256, 5696-5701 (1981).

22. Coller, B.S., Peerschke, E.I., Scuder, L.E., Sullivan, C.A.: J. Clin. Invest. 72, 325-338 (1983).

23. Mustard, J.F., Kinlough-Rathbone, R.L., Packham, M.A., Perry, D.W., Harfenist, E.J., Pai, K.R.M.: Blood 54, 987-993 (1979).

24. McEver, R.P., Baenziger, N.L., Majerus, P.W.: J. Clin. Invest. 66, 1311-1318 (1980).

25. Nachman, R.L., Leung, L.L.K.: J. Clin. Invest. 69, 263-269 (1982).

26. Graber, S., Hawiger, J.: J. Bio. Chem. 257, 14606-14609 (1982).

27. Niewiarowski, S., Budzynski, A.Z., Lipinski, B.: Blood 49, 635-644 (1977).

28. Holt, J.C., Mahmoud, M., Gaffney, P.J.: Thromb. Res. 16, 427-435 (1979).

29. Tomikawa, M., Iwmoto, M., Soderman, S., Blomback, B.: Thromb. Res. 19, 841-855 (1980).

30. Marguerie, G.A., Ardaillou, N., Cherel, G., Plow, E.F.: J. Biol. Chem. 257, 11872-11875 (1982).

31. Strong, D.D., Laudano, A.P., Hawiger, J., Doolittle,

R.F.: Biochemistry 27, 1414-1420 (1982).

32. Hawiger, J., Timmons, S., Kloczewiak, M., Strong, D.D., Doolittle, R.F.: Proc. Natl. Acad. Sci. USA 79, 2068-2071 (1982).

33. Kloczewiak, M., Timmons, S., Hawiger, J.: Biochem. Biophys. Res. Comm. 107, 181-187 (1982).

34. Henschen, A., Lottspeich, F., Kehl, M., Southan, C.: N. Y. Acad. Sci. 408, 28-43 (1983).

35. Kloczewiak, M., Timmons, S., Hawiger, J.: Thromb. Res. 29, 249-255 (1983).

36. Kloczewiak, M., Timmons, S., Lukas, T.J., Hawiger, J.: Biochemistry 23, 1767-1774 (1984).

37. Klotz, I.M.: Science 217, 1247-1249 (1982).

38. Laudano, A.P., Doolittle, R.F.: Biochemistry 19, 1013-1019 (1980).

39. Harfenist, E.J., Guccione, M.A., Packham, M.A., Mustard, J.F.: Blood 59, 952-955 (1982).

40. Plow, E.F., Marguerie, G.: Proc. Natl. Acad. Sci. USA 79, 3711-3715 (1982).

41. Olexa, S.A., Budzynski, A.Z.: J. Biol. Chem. 256, 3544-3549 (1981).

42. Chen, R., Doolittle, R.F.: Biochemistry 10, 4486-4491 (1971).

43. Francis, C.W., Marder, V.J., Martin, S.E.: J. Biol. Chem. 255, 5599-5604 (1980).

44. Wolfenstein-Todel, C., Mosesson, M.W.: Proc. Natl. Acad. Sci. USA 77, 5069-5073 (1983).

45. Hoeprich, P.D., Jr., Doolittle, R.F.: Biochemistry 22, 2049-2055 (1983).

BINDING OF FIBRINOGEN TO ADP-TREATED PLATELETS: IMPORTANCE OF THE Aα CHAIN

Ellinor I.B. Peerschke and Dennis K. Galanakis
Department of Pathology, SUNY
Stony Brook, NY USA

During the last 5 years it has become clear that fibrinogen supports the aggregation of human blood platelets by binding to specific glycoprotein receptors (1-6), presumably the GPIIb/IIIa complex (7-14). Recent studies, using isolated fibrinogen chains (15,16), and antibodies directed against specific domains on the fibrinogen molecule (17), have shown that fibrinogen may bind to platelets with relatively high affinity via the carboxy--terminal segment of the γ-chain, and with lower affinity via an as yet undefined region on the Aα-chain. Since there is evidence, albeit controversial, to suggest that the Aα-chain of fibrinogen may be important in supporting platelet aggregation (18-20), we examined the capacity of several fibrinogen preparations lacking intact Aα-chains to support ADP-induced aggregation and fibrinogen binding. This is of particular interest because a major fraction of fibrinogen isolated by standard procedures (21,22) and used in the evaluation of platelet function, as well as up to one third of the fibrinogen in plasma (21), lack intact Aα-chains.

The fibrinogen fractions and derivatives that were tested are designated fraction I-6, fraction I-9, derivative $I\text{-}9D^{88}$, and derivative $I\text{-}9D^{50}$. These preparations have been extensively characterized (23). The relative sizes of their Aα chains are illustrated in Fig. 1. Fibrinogen fraction I-2, an intact fibrinogen preparation, is composed of Aα-chains with relative molecular weights between 68 and 71K. Its Bβ-and γ-chains are also intact. Fraction I-6 lacks segments amounting to approximately one fifth of the Aα-chain, from the COOH-terminal end of the chain, and possesses Aα-chains with a molecular weight of 57.7K. Fraction I-9 lacks larger COOH-terminal

Fibrinogen - Structural Variants and Interactions
© 1985 Walter de Gruyter & Co., Berlin · New York - Printed in Germany

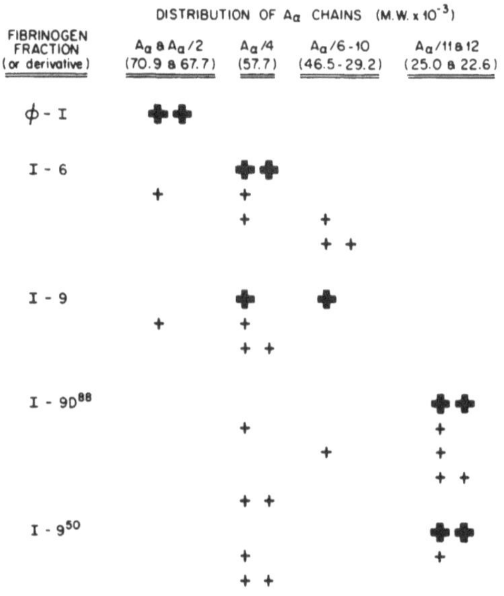

Fig. 1. Diagramatic representation of the Aα-chain composition of molecular species in fibrinogen preparations: I-2, I-6, I-9, I-9D88, and I-9D50. Pairs of crosses: dimeric fibrinogen molecule; heavy crosses: most abudant or major molecular species; light crosses: minor species (Reprinted by permission, C.V. Mosby Co.).

segments and possesses Aα-chains with molecular weights between 31.8K and 57.7K. Derivative I-9D^{88} is composed of Aα-chain residues with molecular weights between 25K and 57.7K, with two-thirds of Aα-chains having a molecular weight of 25K. Finally, fibrinogen derivative I-9D^{50} possesses Aα-chains with a relative MW of 22K. This fraction differs from all the others in that it also lacks amino terminal residues from approximately one half of its Bβ-chains. The molecular weights of its fragments were between 51.2K and 54.7K.

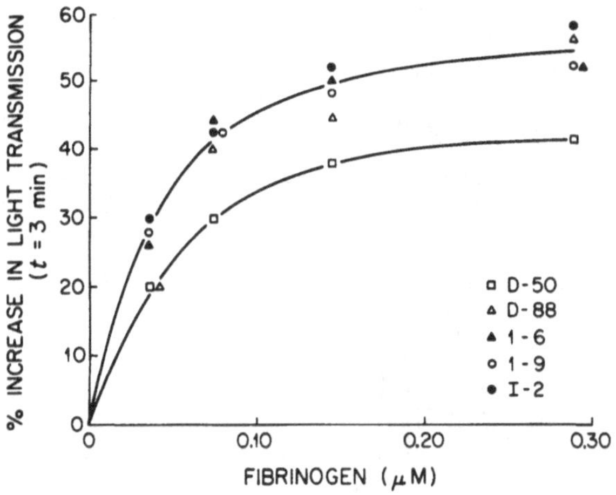

Fig. 2. Effect of increasing concentration of intact fibrinogen (I-2), fractions I-6 and I-9, and derivatives I-9D^{88} (D-88) and I-9D^{50} (D-50) on primary ADP-induced aggregation. One of five similar experiments (Reprinted by permission, C.V. Mosby Co.).

All of these fibrinogen preparations were iodinated with 125-iodine using chloramine T(3). Their ability to bind to ADP-treated platelets and to support aggregation was assessed at room temperature using gel filtered platelets in 0.01 M HEPES-buffered modified Tyrode's solution, pH 7.5 (HBMT) as described elsewhere (3,23). Fig. 2 depicts the ability of increasing concentrations of various fibrinogen preparations to support ADP-induced aggregation. Only fraction I-9D^{50} demonstrated consistently diminished aggregation support, about 25-30% less than that of control fibrinogen.

All fibrinogen preparations tested, bound to platelets in a specific and saturable manner, and bound to the same receptor as shown in Fig. 3. Illustrated is the ability of excess unlabeled fraction I-2 to displace bound labeled fractions I-6, I-9 and derivatives I-9D^{88} and I-9D^{50}.

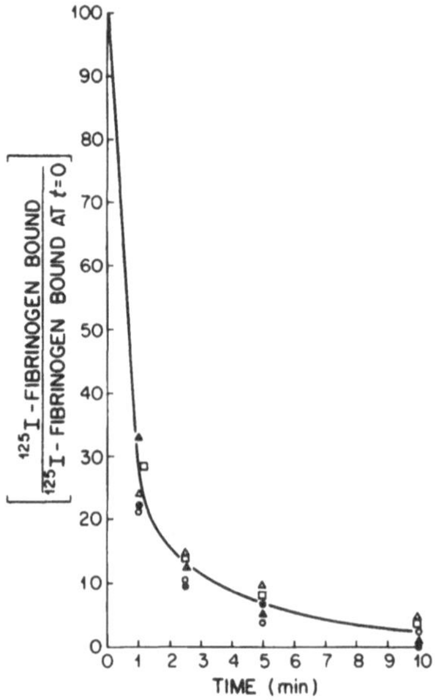

Fig. 3. Displacement of bound iodinated intact fibrinogen ( ), bound iodinated fractions I-6 ( ) and I-9 ( ), and derivatives I-9D^{88} ( ) and I-9D^{50} ( ) by excess unlabeled intact fibrinogen (Reprinted by permission, C.V. Mosby Co.).

The binding characteristics of each of the fibrinogen preparations as calculated by Scatchard analysis are summarized in the Table. The binding of fractions I-6, I-9 and I-9D^{88} was similar to that of intact fibrinogen except that the apparent dissociation constants for high affinity binding of fractions I-6 and I-9 were slightly but significantly higher. Because the affinity returned to normal with further loss of peptides from the Aα chain, as in fraction I-9D^{88}, these changes most likely reflect conformational changes. The apparent dissociation constants for low affinity binding of fractions I-6 and I-9, as well as for derivative I-90^{88} were not significantly different from that of fraction I-2. Changes in binding characteristics were noted, however, with additional losses of interior segments of the Aα chain, and simultaneous losses of amino terminal segments from the Bβ chain. This is illustrated by derivative I-9D^{50}. Compared to the binding of intact fibrinogen, binding of derivative I-9D^{50} reached saturation at a lower fibrinogen concentration (Fig. 4).

Table. Comparison of the Interaction of ADP-Treated GFP with intact Fibrinogen and Fibrinogen Fractions Lacking Increasingly Large Segments from the Carboxy-Terminus of Their Aα Chains

Fibrinogen Fraction or Derivative	MW of major Aα chain population ($x10^{-3}$)	Kd app from aggregation (uM) (means ± S.D.)*	Kd of high/low affinity binding from Scatchard analysis (uM) (mean ± S.D.)*	Fibrinogen bound at maturation molecular/10^8 platelets (mean ± S.D.)*
Intact fibrinogen	70.9, 67.7	0.08±0.04	0.11±0.04/1.70±0.86	18,000±5126
Fraction I-6	57.7	o.33±0.09	0.40±0.05/1.30±0.74	15,000±1686
Fraction I-9	57.7, 46.5-31.8	0.40±0.17	0.39±0.16/1.21±0.53	13,500±3152
Derivative $I-9D^{88}$	46.5-31.8, 25.0, 22.6	0.09±0.04	0.12±0.07/2.09±0.88	13,950±4040
Derivative $I-9D^{50}$	25.0, 22.6	0.12±0.04	0.12±0.06/none	14,330±1058

*n=5

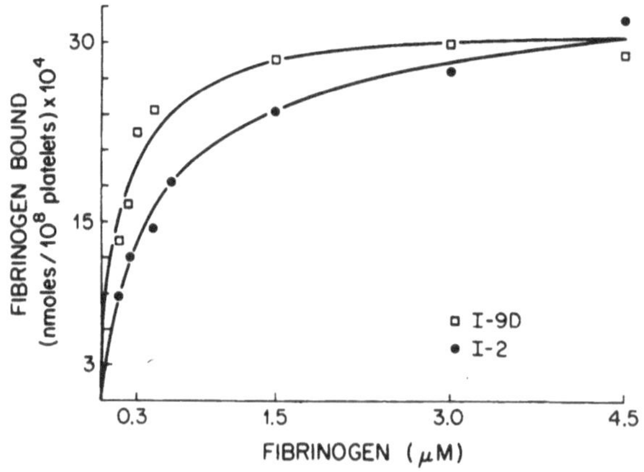

Fig. 4. Specific binding of fibrinogen derivative I-9D^{50} and intact fibrinogen (I-2) at increasing fibrinogen concentrations. One of five similar experiments (Reprinted by permission, C.V. Mosby Co.).

Saturation of fibrinogen receptors occured near 1.5 μM fibrinogen for I-9D^{50}, versus 3.0 μM for control fibrinogen. Scatchard analysis of the binding of fraction I-9D^{50} yielded a more linear plot (Fig. 5). The apparent dissociation constant calculated from the slope of this plot was not significantly different from that obtained for high affinity binding of fraction I-2.

Similar results were recently obtained with an abnormal fibrinogen preparation, fibrinogen Stony Brook. The abnormality was the result of a heterozygous defect in the Aα chain, namely an Arg-Cys substitution at position 16 (26). This fibrinogen was isolated from plasma (21), and the abnormal moiety concentration by removing normally clottable fibrinogen with reptilase (26, 27). At all fibrinogen concentrations tested (0.01-0.20 mg/ml), fibrinogen Stony Brook supported aggregation less well

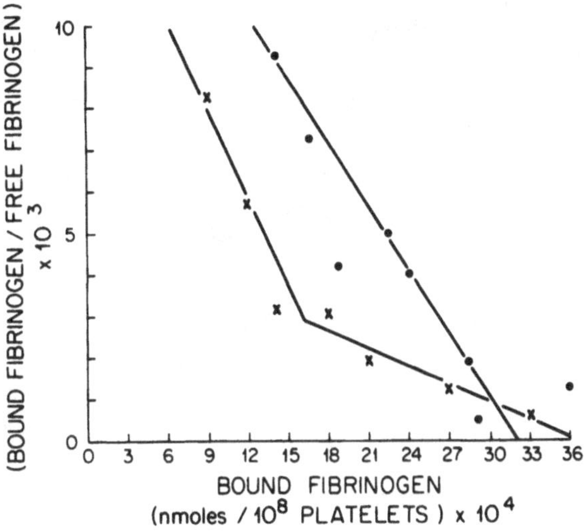

Fig. 5. Scatchard analysis of specific binding of fibrinogen derivative I-9D[50] (o) and intact fibrinogen (x). One of five similar experiments (Reprinted by permission, C.V. Mosby Co.).

than control fibrinogen (Fig. 6), adding further support to the role of the Aα chain in platelet aggregation. A typical aggregometer tracing comparing the ability of ADP treated, gel-filtered platelets to aggregate in the presence of 0.20 mg/ml normal fibrinogen and fibrinogen Stony Brook is depicted in Fig. 6. Evaluation of the specific binding of fibrinogen Stony Brook to ADP-treated platelets is presently underway.

These results as well as those obtained with fibrinogen derivative I-9D[50], suggest that the Aα chain of fibrinogen, particularly, its amino-terminal third, contributes to the capacity of fibrinogen to support platelet aggregation, and may influence the binding characteristics of the molecule (23). Although Bβ chains in fibrinogen Stony Brook are intact, and direct inter-

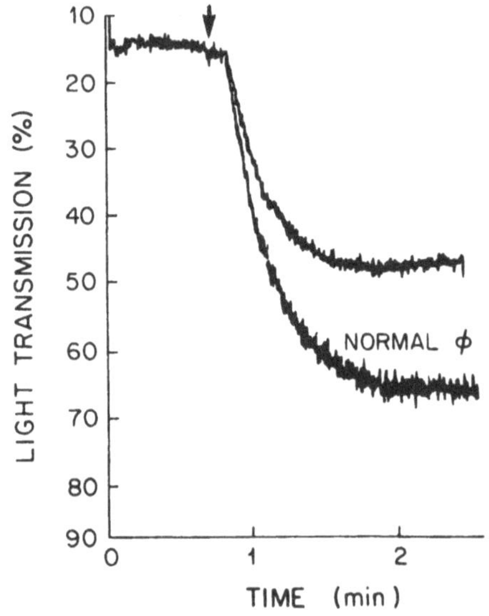

Fig. 6. Ability of 0.20 mg/ml of intact single donor fibrinogen (lower tracing) and Stony Brook fibrinogen (upper tracing) to support 10 M ADP-induced; primary aggregation of gel-filtered platelets in HBMT.

actions between the Bβ chain and platelets in a purified system have not been described (15), a possible role by the Bβ chain cannot be completely excluded as fraction $I\text{-}9D^{50}$ also possesses partially degraded Bβ chains.

Supported by United States Public Health grants HL28183 and HL27196.

References

1. Marguerie, G.A., Edgington, T.S., Plow, E.F.: J. Biol. Chem. 255, 154-161 (1980).

2. Bennett, J.S., Vilaire, G.: J. Clin. Invest. 64, 1393-1401 (1979).

3. Peerschke, E.I., Zucker, M.B., Grant, R.A., Egan, J.J., Johnson, M.M.: Blood 55, 841-847 (1980).

4. Hawiger, J., Parkinson, S., Timmons, S.: Nature 283, 195-197 (1980).

5. Kornecki, E., Niewiarowski, S., Morinelli, T.A., Kloczewiak, M.: J. Biol. Chem. 256, 5696-5701 (1981).

6. Mustard, J.F., Kinlough-Rathborne, R.L., Packham, M.A., Perry, D.W., Harfenist, E.J., Pai, K.R.M.: Blood 54, 987-993 (1979).

7. Nachman, R.L., Leung, L.L.K.: J. Clin. Invest. 69, 263-269 (1982).

8. DiMinno, G., Thiagarajan, P., Perussia, B., Martinez, J., Shapiro, S., Trinchieri, G., Murphy, S.: Blood 61, 140-148 (1983).

9. Kornecki, E., Lee, H.O., Tuszynski, G.P., Niewiarowski, S.: Fed. Proc. 41, 528 (Abstr.) (1982).

10. Bennett, J.S., Hoxie, J.A., Leitman, S.F., Vibire, G., Cines, D.B.: Proc. Natl. Acad. Sci. 80, 2417-2421 (1983).

11. McEver, R.P., Bennett, E.B., Martin, M.N.: Blood 60 (Suppl. 1) 202a, (Abstr.) (1982).

12. Pidard, D., Montgomery, R.R., Kunicki, T.J.: Blood 60 (Suppl. 1) 203a, (Abstr.) (1982).

13. Woods, V.L., Jr., Ginsberg, M.H., Kunicki, T.J., Oh, E.H., Jacques, B.C.: Blood 60 (Suppl. 1) 205a, (Abstr.) (1982).

14. Coller, B.S., Peerschke, E.I., Sudder, L.E., Sullivan, C.A.: J. Clin. Envest. 72, 325-338 (1983).

15. Hawiger, J., Timmons, S., Kloczewiak, M., Strong, D.D., Doolittle, R.F.: Proc. Natl. Acad. Sci. USA 79, 2068-2071 (1982).

16. Kloczewiak, M., Timmons, S., Hawiger, J.: Biochem. Biophys. Res. Comm. 107, 181-187 (1982).

17. Marguerie, G.A., Ardaillon, N., Cherel, G., Plow, E.F.: J. Biol. Chem. 257, 11872-11875 (1982).

18. Jerushalmy, Z. Zucker, M.B.: Thromb. Diath. Haemorrh. 15, 413-419 (1966).

19. Niewiarowski, S., Budzynski, A.Z., Lipinski, B.: Blood 49, 635-644 (1977).

20. Holt, J.C., Mahmoud, M., Gaffney, P.J.: Thromb. Res. 16, 427-435 (1979).

21. Galanakis, D.K., Mosesson, M.W., Stathakis, N.E.: J. Lab. Clin. Med. 92, 376-386 (1978).

22. Mosesson, M.W., Finlayson, J.S., Umfleet, R.A., Galanakis, D.K.: J. Biol. Chem. 247, 5210-5219 (1972).

23. Peerschke, E.I., Galanakis, D.K.: J. Lab. Clin. Med. 101, 453-460 (1983).

24. Plow, E.F., Edgington, T.S.: Thromb. Haemost. 46, 151 (Abstr.) (1981).

25. Goldstein, A., Arnow, L., Kaetian, S.M.: Principles of drug action In: The Basis of Pharmacology, ed. 2, New York, 1974, John Wiley & Sons Inc., p.82.

INHIBITION OF FIBRINOGEN BINDING TO ACTIVATED PLATELETS BY OLIGOAMINES

G. Agam, R. Luria, L. Shoshani, A. Livne
Department of Biology, Ben-Gurion University of the Negev
Beer Sheva, Israel

Introduction

Fibrinogen binding to activated platelets is believed to be indispensable for platelet aggregation. Various compounds inhibit both fibrinogen binding and aggregation (1-6). Amino sugars and basic amino acids inhibit a series of platelet activities including thrombin-, ADP- and A23187-induced platelet aggregation (7,8) and binding of fibrinogen to ADP-activated platelets (9). However, relatively high concentrations of the amines are necessary for the inhibition of the various platelet related processes (7-13). Series of oligomers of primary amines were recently found to be more potent inhibitors of platelet aggregation and platelet lectin activity than the individual component amines (8). The increase in potency was much more than could be accounted for by an additive effect. It is possible that fibrinogen binding is involved in the inhibition of platelet aggregation and lectin activity by the oligoamines. Therefore, it was of interest to compare the pattern of inhibition of these platelet functions by the oligoamines. The present study indeed presents a pattern of inhibitory effect of oligoamines on fibrinogen binding similar to that recently described for the inhibition of platelet aggregation and lectin activity (8).

Materials and Methods

Materials

Albumin (bovine) was purchased from Hyland, Costa Mesa, Ca; the Ca^{2+}

Fibrinogen - Structural Variants and Interactions
© 1985 Walter de Gruyter & Co., Berlin · New York - Printed in Germany.

ionophore A23187 from Calbiochem AG, Luzern; Sepharose 2B from Pharmacia Fine Chemicals, Uppsala; fibrinogen (human, 97% clottable) from Cutter, Berkeley, CA; and sodium iodide - ^{125}I, 13-17 mCi/µg iodide and 100 mCi/ml from Nuclear Research Centre, Negev. L-lysine and L-histidine were obtained from Sigma, St. Louis, MO; histidyl-lysine (abbr. his·lys) and $(lys)_5$ from Serva, Heidelberg, and $(lys)_4$ from Miles-Yeda, Rehovot.

Preparation of platelet suspension

Blood from healthy human volunteers was collected into acid citrate dextrose. Platelet rich plasma (PRP) was obtained by centrifugation at 120 x g for 10 min. Platelets were separated by gel filtration as follows: first, the concentration of platelets in PRP was increased threefold following centrifugation at 200 x g for 10 min; then 1 ml of concentrated PRP was passed through a Sepharose 2B column (0.8 x 12 cm). The column was equilibrated with Ca^{2+}- and Mg^{2+}-free Tyrode solution, pH 7.4, supplemented with 3.6 mg/ml albumin. Platelets were eluted with the same solution.

Radioactive labeling

Fibrinogen was iodinated with ^{125}I-iodine according to Knight et al. (14).

Fibrinogen binding

The binding of ^{125}I-fibrinogen to washed platelets was measured as follows: platelets (2.5×10^7 to 7.5×10^7) were suspended in Tyrode's buffer, supplemented with 0.625 mg ^{125}I-fibrinogen, 0.9 mg albumin and either 2.5 mM $CaCl_2$ or 0.2 mM EDTA in a final volume of 0.25 ml, and activated by 2.5 µg A23187. The suspensions were incubated at 23°C, without stirring, for 10 min, and then aliquots (100 µl, in duplicates) were layered over 1 ml of 20% sucrose in a 1.5 ml conical tube and spun for 1.5 min at 12000 rpm in a Beckman microfuge. The ^{125}I-fibrinogen bound was determined by counting the radioactivity in the tip of the centrifuge tube, following the removal of the supernatant and the detachment of the tip. The total ^{125}I-fibrinogen binding was corrected by subtracting the counts recorded in the absence of Ca^{2+}.

Equilibrium dialysis

Dialysis membranes (Sigma) were treated by immersion in a solution of 10 mM EDTA and 0.1 M $NaHCO_3$ for 30 min at 95°C, then washed thoroughly with distilled water. The apparatus for dialysis was composed of two cells (1 ml each) separated by the membrane (2.7 cm^2). The cells contained 155 mM NaCl-50 mM phosphate buffer pH 8.0, with or without 1 mM $CaCl_2$, 2 mM NaN_3 and either L-lysine at the range of 150 μM to 20 mM or $(lys)_5$ at the range of 25 μM to 1 mM. Fibrinogen (37 μM) was added to one of the cells and the apparatus rotated vertically at 33 rpm at 23°C for 24 hours. Control experiments showed that under these conditions equilibrium between the two cells was fully achieved. Samples from the compartments devoid of fibrinogen were analyzed for amine content at the beginning and at the end of incubation according to the manual method of Böhlen et al. (15). A Perkin-Elmer fluorescent spectrophotometer was used. The assay system enabled the detection of binding of 1.5 n moles $(lys)_5$.

Results

Fig. 1 shows that lysine or histidine inhibit fibrinogen binding to A23187-activated platelets. The extent of inhibition was linearly proportional to the amino acid concentration on a semilogarithmic plot. 50% inhibition was caused by 25-40 mM of either lysine or histidine. The dipeptide his•lys was much more potent than the individual amino acids, and the pentapeptide $(lys)_5$ - even more so: concentrations of only 7 mM and 1 mM, respectively, were required to inhibit fibrinogen binding by 50%.

The mode of action of the oligoamines may involve competition with charged groups on either the platelet membrane or the fibrinogen molecule. To explore such possible competition, the inhibitory effect of a given dose of $(lys)_4$ was tested at different fibrinogen concentrations (fig. 2) as well as different platelet concentrations (fig. 3).

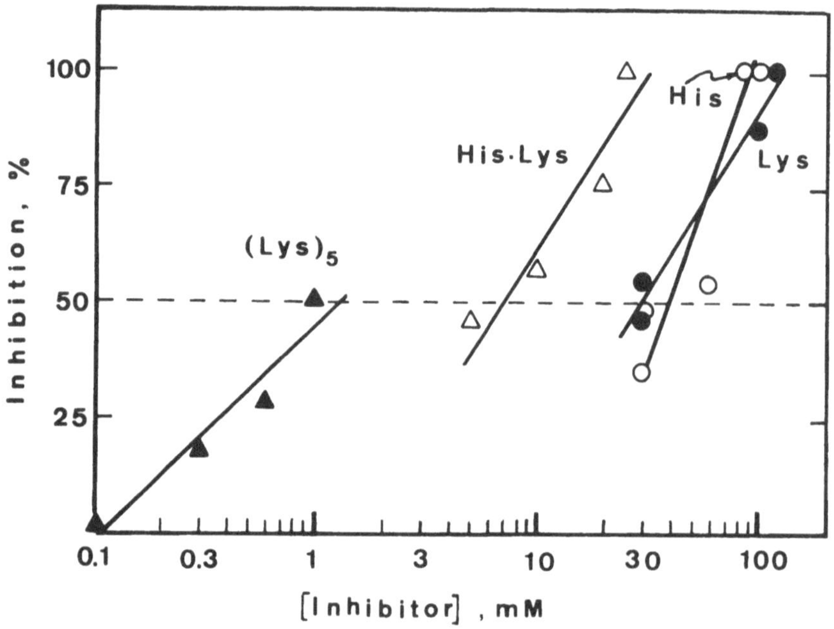

Fig. 1. Inhibition of fibrinogen binding to A23187-activated platelets by L-histidine, L-lysine, his-lys and $(lys)_5$. The binding of fibrinogen, in the absence of inhibitor (750 ng/10^8 platelets) was referred to as 100%. Results are the means of 3 experiments, each point in duplicates.

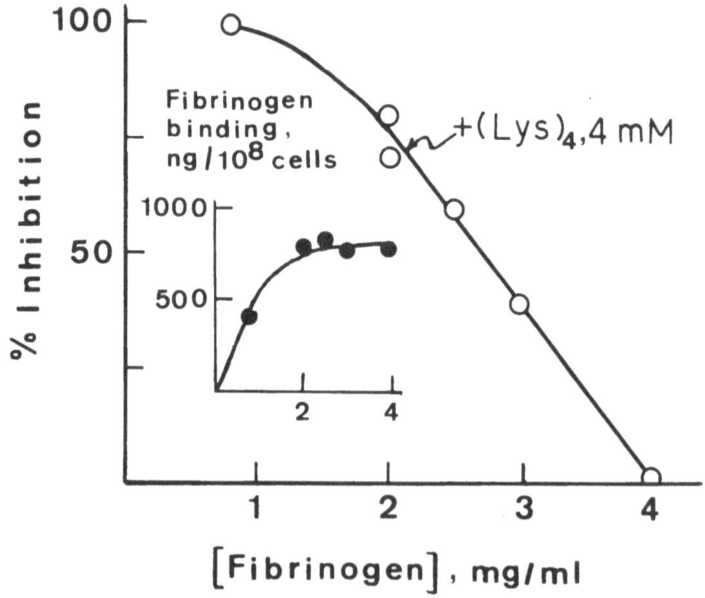

Fig. 2. Inhibition of fibrinogen binding by a constant dose (4 mM) of $(lys)_4$ at different fibrinogen concentrations. Inset: fibrinogen binding as a function of fibrinogen concentration. Results are the means of 5 experiments, each point in duplicates.

The inset in fig. 2 describes the dependence of fibrinogen binding on fibrinogen concentration. It demonstrates that the binding was not limited by fibrinogen concentration above 2 mg/ml. Under these conditions, the figure shows that the inhibitory effect of $(lys)_4$ on the binding is gradually diminished with increasing concentration of fibrinogen. The inhibition, which was as high as 100% at 0.75 mg/ml fibrinogen, was completely eliminated at 4 mg/ml of fibrinogen. In a similar fashion, fig. 3 shows that platelet concentration above 4×10^7/assay obviates the inhibitory effect of 4 mM $(lys)_4$. The inhibition of fibrinogen binding by $(lys)_4$ was also observed using platelets activated by 10 μM ADP (table I). It is noteworthy that platelets activated by 1 μM ADP do bind fibrinogen; however, the binding is not inhibited by the oligoamine.

The possibility that the oligoamines associate with fibrinogen and thereby inhibit fibrinogen binding was tested by equilibrium dialysis measurements. It was found that fibrinogen (37μM) failed to bind either L-lysine or $(lys)_5$ at a wide concentration range of the amines (150 μM - 20 mM and

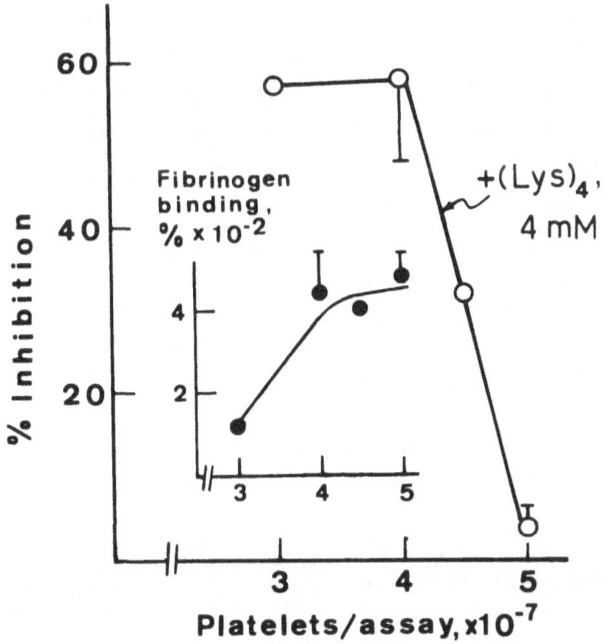

Fig. 3. Inhibition of fibrinogen binding by a constant dose (4 mM) of $(lys)_4$ at different platelet concentrations. Inset: fibrinogen binding as a function of platelet concentration. Results are the means of 3 experiments, each point in duplicates.

25 μM - 1 mM, respectively). Absence of association between the amine and fibrinogen was found both when Ca^{2+} was present in the system or absent.

Discussion

The study of fibrinogen binding to activated platelets has primarily been based on ADP-activated platelets (16-18). Fibrinogen binding to thrombin-, epinephrine-, collagen- and arachidonic acid-induced platelets has also been documented (18-20). The present study is, apparently, the first one which reports the use of the Ca^{2+} ionophore A23187 for fibrinogen binding. The value of maximal binding obtained with the A23187-activated platelets (750 ng/10^8 platelets), as well as the pattern of the dependence of the binding upon fibrinogen concentration are similar to those reported for ADP-activated platelets (17).

The objective of the present study was to test the possibility that the inhibition of aggregation by oligoamines involves fibrinogen binding. Several considerations support this supposition. In analogy to the inhibition of aggregation (8), the oligoamines do inhibit fibrinogen binding. Furthermore, a similar pattern of inhibitory potency was observed, both qualitatively and quantitatively, in the two processes. For example, fig. 1 in the present study is almost superimposable on the data concerning platelet aggregation (8). The results in the present work corroborate

Table I. Effect of ADP concentration on the extent of inhibition of fibrinogen binding by $(lys)_4$.

ADP	fibrinogen binding	inhibition by 1.6 mM $(lys)_4$
μM	ng/10^8 platelets	%
1	1180	0
10	2230	86

Results are the means of two experiments, each carried out in quadruplicates.

earlier reports that fibrinogen binding to activated platelets is inhibited by primary amines (9).

Fibrinogen binding to activated platelets may occur through interaction with the surface glycoproteins IIb - IIIa (21, 22), and/or with thrombospondin released from the α granules and attached onto the platelets (12, 13, 23). Fibrinogen binding to platelets activated by 10 μM ADP was inhibited by $(lys)_4$, whereas the binding to platelets activated by 1 μM ADP was not (table I). Since no release takes place with the lower ADP concentration (24), the differential inhibitory effect may indicate that the inhibition by the oligoamines is directly related to thrombospondin-fibrinogen interaction. Therefore, it is of interest to further extend the study of the impact of oligoamines on thrombospondin-fibrinogen complex formation (13).

The mode of action of the oligoamines may involve their interaction with the fibrinogen molecule and/or the platelet surface. The apparent competitive relationship reflected by figures 2 and 3 is in line with these possibilities. Negatively charged regions on the fibrinogen molecule (25) may facilitate an electrostatic association with the positively charged oligoamines. However, two lines of evidence indicate that a direct interaction of the oligoamines with fibrinogen is unlikely:

a) No binding of either L-lysine or its oligopeptide $(lys)_5$ to fibrinogen could be detected by equilibrium dialysis.

b) At 2.5 μM fibrinogen, 4 mM $(lys)_4$ exerted 100% inhibition of fibrinogen binding, while a change in the fibrinogen concentration to about 12 μM completely abolished the inhibitory effect (fig. 2).

It is thus concluded that the mode of action of the oligoamines probably involves their interaction with the platelet surface.

The synthetic oligoamines may serve as a model for the study of the natural fibrinogen fragments which comprise the active site for interaction with and between platelets.

References

1. Mustard, J.F., Packham, M.A., Kinlough-Rathbone, R.L., Perry, D.W., Rogoeczi, E.: Blood 52, 453-466 (1978).

2. Peerschke, E.I., Zucker, M.B.: Blood 57, 663-640 (1981).

3. Hawiger, J., Parkinson, S., Timmons, S.: Nature 283, 195-197 (1980).

4. Marguerie, G.A., Plow, E.F.: Biochemistry 20, 1074-1080 (1981).

5. Peerschke, E.I., Zucker, M.B.: Thromb. Haemost. 43, 58-60 (1980).

6. Harfenist, E.J., Packham, M.A., Kinlough-Rathbone, R.L., Mustard, J.F.: J. Lab. Clin. Med. 97, 680-688 (1981).

7. Gartner, T.K., Williams, D.C., Minion, F.C., Phillips, D.R.: Science 200, 1281-1283 (1978).

8. Agam, G., Gartner, T.K., Livne, A.: Submitted for publication.

9. Kinlough-Rathbone, R.L., Chahil, A., Perry, D.W., Packham, M.A., Mustard, J.F.: Blood 54 (suppl.), 249a (1979).

10. Gartner, T.K., Williams, D.C., Phillips, D.R.: Biochem. Biophys. Res. Commun. 79, 592-599 (1977).

11. Gartner, T.K., Gerrard, J.M., White, J.G., Williams, D.C.: Nature 289, 688-690 (1981).

12. Jaffe, E.A., Leung, L.L.K., Nachman, R.L., Levin, R.I., Mosher, D.G.: Nature 295, 246-248 (1982).

13. Leung, L.L.K., Nachman, R.L.: J. Clin. Invest. 70, 542-549 (1982).

14. Knight, L.C., Budzynski, A.Z., Olexa, S.A.: Thromb. Haemost. 46, 593-596 (1981).

15. Böhlen, P., Stein, S., Dairman, W., Udenfriend, S.: Arch. Biochem. Biophys. 155, 213-220 (1973).

16. Marguerie, G., Plow, E.F., Edgington, T.S.: J. Biol. Chem. 254, 5357-5363 (1979).

17. Peerschke, E.I., Zucker, M.B., Grant, R.A., Egan, J.J., Johnson, M.M.: Blood 55, 841-847 (1980).

18. Bennett, J.S., Vilaire, G.: J. Clin. Invest. 64, 1393-1401 (1979).

19. Plow, E.F., Marguerie, G.A.: Blood 56, 553-555 (1980).

20. Di Minno, G., Thiagarajan, P., Perussia, B., Martinez, J., Shapiro, S., Trinchieri, G., Murphy, S.: Blood 61, 140-148 (1983).

21. Nachman, R.L., Leung, L.L.K.: J. Clin. Invest. 69, 263-269 (1982).

22. Bennett, J.S., Vilaire, G., Clines, D.B.: J. Biol. Chem. 257, 8049-8054 (1982).

23. Phillips, D.R., Jennings, L.K., Prasanna, H.R.: J. Biol. Chem. 255, 11629-11632 (1980).

24. Charo, I.F., Feinman, R.D., Detwiler, T.C.: J. Clin. Invest. 60, 866-873 (1977).

25. Hermans, J., Mcdonagh, J.: Sem. Thromb. Hemost. 8, 11-24 (1982).

SPECIFIC INTERACTION BETWEEN FIBRINOGEN-FIBRIN AND ENDOTHELIAL CELLS

E. Dejana, L.R. Languino, Vergara M. Dauden, G. Balconi

Istituto di Ricerche Farmacologiche Mario Negri
Via Eritrea 62,
20157 Milano

G. Marguerie

Inserm U143, Institut de Pathologie Cellulaire, Hopital Bicêtre
94270 Le Kremlin-Bicêtre

Fibrinogen is a major plasma protein involved in thrombotic disease and blood vessel injury. Platelets require fibrinogen to aggregate in response to ADP and other aggregating agents (1). After stimulation, platelets undergo morphological changes and acquire the capacity to interact with plasma fibrinogen (2). Fibrinogen binding to platelets is the prerequisite for platelet aggregation, and dissociation of platelet-bound fibrinogen correlates with platelet disaggregation. The interaction is time-dependent and mediated by specific, saturable receptors (3).

Platelets can also interact with polymerizing fibrin and induce

Fibrinogen - Structural Variants and Interactions
© 1985 Walter de Gruyter & Co., Berlin · New York - Printed in Germany

retraction of a fibrin clot. This property is shared by other cell types such as cultured normal human and mouse fibroblasts from different sources (4) and cultured human and bovine endothelial cells (5).

These experiments are summarized in fig. 1. Fibrin clot retraction (FCR) was studied using 200 μl of single-cell suspension, 100 μl of human, platelet-free, citrated plasma and 7 U of thrombin, added in rapid succession.

After addition of the clotting enzyme, the diameter of the clot was measured at intervals with microcalipers. Retraction values were expressed as a percentage of complete clot retraction according to a procedure previously described (4). As shown in fig. 1, cultured human umbilical vein and bovine aorta endothelial cells induced FCR comparable to that induced by the same number of platelets. FCR was time-dependent and reached a plateau about 6 hours after addition of thrombin. With the three types of cells used no retractile activity could be detected at 22°C, or in the presence of Na_2-EDTA, or if the cells were mechanically disrupted. Moreover, within the same culture of endothelial cells, FCR activity appeared to be dependent to the functional state of the cells, being lower in cells with higher density and rising as the number of cell doublings increased.

It is known that platelet-induced FCR requires binding of thrombin to platelets with subsequent activation of the platelet contractile protein(s) and binding of platelets to fibrin through specific receptor sites (6). It is not yet known whether the fibrin receptor on platelets is the same as or closely related to the fibrinogen receptor. The existence of similar receptor systems in endothelial cells may be hypothesized to explain their functional interaction with

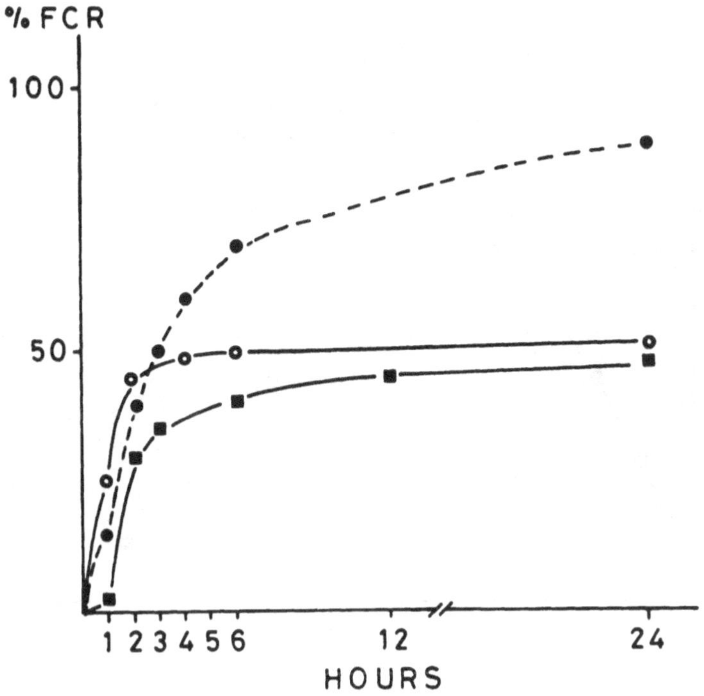

Figure 1 : Course of fibrin clot retraction (FCR) activity of human endothelial cells (●--●) bovine endothelial cells (o--o) and human platelets (■--■). The final cell concentration was $2x10^6$ cells/ml.

fibrin or fibrinogen. To assess this concept we initiated an investigation on the direct interaction between native fibrinogen and cultured human and bovine endothelial cells. Endothelial cells in single-cell suspension were incubated with ^{125}I-fibrinogen. To separate the free from the bound ligand the cell suspension was centrifuged on 20% sucrose gradient and the radioactivity associated with the cell pellet was measured. Binding was time-dependent and reached a plateau within 15 min. (Fig. 2). When a 100-fold molar excess of unlabelled fibrinogen was added to the system, very little binding of ^{125}I-fibrinogen was observed (10-

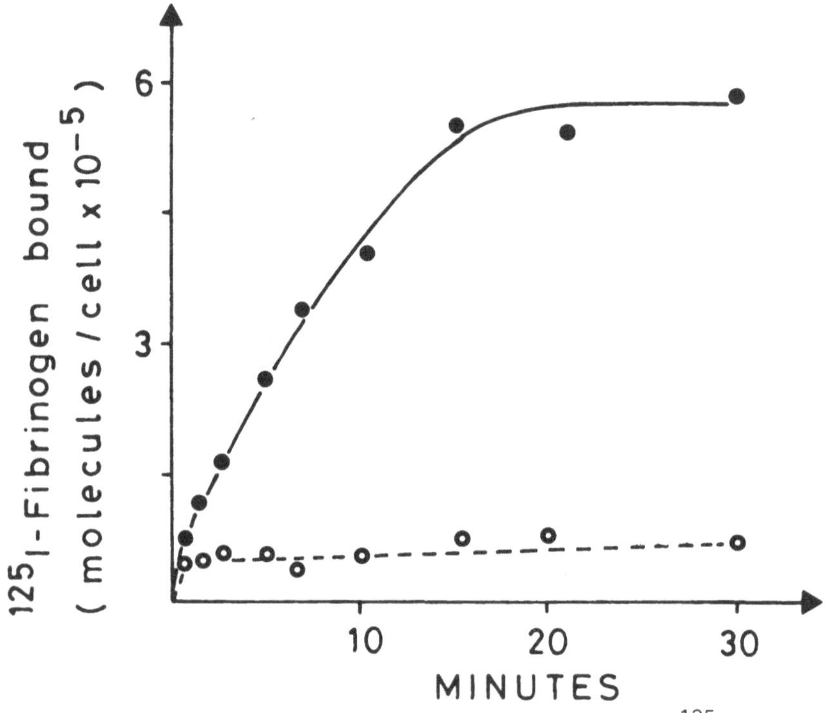

Figure 2 : Time course of the association of ^{125}I-fibrinogen with bovine endothelial cells. Single-cell suspensions at a concentration of $5x10^6$/ml were incubated at 37°C with 0.04 µM ^{125}I-fibrinogen in phosphate buffer saline containing 1% bovine serum albumin and 1 mM Ca^{2+} and Mg^{2+} total binding (●--●) and binding obtained in the presence of 100-fold molar excess of unlabelled fibrinogen (o--o).

~15% of total binding). This binding was very fast and peaked at about 1 min consistent with kinetics observed for non specific interaction.

The concentration-dependence of the specific ^{125}I-fibrinogen binding to

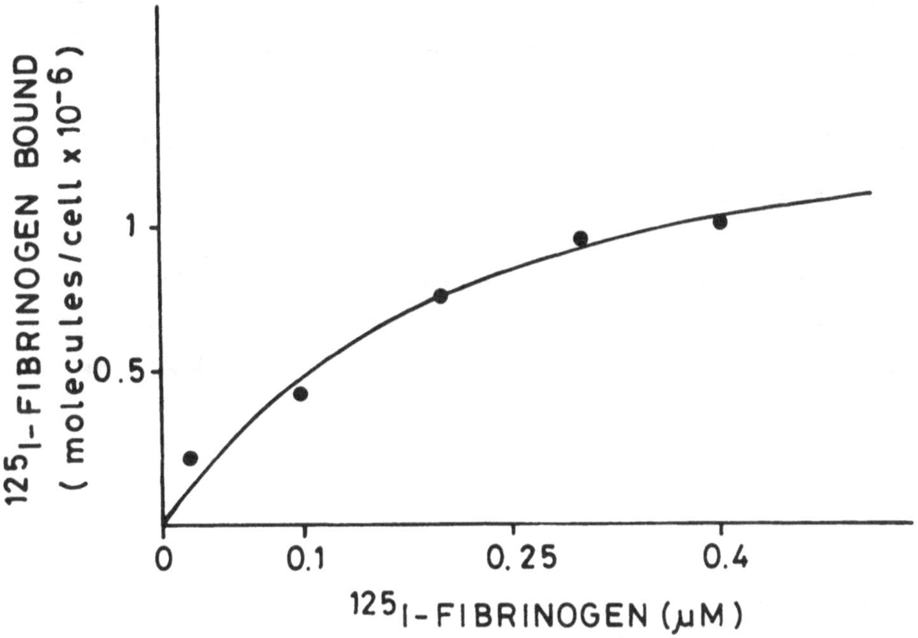

Figure 3: Specific binding of ^{125}I-fibrinogen to bovine endothelial cells as a function of ^{125}I-fibrinogen concentration. Experimental conditions were as in Fig. 2. Specific binding was obtained for each point by subtracting the values obtained in the presence of a 100-fold molar excess of unlabelled fibrinogen from total binding.

endothelial cells is shown in fig. 3. The cells were incubated for 30 min with ^{125}I-fibrinogen at concentrations ranging from 0.01 to 0.4 μM. Binding parameters were derived using a non-linear regression program (7) and were B max=$1.7x10^{3}$ molecules/cell and Kd=$0.23x10^{-6}$ M.

The specificity of this interaction for fibrinogen was established by competitive inhibition of binding by an excess of unlabelled fibrinogen in comparison to other proteins. A 100-fold molar excess of unlabelled fibrinogen competitively inhibited ^{125}I-fibrinogen binding to the cells by 80-90% of total binding. The same molar concentration of egg albumin or myoglobin had very little effect.

These results indicate that endothelial cells may specifically interact with fibrinogen and that this interaction exhibits the same characteristics as a receptor system, i.e. saturability, time-dependence, specificity.

Table 1

Binding of ^{125}I-fibrinogen and thrombin treated ^{125}I-fibrinogen to human endothelial cells

	cpm/10^5 cells
^{125}I-fibrinogen	30.900 ± 4900
Thrombin treated ^{125}I-fibrinogen	108.300 ± 12300

We further investigated if endothelial cells may interact also with fibrin ^{125}I-fibrinogen (0.02 μM) was treated with 0.04 NIH U/ml thrombin for 15 min at 37°C. Thrombin was then neutralized by addition of 4 NIH U/ml hirudin and the mixture added to endothelial cells in monolayer. After 30 min of incubation the cells were washed four times with phosphate-buffer-saline containing 1% bovine serum albumin 1 mM Ca^{2+} and Mg^{2+} and the radioactivity associated to the cells extracted with 2% sodium-dodecyl-sulphate+ 0.1 M NaOH and counted. As shown in table 1 thrombin-treated ^{125}I-fibrinogen bound to endothelial cells and this binding was higher than the binding of untreated ^{125}I-fibrinogen. Since fibrin aggregates these results are difficult to interpret in terms of affinity nevertheless they indicate that endothelial cells may associate with polymerizing fibrin.

In summary we report here that human or bovine endothelial cells can retract a fibrin clot and that this biological activity may be mediated via specific interactions.

In support of this concept are observations from others that fibrinogen and its physiological derivatives can interact and participate to endothelial cell functions. Witte (8) demonstrated association of fluorescent fibrinogen to the endothelial cells in the mesenteric microcirculation <u>in vivo</u> in the rat. Kadish et al (9), demonstrated that interaction between fibrin and endothelial cells leads to a complete disorganization of the cell monolayer. Nicosia et al (10), showed a fibrin induced migratory and proliferative effect on capillary endothelial cells, in an <u>in vitro</u> system of angiogenesis. Plasmin degradation fragments of fibrinogen are reported to increase vascular cell permeability (11).

It remains to be established if the specific interaction of fibrinogen with the endothelium may be correlated with the modifications induced by this molecule, and its derivative, on functions of the EC.

Acknowledgements: The authors wish to thank Ms Judith Baggott, Vincenzo and Felice de Ceglie, Ivana Garimoldi and Vanna Pistotti who helped prepare the manuscript. This work was partially supported by grants from NATO No 0572-82 and Italian National Research Council No 82-02384-51. L.R.L. is Fellow of Pepsico Foundation, Purchase, N.Y., U.S.A.

REFERENCES

1. Mustard, J.F., Perry, D.W., Kinlough-Rathbone, R.L., Packham, M.A.: Am. J. Physiol. 228, 1757-1765 (1975).
2. Marguerie, G.A., Plow, E.F., Edgington, T.S.: J. Biol. Chem. 254, 5357-5363 (1979).
3. Marguerie,G.A., Edgington, T.S., Plow, E.F.: J. Biol. Chem. 255, 154-161 (1980).
4. Curatolo, L., Balconi, G., Borgia, R., Morasca, L., Donati, M.B.: In Vitro 16, 731-737 (1980).
5. Barbieri, B., Balconi, G., Dejana, E., Donati, M.B.: Proc. Soc. Exp. Biol. Med. 168, 204-207 (1981).
6. Niewiarowski, S., Regoeczi, E., Stewart, G.J., Senyi, A.F., Mustard, J.F.: J. Clin. Invest. 51, 685-700 (1972).
7. Sacchi Landriani, G., Guardabasso, V., Rocchetti, M.: Comput. Programs Biomed. 16, 35-42 (1983).
8. Witte, S.: Thromb. Res. suppl.V, 93-104 (1983).
9. Kadish, J.L., Butterfield, C.E., Folkman, J.: Tissue Cell 11, 99-108 (1979)
10. Nicosia, R.F., Tchao, R., Leighton, J.: In Vitro 18, 538-549 (1982).
11. Gerdin, B., Saldeen, T.: Thromb. Res. 13, 995-1006 (1978).

INFLUENCE OF FIBRIN, FIBRINOGEN AND FIBRINOGEN DEGRADATION PRODUCTS ON CULTURED ENDOTHELIAL CELLS

Kenzo Tanaka, Kentaro Watanabe

Department of Pathology, Faculty of Medicine, Kyushu University,
Fukuoka 812, Japan

Summary

Thin films of fibrin-agar mixture induced significant morphological changes in porcine aortic endothelial cells in culture. These morphological changes became more significant with higher concentration of fibrin. Films of fibrinogen-agar mixture or thrombin-agar mixture induced no such morphological changes, thereby suggesting that fibrin itself injures the endothelial cells. Lower molecular weight fibrinogen degradation products also induced injury in endothelial cells. The incubation of cultured aortic smooth muscle cells for 24h with fibrinogen degradation product D, E and lower molecular weight fibrinogen degradation products decreased prostaglandin I_2 synthesis. After 24h incubation with fibrinogen degradation products, exogenous arachidonic acid-induced prostaglandin I_2 synthesis by endothelial cells was diminished. These data suggest that the synthesis of prostaglandin I_2 of vascular wall may be affected by fibrinogen degradation products.

As damage to the endothelium is a key event in initiation of atherosclerosis, fibrinogen and its metabolites probably play important roles.

Introduction

Many investigators noted the importance of the coagulation-fibrinolysis system in atherosclerosis (1, 2, 3). Recently the roles of prostaglandins, particularly prostaglandin I_2 (PGI_2)(4, 5), has been discussed in atherogenesis. We investigated the morphological changes of endothelial cells

Fibrinogen - Structural Variants and Interactions
© 1985 Walter de Gruyter & Co., Berlin · New York - Printed in Germany

in contact with fibrin, fibrinogen and fibrinogen degradation products (FDPs). And we also investigated the influence of FDPs on the synthesis of PGI_2 by endothelial and smooth muscle cells using culture technique.

Materials and Methods

1) Isolation and culture. Porcine aortic endothelial cells were isolated and cultured as previously described (6). Porcine aortic medial smooth muscle cells were isolated and cultured from explants by the method of Ross (7). Medium of both cultures was Medium 199 (M199) (GIBCO) supplemented with 10% fetal calf serum (FCS) and antibiotics (penicillin G 100U/ml, streptomycin 100µg/ml).
2) Purification of porcine plasminogen-free fibrinogen, FDP D, E and lower molecular weight fibrinogen degradation products (LMWFDPs). The details of the purification procedures were previously described (6). Porcine plasminogen-free fibrinogen was purified by applying plasma to lysine-Sepharose column, followed by ethanol fractionation described by Blombäck, B and Blombäck,M (8). FDP D and E were purified by plasmin digestion, Sephacryl S-300 column chromatography and Pevicon C-870 block electrophoresis (9). LMWFDPs were purified as follows. After plasmin digestion for 24h, the digests were passed through a PM10 filter (Amicon, Lexington, MA, USA). The effusate with molecular weights less than 10,000 was chromatographed on a Sephadex G-25 column equilibrated with 0.1M ammonium acetate pH 7.4 (10). The first, second and third peaks were separated and lyophi-

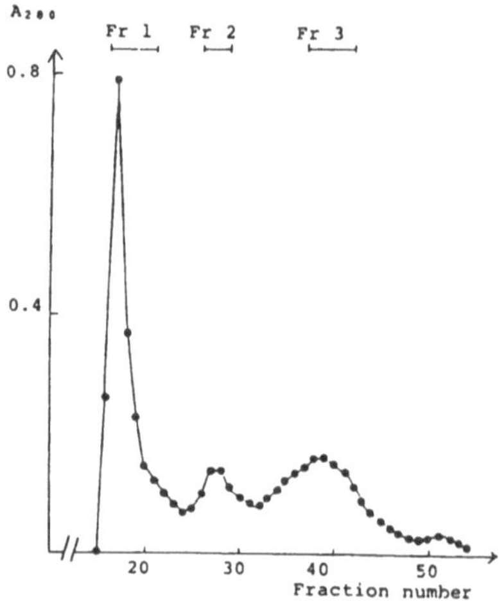

Fig.1. LMWFDPs "Fraction 1" (Fr 1), "Fraction 2" and "Fraction 3" Chromatographed on a Sephadex G-25.

lized (Fig.1). Protein concentration was determined by the method of Lowry et al (11).

3) Effects of fibrin,fibrinogen, thrombin and FDPs. Two ml of plasminogen-free fibrinogen (2mg/ml) in M199 were clotted on cultured endothelial cells in Falcon 3002 dishes by adding 150μl of 12.5U bovine thrombin (Mochida Pharmaceutical Co., Ltd., Tokyo, Japan) and morphological changes in endothelial cells were continuously observed until 24h. Fibrinogen-agar films, fibrin-agar films, thrombin-agar films and agar films were made as previously described (6). These thin films of various concentrations were gently placed on the monolayers of cultured endothelial cells. Morphological changes were observed after 24h. The effects of FDP D or E, LMWFDP Fr 1, 2 or 3 without agar on endothelial cells were also examined.

4) Endothelial injury by FDPs measured by ^{51}Cr release. Confluent endothelial cells cultured in Falcon 3047 dishes (24-well plates) were incubated with $Na[^{51}Cr]O_4$ (spec. act. 3.0GBq/mg Cr, JRIA) in M199 with 10% FCS, 5μCi/well, for 24h. Each well was then washed with serum-free M199 and fresh M199 with 5% FCS and various concentrations of FDP fractions were added. ^{51}Cr release after 24h was calculated by dividing ^{51}Cr radioactivity of the floating cells and medium by total ^{51}Cr radioactivity (6).

5) Effects of FDPs on PGI_2 synthesis by cultured endothelial and smooth muscle cells. When confluent, cultures were washed twice with serum-free M199 and incubated with various concentrations of FDPs in 5% CO_2 in air at 37°C. After 24h, the media were aspirated and the levels of 6keto-$PGF_{1\alpha}$ were determined by radioimmunoassay later. Then another 1ml of M199 was poured on each culture. After unlabeled arachidonic acid (AA) was added (final concentration 25μM), the cultures were incubated on a rocker platform at room temperature for 30 min. The media was aspirated for assay of PGI_2 and each well was treated with trypsin for counting cell numbers. These samples were assayed for 6keto-$PGF_{1\alpha}$ using specific antibody developed in our department. All these procedures were duplicated and repeated four (endothelial cells) or three (smooth muscle cells) times.

Results

1) Effects of fibrin, fibrinogen, thrombin and FDPs. Figure 2a shows porcine aortic endothelial cells in culture at 5th passage. Within 2h of the contact with fibrin, intercelullar spaces became apparent between the endothelial cells that had been adherent (Fig.2b). After 24h, the endothelial dissociation was prominent (Fig.2c). Fig.3a shows the endothelial cells in culture after contact with a fibrin-agar film with a concentration of fibrinogen 1mg/ml. Fig.3b 5mg/ml, and Fig.3c 10mg/ml. These morphological changes were more prominent with higher concentration of fibrin. Fibrinogen-agar films, thrombin-agar films or agar films without fibrinogen or thrombin did not induce any morphological change in the cultured endothelial cells, even at concentrations up to 10mg/ml, 10U/ml or 2%, respectively. FDP D or E in serum-free M199 without agar induced no morphological changes at concentrations up to 1400μg/ml or 700μg/ml, respectively. LMWFDP Fr 1-3 in serum-free M199 were also poured onto the cultures. Endothelial cells exposed to LMWFDP Fr 3 were damaged and many cells became detached from the wells (Fig.4).
2) Endothelial injury by FDPs measured by ^{51}Cr release. There was an increase in ^{51}Cr release with increasing concentration of LMWFDP Fr 3 (Fig.5). Correlation between ^{51}Cr release and the concentration of Fr 3 was statistically significant. LMWFDP Fr 1, 2 and FDP D, E showed no increase of the release of ^{51}Cr.
3) Effects of FDPs on PGI_2 synthesis by cultured smooth muscle and endothelial cells. PGI_2 synthesis by smooth muscle cells without exogenous AA is higher than that by endothelial cells. On the contrary, AA-induced PGI_2 synthesis by smooth muscle cells is lower than that by endothelial cells. Fig.6a shows PGI_2 synthesis by cultured smooth muscle cells and endothelial cells during 24h incubation with FDP D. The levels of PGI_2 synthesis by smooth muscle cells are decreased with increasing concentration of FDP D. Similar data were obtained in FDP E, LMWFDP Fr 1 and 2. In LMWFDP Fr 3, however, PGI_2 synthesis per cell by endothelial cells is remarkably increased with increasing concentration of Fr 3 (Fig.6b). Fig.7a shows AA-induced PGI_2 synthesis after incubation with FDP E. PGI_2 synthesis by endothelial cells was decreased with increasing concentration of FDP E. Similar data were obtained in FDP D,

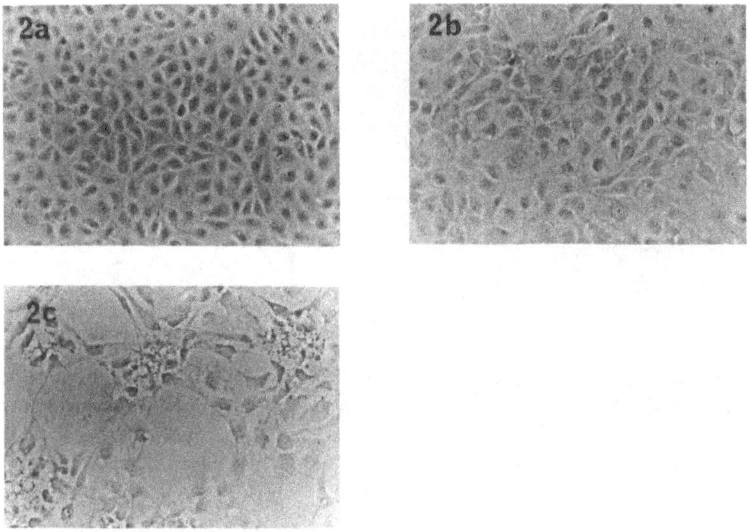

Fig.2. Phase contrast micrographs of the cultured endothelial cells at 5th passage in response to contact with fibrin (2mg/ml fibrinogen).
a : before treatment, b : 2h after the contact with fibrin, c : 24h after the contact with fibrin. The cells became spindle-shaped, piled up and intercellular spaces were widened.

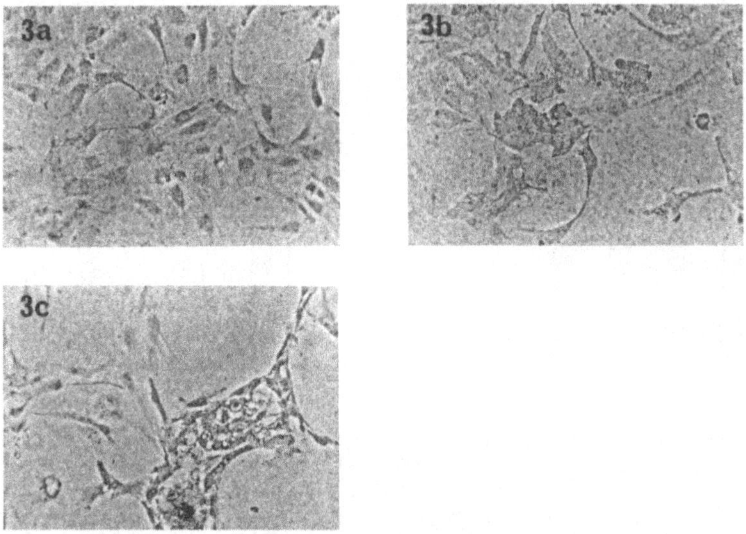

Fig.3. Phase contrast micrographs of the cultured endothelial cells at 4th passage in response to contact with fibrin-agar films (after 24h).
a : 1mg/ml fibrinogen, b : 5mg/ml fibrinogen, c : 10mg/ml fibrinogen.

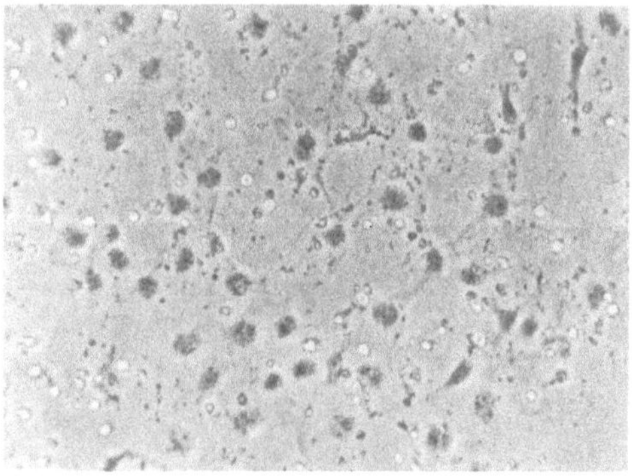

Fig.4. Phase contrast micrographs of cultured endothelial cells at 4th passage exposed to 100μg/ml LMWFDP Fr 3 (after 24h).

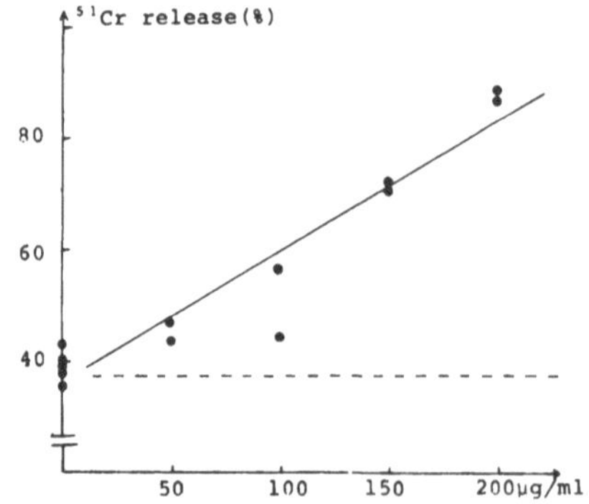

Fig.5. The effect of LMWFDP Fr 3 on the release of ^{51}Cr from endothelial cells. The correlation between ^{51}Cr release with respect to the concentration of Fr 3 is given by $Y=35.9 + 0.237X$ ($r=0.961$, $p<0.005$).

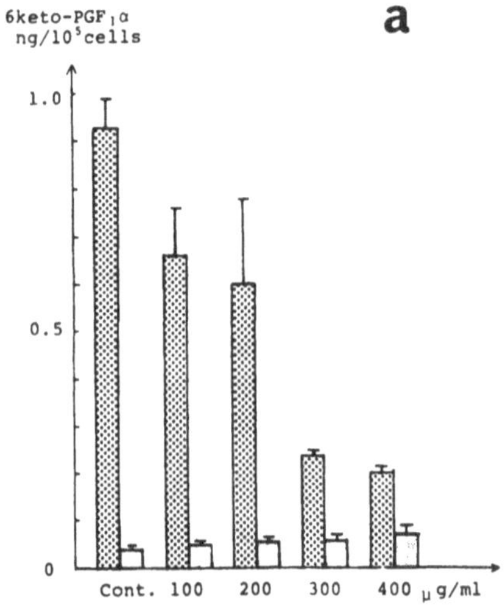

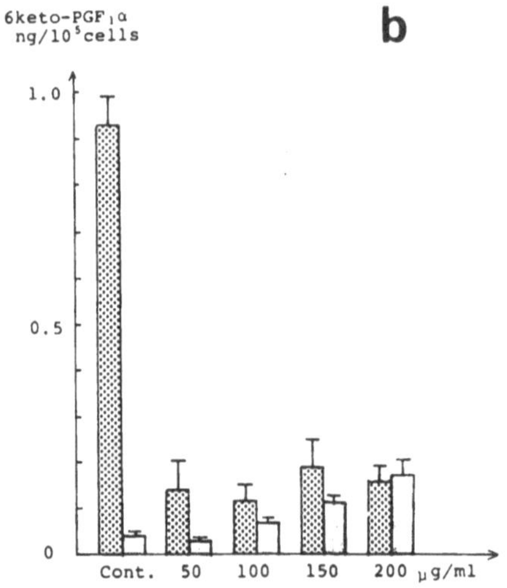

Fig.6. PGI_2 synthesis by cultured smooth muscle cells (columns with large dots) and endothelial cells (columns with small dots) at 4th passage during 24h incubation with FDP. a : incubation with FDP D, b : incubation with LMWFDP Fr 3. Data represent the mean value ± SEM of three duplicated dishes (smooth muscle cells) and four duplicated dishes (endothelial cells).

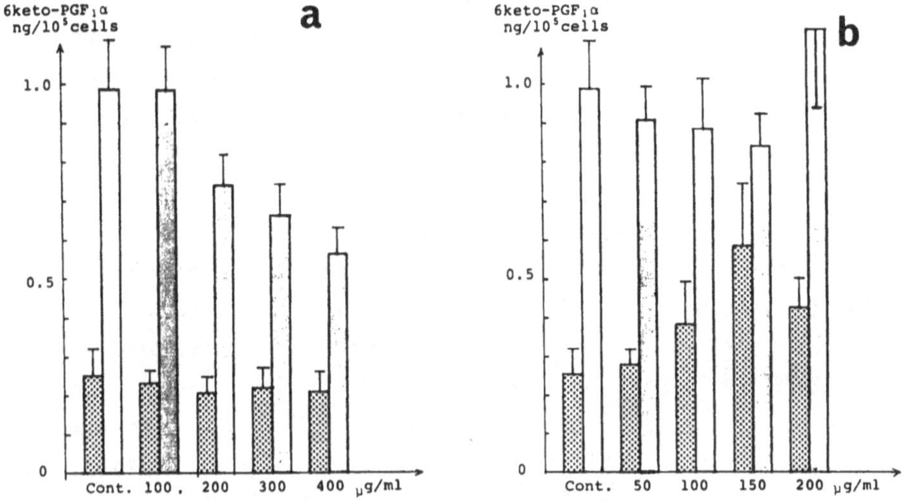

Fig.7. AA-induced PGI_2 synthesis by cultured smooth muscle cells (columns with large dots) and endothelial cells (columns with small dots) at 4th passage after incubation with FDP for 24h. a : after incubation with FDP E, b : after incubation with LMWFDP Fr 3. Data represent the mean value ± SEM of three duplicated dishes (smooth muscle cells) and four duplicated dishes (endothelial cells).

LMWFDP Fr 1 and 2. In LMWFDP Fr 3, however, PGI_2 synthesis by endothelial cells after incubation with 200μg/ml of Fr 3 was higher than control (incubation with serum-free M199)(Fig.7b).

Discussion

In 1976. Ross and Glomset (12) stressed the role of endothelial injury and subsequent proliferation of smooth muscle cells in initiation of atherosclerosis. We have examined the role of fibrin, fibrinogen and their degradation products as the cause of endothelial injury in initiation of atherosclerosis. Fibrin is in an insoluble state in vivo, and mechanical pressure by insoluble material, as well as fibrin, might induce morphological changes in endothelial cells. In an attempt at elucidation we used agarfilms as a control. Thin films of 1% agar without fibrin, fibrinogen or thrombin, produced no morphological changes such as were produced by fibrin in agar film. Fibrinogen or thrombin with or without agar also produced no morphological changes. The degree of morphological changes correlated with the concentration of fibrin. Thus fibrin does injure endothelial cells. Kadish et al (3) support our study. This injury, in vivo, leads to the exposure of the subendothelium, followed by insudation of blood constituents, platelet adhesion, further fibrin deposition and probably myointimal proliferation. Though FDP D or E, LMWFDP Fr 1 or 2 had no effects on the cultured endothelial cells morphologically, AA-induced PGI_2 synthesis by endothelial cells was decreased. And PGI_2 synthesis by smooth muscle cells incubated with FDP D or E, LMWFDP Fr 1, 2 or 3 for 24h also decreased. The increase of PGI_2 synthesis by endothelial cells incubated with LMWFDP Fr 3 for 24h may result from the release of stocked PGI_2 in the endothelial cells by injury. From our data, FDPs may affect the synthesis of PGI_2 of vascular wall. That is, FDPs may depress PGI_2 synthesis by smooth muscle cells and endothelial cells. These phenomena, in turn, may facilitate platelet to aggregate in vivo. But AA-induced PGI_2 synthesis of non-detached endothelial cells after injury by LMWFDP Fr 3 was rather increased. This seems to be the defense mechanism of endothelial cells after injury. Ishida and Tanaka (13) reported that fibrin enhanced the growth of smooth muscle cells in culture and FDPs yielded by fibrinolysis inhibited such growth.

Thus the initiation of atherosclerosis may be affected by the contact of fibrin, fibrinogen and the degradation products with vascular endothelial cells and smooth muscle cells.

References

1. Astrup, T.: Lancet II , 565-568 (1956).
2. Kwaan, H.C.: Artery 5, 285-290 (1979).
3. Kadish, J.L., Butterfield, C.E.: Tissue and Cell 11, 99-108 (1979).
4. Dembinska-Kiec, A., Gryglewska, T., Zmuda, A., Gryglewski, R.J.: Prostagladins 14, 1025-1034 (1977).
5. Larrue, J., Rigaud, M., Daret, D., Demond, J., Durand, J., Bricaud, H.: Nature 285, 480-482 (1980).
6. Watanabe, K., Tanaka, K.: Atherosclerosis, 48, 57-70 (1983).
7. Ross, R.: J. Cell Biol. 50, 172-186 (1971).

8. Blomback, B., Blomback, M.: Ark. Kem. 10, 415-443 (1956).

9. Marder, V.J., James, H.L., Sherry, S.: Throm. Diath. Haemorrh. 22,234-239 (1969).

10. Busch, C., Gerdin, B.: Thromb. Res. 22, 33-39 (1981).

11. Lowry, O.H., Rosebrough, N.J., Farr, A.L., Randall, R.J.: J. Biol. Chem. 193, 265-275 (1951).

12. Ross, R., Glomset, J.A.: N. Engl. J. Med. 295, 369-377 (1976).

13. Ishida, T., Tanaka, K.: Atherosclerosis 44, 161-174 (1982).

INTERACTION OF FIBRIN WITH MALIGNANT MELANOMA CELLS IN LONG TERM CULTURE

Maria Jeleńska and Maria Kopeć

Institute of Nuclear Chemistry and Techniques
Warsaw, Poland

Maria Rochowska and Henryk Skurzak

Institute of Oncology
Warsaw, Poland

Introduction

It has been reported recently that fibrin retraction can be mediated not only by platelets and fibroblasts (1) but also by epithelial and endothelial cells (2,3). Transformation of fibroblasts, spontaneous as well as induced by various agents appeared to be accompanied by a loss of fibrin retracting cell capacity (4). In comparative study on normal, established and tumorigenic epithelial cells in cultures Azzarone et al (2) found that in contrast to normal cells malignant ones do not retract fibrin. Hence the loss of fibrin retracting potency has been suggested to belong to the indicators of tumoral cell transformation (2, 5).

This paper presents the results of studies on fibrin retraction under the influence of human melanoma cell lines in long term cultures. Human fetal fibroblasts are included into investigation as a control.

Results

Fibroblasts and melanoma cells grow as monolayers at 37°C in 5% CO_2 in air

Fibrinogen - Structural Variants and Interactions
© 1985 Walter de Gruyter & Co., Berlin · New York - Printed in Germany

atmosphere, were maintained in Eagles'1959 medium supplemented with 10% fetal calf serum, 100 Ul/ml penicillin and 100 UG/ml streptomycin. Single-cell suspensions were prepared by treatment of cultures with 0.25% trypsin for 5-10 min. Detached cells were suspended in complete medium, centrifuged at 200 x g for 10 min resuspended and counted in Coulter Counter.

TABLE 1

Tested human cell lines*

Line	Origin and cell type	Source
FLWp-1 p.5-10	skin fibroblast from 12 weeks fetus	Institute of Oncology, Warsaw
MeW-132	primary skin melanoma fibroblast like	Institute of Oncology, Warsaw
Me-81	skin melanoma fibroblast like	Della Porta, Milano
Me-22	skin melanoma fibroblast like	Della Porta, Milano
BCOA	skin melanoma fibroblast like	NCI, Bethesda
Me-6278	skin melanoma fibroblast like	NCI, Bethesda
MeW-18	primary skin melanoma fibroblast like	Institute of Oncology, Warsaw
MeW-163 p.10-13	primary skin melanoma fibroblast like	Institute of Oncology, Warsaw
FT-SLC	metastatic skin melanoma epithelial like	Cancer Institute, Wien
GTBSB	metastatic skin melanoma epithelial like	Cancer Institute, Wien
Me-4143	skin melanoma epithelial like	NCI, Bethesda
MeW-151 p.10-13	metastatic skin melanoma fibroblast like	Institute of Oncology, Warsaw

*All tumour cell lines, except MeW 163, MeW 151, are established in culture and maintained from 2-10 years.

Fibrin clot retraction (FCR) was estimated as described previously (6) in the following system: cells in Eagles'(MEM) medium 50-200 µl, human serum 100 µl, human platelet free plasma 100 µl, Eagles'(MEM) medium to 900 µl, thrombin (20 u/ml) 100 µl or Reptilase (R) (20u/ml) 100 ul.

Hirudin (20 u/ml) 100 µl (instead of 100 µl of MEM) was added to samples clotted by Reptilase. Results are expressed as volumes of extruded fluid in per cent of the initial volume.

A majority of the investigated melanoma cell lines namely seven out of eleven retracted fibrin clots. However, numbers of cells required for maximal FCR (ca 80%) were 2.5 to 8 fold higher for all but two of melanoma cell lines as compared with control fibroblasts.

TABLE 2

FCR mediated by various human melanoma cell lines (maximal values)

Cell line	FCR	Cells/ml 1×10^6	Tumorigenicity*
Control-HFF	82-88	0.2	
MeW-132	84-90	0.2	
Me-81	82-90	0.1	+
Me-22	34-36	0.2	+
BCCA	80-82	0.5	+
Me-6278	64-68	1.0	+
MeW-18	78-82	0.5	
MeW-165	82-84	4.0	
FT-SLC	Lysis		+
GTBSB	Lysis		+
Me-4143	Lysis		+
MeW-151	Lysis		

*Assessed in athymic nude mice

Four other melanoma cell lines provoked complete lysis of fibrin clots containing normal human serum and plasma as a source of fibrinogen. Fibrinolytic activity of these cells was mediated by plasminogen activator. It could be inhibited by EACA or Trasylol (not shown) and prevented by removal of plasminogen from the serum and plasma by filtration on columns of Sepharose-bound lysine. Under such conditions fibrin containing the same number of melanoma cells (10×10^6/ml) remained undigested during

24 h. Three melanoma cell lines producing plasminogen activator have been tested in such plasminogen free system (Plg-). Two did not retract fibrin at all while the third showed low activity in FCR (30%, Table 3).

TABLE 3

Dependence of fibrin lysis on plasminogen (Plg)

Cell line	FCR (%)	
	Plg^+	Plg^-
Control-HFF	85	84
GTBSB	Lysis	30
Me-4143	Lysis	0
MeW-151	Lysis	0

Tumorigenicity in athymic, nude mice could be estimated in not all studied melanoma cell lines. The data presented in Tables 2 and 3 indicate however no relation between the tumorigenicity of melanoma cells in vivo and their potency to retract fibrin in vitro. Some cell lines completely inactive as well as some highly active in FCR expressed the malignant behaviour in vivo.

It is known that activation of platelets and fibroblasts is a necessary requirement for their FCR activity (7,8,9). These cells do not retract fibrin formed by thrombin-like enzymes from snake venoms, which do not induce platelet activation and release reaction (1, 8, 9). The data presented in Table 4 show that activation with thrombin is also a prerequisite for the full expression of fibrin retracting capacity of melanoma cells. Some cell lines that provoked strong retraction of clots formed by thrombin did not induce FCR when Reptilase had been used for conversion of fibrnogen into fibrin. One cell line produced only partial retraction of Reptilase-fibrin but full FCR in the system clotted with thrombin. (Table 4).

The results of our investigation allow to include human malignant melanoma cells to the list of cells expressing the fibrin clot retractile activity.

Table 4

Requirement of cell stimulation by thrombin for FCR

Cell line	Retraction of fibrin (%) induced by	
	Thrombin	Reptilase
MeW-132	87	0
Me-81	86	50
Me-6278	66	0
MeW-18	80	0
MeW-165	83	0

Previous studies of several groups of authors (2, 4, 5, 7, 10) on normal, defective and malignant fibroblasts and epithelial cells have demonstrated a great variability in FCR activity depending on aging, proliferative capacity and phenotypic characteristics. On the basis of the obtained results unlimited proliferative life span in vitro and tumorigenicity in vivo have been postulated to be associated with the loss of the cells ability for interactions with fibrin leading to retraction (2, 4, 5).

Impairment of fibrin retractile activity has been also discovered in fibroblasts and fibroblast-like cells from some patients with malignancies and from cases of genetic defects such as Glanzmann´s thrombasthenia (12), premature aging syndromes (7) and Cockayne syndrome (10).

Proper anchorage of fibrin on the cell surface and functionally efficient cellular contactile system are the necessary requirement for FCR.

The lack or defect in fibrin(ogen) receptors is the most probable reason of inefficiency of thrombasthenic platelets in FCR. Thrombasthenic platelets are fully active in retraction of collagen gels, what indicates functionally normal contractile machinery (6). For other cells the reasons of loss of FCR activity remain unclarified. Changes in composition of cell membrane components as well as in structure of contractile system elements are known to be associated with tumoral cell transformation (13,

14). Contribution of these two types of changes to the loss of FCR potency is however still unknown. In preliminary experiments we did not found differences in ^{125}J-fibrinogen binding to melanoma cells nonactive and active in FCR.

Our investigation demonstrated a great variability of malignant melanoma cell lines for FCR capacity. It should be stressed that the investigated cell lines differed also considerably in other properties including morphology. It can be however concluded that for melanoma cells in cultures the notion that the loss of activity in FCR is an indicator of tumorigenicity in vivo and infinite life span in vitro is not valid.

It has been suggested that plasminogen activator production may be relevant for adhesive cell properties and the pattern of growth. Pollack et al (15) reported that plasminogen activator production accompanies the loss of anchorage regulation of growth in rat embryo cells transformed by SV40 virus. The few examined melanoma cell lines producing plasminogen activator were the only ones inactive or hardly active in FCR. Curatolo et al (16) found however complete dissociation between plasminogen activator production and loss of FCR capacity in hybridomas of normal diploid human fibroblasts and heteroploid transformed mouse cell line. Hence this problem merits further studies.

This study was supported by a grant from Polish Government Research Program of Control of Neoplastic Diseases, PR-6 No 2118.

References

1. Niewiarowski, S., Regoeczi, E., Mustard, J.G.: Proc. Soc. Exp. Biol. Med. 250, 4007-4021 (1972).

2. Azzarone, B., Curatolo, L., Vanier, O., Donati, MB., Macieira-Coelho, A., Morasca, L.: J.Natl. Cancer Inst. 71, 879-885 (1983).

3. Barbieri, B., Balconi, G., Dejana, E., Bonati, MB.: Thrombos. Haemost. (abstr) 46, 72 (1981).

4. Azzarone, B., Curatolo, L., Carloni, G., Donati, MB., Morasca, L., Macieira-Coelho, A.: J. Natl. Cancer Inst. 67, 89-94 (1981).

5. Azzarone, B., Brouty-Boye, D., Macieira-Coelho, A.: Int J. Cancer 28, 799-803 (1981).

6. Jeleńska, M., Kopeć, M.: Thrombos. Haemost. 44, 161-165 (1980).

7. Goldstein, S., Niewiarowski, S., Singal, D.P.: Federation Proceedings 34, 56-63 (1975).

8. De Gaetano, G., Bottecchia, D., Vermylen, J.: Thrombos. Res. 2, 71-84 (1973).

9. Niewiarowski, S., Goldstein, S.: Proc. Soc. Exp. Biol. Med. 151, 253-255 (1976).

10. Curatolo, L., Borgia, R., Donati, MB., Morasca, L.: Cell Biol. Int. Rep. 6, 609-618 (1982).

11. Niewiarowski, S., Goldstein, S: J.Lab. Clin. Med. 82, 605-610 (1973).

12. Donati, MB., Balconi, G., Renauzzi, G., Borgia, R., Morasca, L., de Gaetano, G.: Thromb Res 10, 173-174 (1977).

13. Tarone, G., Comoglio, P.M.: Exp. Cell Res. 110, 143-152 (1977).

14. Pollack, R., Osborn, M., Weber, K.: Proc. Natl. Acad. Sci. USA 72, 994-998 (1975).

15. Pollack, R., Rissner, R., Conlon, S., Rifkin, D.: Proc. Natl. Acad. Sci. USA 71, 4792-4796 (1974).

16. Curatolo, L, Azzarone, B., Fally-Crepin, C., Morasca, L., Macieira-Coelho, A.: Int. J. Cancer 31, 249-255 (1983).

AUTHOR INDEX

SUBJECT INDEX

Fibrinogen

Recent Biochemical and Medical Aspects
Volume 1

Proceedings · International Symposium, Max-Planck-Institute for Biochemistry, February 17–18, 1981, Martinsried/Munich, Germany

Editors *A. Henschen · H. Graeff · F. Lottspeich*

1982. 17 x 24 cm. X, 400 pages. Numerous illustrations.
Hardcover. DM 135,–; approx. US $52.00 ISBN 3 11 008543 7

Fibrinogen

Structure – Functional Aspects – Metabolism
Volume 2

Proceedings · Workshop on Fibrinogen
May 12–14, 1982, Leiden, The Netherlands

Editors *F. Haverkate · A. Henschen · W. Nieuwenhuizen · P. W. Straub*

1983. 17 x 24 cm. XVI, 343 pages. Numerous illustrations.
Hardcover. DM 145,–; approx. US $55.75 ISBN 3 11 008986 6

Fibrinogen

Structural Variants and Interactions
Volume 3

Proceedings · Workshop on Fibrinogen
July 9–10, 1983, Stockholm, Sweden

Editors *A. Henschen · B. Hessel · J. McDonagh · T. Saldeen*

1985. 17 x 24 cm. XII, 418 pages. Numerous illustrations.
Hardcover. DM 270,–; approx. US $103.90 ISBN 3 11 009806 7

Forthcoming issue:

Fibrinogen

Fibrin Formation and Fibrinolysis · Volume 4

Proceedings · Workshop on Fibrinogen
April 2–3, 1985, London, England

Editors *D. A. Lane · A. Henschen · M. K. Jasani*

1986. 17 x 24 cm. XVIII, 398 pages. Numerous illustrations.
Hardcover. DM 250,–; approx. US $96.20 ISBN 3 11 010597 7

Prices are subject to change without notice

Walter de Gruyter · Berlin · New York

Biological Chemistry Hoppe-Seyler

Editors: K. Decker, W. Stoffel, H. G. Zachau
Editorial Office: A. Dillmann, G. Peters

This Journal was founded in 1877 as *Zeitschrift für Physiologische Chemie* by F. Hoppe-Seyler and was continued after his death under the editorship of A. Kossel, F. Knoop, K. Thomas, F. Lynen, A. Butenandt, and G. Weitzel as *Hoppe-Seyler's Zeitschrift für Physiologische Chemie*
[Volume 21 (1875) – Volume 365 (1984)]

Subscription Information: 1986, Vol. 367 (12 issues): DM 880,–; approx. US $ 338.50. Single issue: DM 88,–; approx. US $ 33.90

Journal of Clinical Chemistry and Clinical Biochemistry Zeitschrift für Klinische Chemie und Klinische Biochemie

Gemeinsames Organ der Deutschen, der Niederländischen, der Österreichischen und der Schweizerischen Gesellschaft für Klinische Chemie

This journal publishes all IFCC Recommendations regularly

Editors-in-Chief: Johannes Büttner, Hannover – Walter Guder, München
Managing Editor: Friedrich Körber, Berlin
Special Editor for IFCC Recommendations: Nils-Erik Saris, Helsinki

Subscription Information: 1986, Volume 24 (12 issues): DM 690,–; approx. US $ 265.40. Single issue: DM 65,–; approx. US $ 25.00

Prices are subject to change without notice

Walter de Gruyter · Berlin · New York

FSC
www.fsc.org
MIX
Papier | Fördert
gute Waldnutzung
FSC® C083411

Kolibri 360 GmbH
Ferdinand-Jühlke-Straße 7
99095 Erfurt, Deutschland
produktsicherheit@kolibri360.de